SAFETY AT THE SHARP END

Safety at the Sharp End

A Guide to Non-Technical Skills

RHONA FLIN, PAUL O'CONNOR and MARGARET CRICHTON

CRC Press is an imprint of the
Taylor & Francis Group, an informa business

CRC Press
Taylor & Francis Group
6000 Broken Sound Parkway NW, Suite 300
Boca Raton, FL 33487-2742

CRC Press is an imprint of Taylor & Francis Group, an Informa business

Printed on acid-free paper
Version Date: 20160226

International Standard Book Number-13: 978-0-7546-4600-6 (Paperback) 978-0-7546-4598-6 (Hardback)

Visit the Taylor & Francis Web site at
http://www.taylorandfrancis.com

and the CRC Press Web site at
http://www.crcpress.com

Contents

List of Figures

List of Tables

Acknowledgements

We would like to thank the following people for providing material or commenting on draft chapters: Bob Hahn, Matthew Hilscher, Terry Kelly, Bill Little, Mike Lodge, Jean McLeish, Richard Miller, Ed Naggiar, Simon Paterson-Brown, Tanja Pullwitt, Cameron Ramsay, Tom Reader, Dick Stark, John Tse, George Youngson, and Steven Yule.

In particular, our special thanks go to John Thorogood for his full and frank but always constructive comments on a number of chapters.

We would also like to acknowledge the contribution of Guy Loft and John Hindley of Ashgate for their support and advice, and Gemma Lowle for her help with editing and production.

Any remaining errors are our own and we would be grateful to hear from any readers who can provide corrections.

Chapter 1

Introduction

When things go wrong in high-risk organisations, the consequences can result in damage to humans, equipment and the environment. Significant levels of protection and redundancy are built into modern technical systems, but as the hardware and software have become increasingly reliable, the human contribution to accidents has become ever more apparent. Analyses in a number of industrial sectors have indicated that up to 80% of accident causes can be attributed to human factors (Helmreich, 2000; Reason, 1990; Wagenaar and Groenweg, 1987). This means that managers also need to understand the human dimension to their operations, especially the behaviour of those working on safety-critical tasks – the 'sharp end' of an organisation. Psychologists have long been interested in the factors that enhance workers' performance and minimise error rates (Munsterberg, 1913). We know that human error cannot be eliminated, but efforts can be made to minimise, catch and mitigate errors by ensuring that people have appropriate non-technical skills to cope with the risks and demands of their work.

Non-technical skills are the cognitive and social skills that complement workers' technical skills (Flin et al., 2003). The definition of non-technical skills used to underpin this book is 'the cognitive, social and personal resource skills that complement technical skills, and contribute to safe and efficient task performance'. They are not new or mysterious skills but are essentially what the best practitioners do in order to achieve consistently high performance and what the rest of us do 'on a good day'. This book describes the basic non-technical skills and explains why they are important for safe and efficient performance in a range of high-risk work settings from industry, health care, military and emergency services.

The seven skills we discuss are:

- situation awareness (attention to the work environment)
- decision-making
- communication
- teamwork
- leadership
- managing stress
- coping with fatigue.

This skills approach provides a set of established constructs and a common vocabulary for learning about the important behaviours that influence safe and efficient task execution. Methods of identifying, training and assessing key non-technical skills for particular occupations are outlined in the chapters to follow. The skills listed above are required across a range of settings. Much of the background material is drawn from the aviation industry but our aim is to demonstrate why these non-technical skills are critical for many different tasks, from operating a control room on a power plant, to operating on a surgical patient. Human behaviour is remarkably similar across all kinds of workplaces.

Before examining the non-technical skills, it should be emphasised that this focus on workers' behaviours is only one component of an effective safety management strategy. Organisational safety is influenced by regulatory and commercial pressures, the working environment and management demands. So, while this book concentrates on the skills of those operating the system, it is acknowledged that their behaviour is influenced by the conditions they work in and by the behaviours of others, particularly those in managerial positions (Flin, 2003; Hopkins, 2000).

As Reason (1997) illustrated in his 'Swiss Cheese' model (see Figure 1.1), accidents are usually caused by a sequence of flaws in an organisation's defences. These can be attributed to a combination of errors and violations by the operational staff (active failures) and the latent unsafe conditions ('resident pathogens') in the work system that are created by managers, designers, engineers and others.

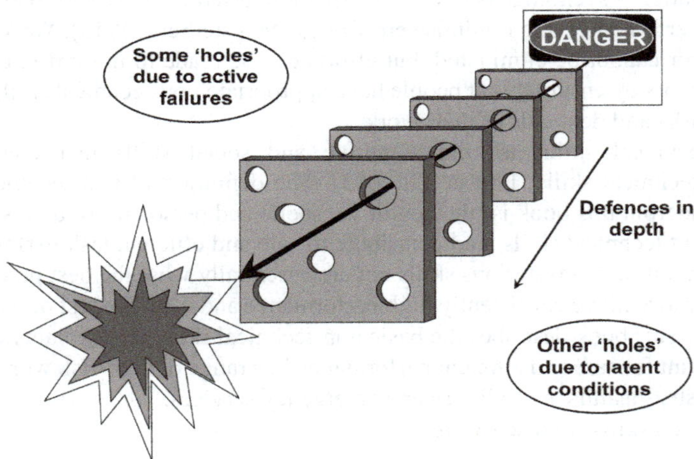

Figure 1.1 An accident trajectory passing through corresponding holes in the layers of organizational defences (Reason, 1997, Fig 1.5)

Reprinted with permission of Ashgate.

The 'front line' personnel represent the last line of protection of the system's defences. They are not only responsible for the active failures that can contribute to losses and injuries, but more importantly, they regularly catch and correct their own and others' errors. Helmreich et al.'s (2003) observations on aircraft flight decks showed that airline pilots make about two errors per flight segment, but that most of these are caught and corrected by the air crew themselves. Moreover, front-line staff can also recognise and remedy technical malfunctions and can cope with a wide variety of risky conditions. Therefore, while human error is inevitable and pervasive, as Reason (1997) pointed out, humans can also be heroes by providing the essential resilience and expertise to enable the smooth operation of imperfect technical systems in threatening environments.

We begin by describing how a series of puzzling aircraft accidents in the 1970s led to the recognition of the importance of non-technical skills in aviation. This is followed by an account of how these skills have since become a focus of attention in other work settings.

Non-technical accidents in aviation and beyond

Thirty years ago, a series of major aviation accidents, without primary technical cause, forced investigators to look for other contributing factors. The best known of these events is the Tenerife crash in 1977, when two jumbo jets crashed on an airport runway, as described below.

Box 1.1 Tenerife Airport Disaster

At 17:06 on 27 March 1977, two Boeing 747 aircraft collided on the runway of Los Rodeos airport on the island of Tenerife. The jets were Pan Am flight 1736 en route to Las Palmas from Los Angeles via New York and KLM flight 4805 from Amsterdam, also heading for Las Palmas. Both had been diverted to Tenerife because of a terrorist incident on Las Palmas. After several hours, the airport at Las Palmas re-opened and the planes prepared for departure in the congested (due to re-routed aircraft), and now foggy, Los Rodeos airport. The KLM plane taxied to the end of the runway and was waiting for air traffic control (ATC) clearance. The Pan Am plane was instructed to taxi on the runway and then to exit onto another taxiway. The KLM plane was now given its ATC clearance for the route it was to fly – but not its clearance to begin take-off. The KLM captain apparently mistook this message for a take-off clearance, released the brakes, and despite the co-pilot saying something, he proceeded to accelerate his plane down the runway. Due to the fog, the KLM crew could not see the Pan Am 747 taxiing ahead of them. Neither jet could be seen by the control tower and there was no runway radar system. The KLM flight deck engineer, on hearing a radio call from the Pan Am jet, expressed his concern that the US aircraft might not be clear of the runway, but was over-ruled by his captain. Ten seconds before collision, the Pan Am crew noticed the approaching KLM plane but it was too late for them to manoeuvre their plane off the runway. All 234 passengers and 14 crew on the KLM plane and 335 of 396 people on the Pan Am plane were killed.

Analyses of the accident revealed problems relating to communication with ATC, team co-ordination, decision-making, fatigue and leadership behaviours.

See Weick (1991) and Box 5.4. for further details.

This was not an isolated incident. Other aircraft accidents had happened which, like the Tenerife crash, did not have primary technical failures. United Airlines suffered a sequence of crashes in the late 1970s, also attributed to what was now being called 'pilot error' rather than technical faults. Due to growing concern, an aviation industry conference was held at NASA in 1979 bringing together psychologists and airline pilots to discuss how to identify and manage the human factors contributing to accidents. The aviation industry was fortunate to have one invaluable source of information, namely the cockpit voice recorders that had been built into modern jet aircraft (CAA, 2006). They revealed what the flight deck crew were saying in the minutes before and during these accidents. Analysis of these conversations suggested failures in leadership, poor team co-ordination, communication breakdowns, lack of assertiveness, inattention, inadequate decision-making and personal limitations, usually relating to stress and fatigue (Beaty, 1995; Wiener et al., 1993).

Accidents involving failures in non-technical skills are certainly not unique to the aviation industry. Table 1.1 gives a sample of such events with failures in non-technical skills at the 'sharp' or operational end of an organisation. In two of the world's most serious nuclear power plant incidents, Chernobyl (Reason, 1987) and Three Mile Island (NRC, 1980), operator error relating to loss of situation awareness and flawed decision-making played a major role. In the military domain, the accidental attack in 1988 by the *USS Vincennes* on a passenger aircraft killing all 290 passengers and crew was caused by decision-making and communication failures (Klein, 1998).

The loss of the *Piper Alpha* oil platform with 167 deaths was caused by poor communication at shift handover, compounded by leadership failures in emergency response (Cullen, 1990). Shipping accidents such as groundings and collisions are frequently characterised by failures in bridge leadership or crew co-ordination (Barnett et al., 2006; Hetherington et al., 2006). In hospitals, rates of adverse events (injuries caused by medical treatment) to patients can be as high as 10% of admissions – many of these can be attributed to failures in communication, breakdowns in teamwork and flawed decision-making (Bogner, 1994; Helmreich, 2000; Vincent, 2006).

As in the aviation sector, high-risk industries, emergency services and military organisations also began to realise that they would not manage to address their safety problems by attending only to the technology or technical skills. Other aspects of workers' competence needed to be considered. The term non-technical skills (NTS) is used in several technical specialties but it was introduced to aviation by the European civil aviation regulator, the Joint Aviation Authorities (JAA).

Table 1.1 Accidents with non-technical skills failures in causation or response*

Year	Industry	Incident	Non-technical skills failures
1979	Nuclear power	*Three Mile Island* nuclear power plant release	Problem-solving, teamwork, situation awareness
1986	Nuclear power	*Chernobyl* nuclear power plant release	Decision-making, situation awareness, personal limitations
1987	Maritime	*Herald of Free Enterprise* Ship sails with bow doors open and capsizes	Team co-ordination, situation awareness
1988	Oil and gas production	*Piper Alpha* Oil platform explosion	Communication, leadership, decision-making, team handover
1988	Military	*USS Vincennes* Warship destroys passenger plane	Team co-ordination, decision-making, situation awareness
1989	Police	*Hillsborough* Police response to football crowd being crushed	Communication, situation awareness, leadership
1989	Aviation	*Kegworth* Plane crash – wrong engine shut down	Situation awareness, decision-making
1990	Maritime	*Scandinavian Star* Response to ship fire	Teamwork, leadership
1994	Health care	Betsy Lehman Chemotherapy overdose	Situation awareness, decision-making
1996	Transport	*Channel Tunnel* Response to fire in tunnel	Communication, stress, team co-ordination
1998	Petrochemical	*Esso Longford* Refinery explosion	Communication (shift handover), situation awareness
2000	Health care	Graham Reeves – wrong kidney removed	Situation awareness, teamwork, leadership
2001	Health care	Wayne Jowett Chemotherapy site error	Decision-making, situation awareness, communication
2005	Petrochemical	*BP Texas City* Refinery explosion	Leadership, decision-making, fatigue, communication

*Source reports are listed at the end of the chapter.

The relationship between non-technical skills and human error is illustrated in Figure 1.2. This diagram is oversimplified but indicates that poor non-technical skills can increase the chance of error, which in turn can increase the chance of an adverse event. Good non-technical skills (e.g. high vigilance, effective communication, leaders who maintain standards) can reduce the likelihood of error and consequently of adverse events.

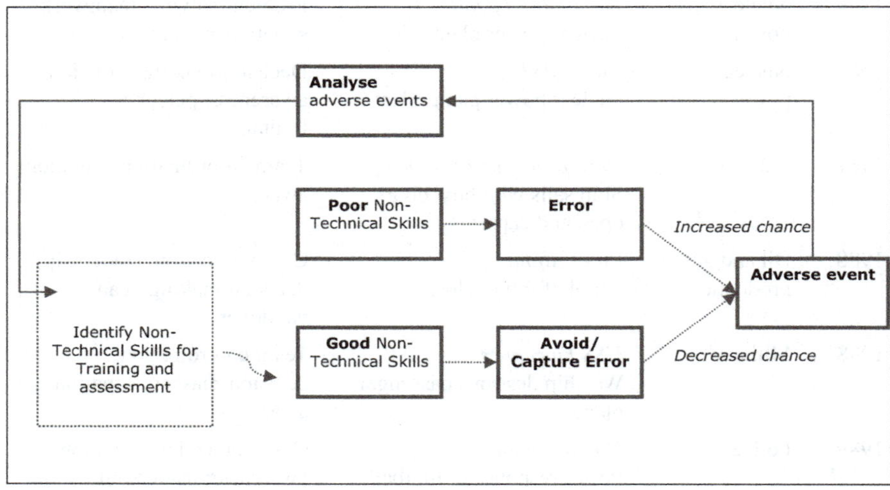

Figure 1.2 Relationship between non-technical skills and adverse events

Adverse events can be costly production failures that do not harm anyone – such as getting the drill pipe stuck when exploring for oil or shutting down a manufacturing process by mistake. More importantly, adverse events can have safety consequences such as injuries to workers or to other humans in the system, e.g. hospital patients or airline passengers. Analysing and learning from adverse events (as well as from other occupational data discussed in Chapter 9) can result in new knowledge about the skills and component behaviours that influence adverse outcomes. As Figure 1.2 shows, this information should be fed back into the training system, as well as being used to direct regular assessments from observations and audits of operational behaviour.

The next section describes how the process shown above was enacted first in aviation and then in other industries, in order to identify the main categories of non-technical skills for a given occupation, so that they could be trained and monitored.

Identifying and training non-technical skills in aviation and beyond

Once the aviation industry realised that maintaining high standards of safety was going to require more than reliable technology and proficient technical skills in the system operators (e.g. pilots, engineers, air traffic controllers), it began several programmes of research to identify the key non-technical skills (NTS). These were not mysterious or rare behaviour patterns; in fact they were well-known skills that pilots regarded as an essential part of good airmanship (Beaty, 1995). But traditionally, non-technical skills had been tacitly rather than explicitly addressed and consequently were taught in an informal and inconsistent manner from one generation of pilots to the next. Studies were commissioned by the airlines and by regulators to enable aviation psychologists in centres such as NASA Ames and the University of Texas at Austin to conduct research with pilots looking at the key behaviours. They began to run experiments in flight deck simulators, to interview pilots, to analyse accident reports, in order to discover which skill components either contributed to accidents or were effective in preventing adverse events (Wiener et al., 1993). Once the core NTS had been identified, then the airlines began to develop special training courses to raise awareness of the importance of these skills, to provide the necessary underpinning knowledge and practice for skill development. These courses were initially called cockpit resource management, later amended to crew resource management (CRM) as other crew members, e.g. cabin attendants, were involved. This CRM approach can be defined as 'a management system which makes optimum use of all available resources – equipment, procedures and people – to promote safety and enhance efficiency of flightdeck operations' (CAA, 2006, p1).

In the last decade, other industries have started to introduce CRM-based training, such as nuclear power generation, Merchant Navy, prison service, emergency services and hospital medicine (Flin et al., 2002; Musson and Helmreich, 2004; Salas et al., 2006). The training of NTS using CRM courses is discussed in Chapter 9, and several general texts on CRM are available: Weiner et al. (1993) give a comprehensive account of the development of the early CRM courses and Salas et al. (2001) provide reports of CRM training in other industries. There are specialist texts on CRM training for aviation (e.g. CAA, 2006; Jensen, 1995; Macleod, 2005; McAllister, 1997; Walters, 2002) and also for other occupations, such as the fire service (Okray and Lubnau, 2004).

Why we have written this book

So why do we need another book? The reason for writing this book is based on our experiences as psychologists, working with industries and professions that are using or introducing non-technical skills/CRM training. For a number of years, we ran a CRM users group meeting at Aberdeen University to share information and expertise across different sectors. This would be attended by experienced CRM trainers from aviation as well as representatives from a wide range of other organisations (e.g. emergency services, energy sector, anaesthesia, surgery, finance, prison service) interested in non-technical skills and CRM training.

These practitioners often asked us to recommend a basic guidebook on non-technical skills. They wanted to know what these skills are; the evidence that they were related to task performance; how to identify them and to design training for their own workplace. They needed to find psychological or other human factors material for the development of their own programmes and sometimes they were looking for methods to assess the performance of non-technical skills.

This book was designed to meet these requirements. The aims are:

- to explain the basic categories of non-technical skills, and to summarise relevant psychological evidence for their importance for workplace safety,
- to describe the basic procedures for identifying, training and assessing non-technical skills.

The material is not concentrated on any one single industry or occupation; indeed, our aim is to show how these skills are appropriate to many organisations; examples have been drawn from aviation and from other sectors in which we are working or have relevant knowledge. We have been involved in research projects to identify non-technical skills and to design CRM training with civil aviation, production and drilling in the oil and gas industry, marine industry, nuclear power generation, military (aviators and special forces) and acute medicine (anaesthetists, scrub nurses, surgeons). This book has been written primarily for use by professionals developing or attending courses in CRM, NTS, safety leadership and human performance limitations. The material would also be suitable for undergraduate and post-experience students studying applied psychology, human factors, industrial or medical safety. References have been given throughout the text in order to give the scientific background and to direct the reader to more detailed published sources or websites.

Where appropriate, examples of actual events where these skills appear to have failed or have proved valuable are provided in boxes. These are often more memorable for trainees than the theoretical explanations of the underlying psychology or physiology.

What this book is about (and not about)

There are four issues we have to clarify in relation to our focus. This book is about:

a. the individual rather than the group
b. behaviour rather than personality
c. routine work, as well as unusual situations
d. technically relevant skills – not 'soft' skills.

a) Individual rather than group

First, it is important to note that the focus of our attention is on the individual rather than the work team. In most safety-critical industries, work is carried out by teams of technical specialists and so the work group will normally provide the context for the individual's behaviour. The social skills we discuss are about team members working collaboratively on overlapping tasks to achieve common goals or about how to lead a team to produce safe and efficient work performance. But our unit of analysis is the individual team member, not the work group as a whole, so for example we do not discuss measures to observe or rate a team as an entity.

The reason for this focus on the individual is because the individual is the basic 'building block' from which teams and larger organisational groupings are formed. Moreover, in our experience (e.g. civil aviation, the energy industry, hospital medicine), people often do not work in the same team every day. In fact, team composition is rarely fixed due to shift and rotation patterns, on-the-job training, organisational constraints and working time restrictions. In the larger airlines, the same pilots rarely fly together and for that reason, the focus in European aviation has been on the individual pilot's technical and non-technical competence, rather than on a crew.

Pilots are taught and examined on the psychological and physiological factors influencing human performance in their initial training programme (human performance limitations, see Campbell and Bagshaw, 2002). This training in human factors is usually a prerequisite for pilots attending CRM training. Therefore, when pilots begin to undertake CRM training, they already possess the basic human factors knowledge. We would advocate that, as in aviation, practitioners should be taught the basic principles of non-technical skills and their impact on human performance for their own profession, before they embark on CRM or other multidisciplinary team training.

b) Behaviour rather than personality

A second caveat is that this book is about behaviour and not about personality. Psychologists define personality as 'internal properties of a person that lead to characteristic patterns of behaviour' (Hampson and Coleman, 1995: pxi). While acknowledging that personality has an influence on behaviour, the interventions described here are about changing patterns of behaviour rather than attempting to modify the individual's underlying personality. Moreover, personality should not be accepted as an excuse for the maintenance of unsafe behaviour at work. Of course, our particular blend of traits and characteristics makes it easier or more difficult for us to adopt less instinctive behaviours. So, for example, the talkative extroverts may have to learn to talk less and to listen more. The quiet introverts may need to start sharing their thoughts about task goals and planned actions. The autocratic leader has to adopt new behaviours that encourage team members to express their views. The shy trainee has to be confident enough to speak up when they think the team leader is making an error. As one airline told its pilots, 'You can have any personality you like, but this is the standard of behaviour we expect on our flight decks.'

c) Routine as well as unusual situations

Third, non-technical skills are important in everyday, routine work conditions; they are not just for managing critical situations or emergencies, although they may be particularly critical in high-demand circumstances. When workers are attentive, make sound decisions, share information and co-operate with fellow workers, then errors and accidents are less likely to occur. We use examples of difficult, unexpected work conditions throughout the book, as they best demonstrate the necessity of good non-technical skills. These situations may have been partly caused by the crew, as in the *Piper Alpha* accident where critical information was not transferred across shifts, or the cause may have occurred independently from a technical failure or external conditions. Major accident investigations usually reveal important information on how people behaved before or during the event, and they tend to be well documented. The investigations can show how non-technical skills have helped or hindered the management of the situation.

Routine job performance is rarely scrutinised to the same degree[1] even though accidents regularly occur in normal conditions, and therefore non-technical skills are for 'everyday' use, not just for special occasions.

d) Technically relevant skills

Finally, the skills discussed in this book are sometimes referred to in industry (often with a disparaging tone) as 'soft' skills. This is misleading because they are as critical as the 'hard' technical skills and, in fact, are essential when competent technical people do technical things in technical work settings. So we would emphasise that the non-technical skills are needed to complement the technical skills possessed by an engineer, a scrub nurse or a pilot. Their decision-making is usually about technical matters, the content of their communications is likely to be technical, they have to co-ordinate technical tasks. When asked about creating high-performing teams, members of an oil industry drilling department said they preferred to select technically competent applicants who could work well in a team due to their non-technical skills, rather than relying solely on technical competence.

It should always be remembered that high technical expertise, although important, does not provide adequate protection from human error (see Box 1.2).

Box 1.2 Technical expertise

The surgeon showed a slide of the site of a gall bladder operation conducted by an experienced surgeon. Each clip was neatly positioned, the dissection had clearly been executed with precision and there had obviously been minimal unnecessary tissue handling or bleeding. In fact, it was a technically perfect procedure.

The only problem was that it had been executed on the wrong structure. The surgeon had excised the common bile duct along with the gall bladder. This is a disastrous error requiring reconstructive surgery and is associated with major morbidity and significant mortality.

Non-Technical Skills for Surgery Course ('Safer Operative Surgery')
Royal College of Surgeons of Edinburgh (2006) (see Flin et al., 2007)

1 Helmreich et al.'s (2003) observational LOSA system is one exception, see Chapter 11.

Overview of the book

The main categories of non-technical skills that are important for safe operations are shown in Table 1.2 below. They are not intended to be a definitive or comprehensive list but have been selected as common categories from CRM courses and non-technical skills lists from various organisations. Thus these categories and their component elements are only representative components of NTS/CRM training and assessment systems. For training and assessment, the target non-technical skills should be customised to suit a particular occupation, whether for a train driver, surgeon, firefighter or nuclear plant control room operator. The list in Table 1.2 is intended only to serve as an organising framework for Chapters 2–8, where each skill category is described in more detail.

Table 1.2 Main categories and elements of non-technical skills

Category	Elements
Situation awareness *Chapter 2*	Gathering information Interpreting information Anticipating future states
Decision-making *Chapter 3*	Defining problem Considering options Selecting and implementing option Outcome review
Communication *Chapter 4*	Sending information clearly and concisely Including context and intent during information exchange Receiving information, especially by listening Identifying and addressing barriers to communication
Team Working *Chapter 5*	Supporting others Solving conflicts Exchanging information Co-ordinating activities
Leadership *Chapter 6*	Using authority Maintaining standards Planning and prioritising Managing workload and resources
Managing stress *Chapter 7*	Identifying symptoms of stress Recognising effects of stress Implementing coping strategies
Coping with fatigue *Chapter 8*	Identifying symptoms of fatigue Recognising effects of fatigue Implementing coping strategies

The specific non-technical skills required to achieve safe and efficient performance within a given occupation and task set have to be determined by a systematic process of identification (techniques for this are described in Chapter 9). Nevertheless, as a number of safety-critical occupations have now begun to identify core NTS in order to design non-technical or CRM training, it is apparent that broad skill categories (e.g. decision-making) are similar across professions. This is hardly surprising as similar kinds of behaviour are conducive to safe and efficient work performance in a wide range of tasks, especially where teams are working in high-risk environments.

The seven chapters listed above constitute the first part of the book, providing a basic description of the core non-technical skills required by professionals working in teams in safety-critical environments and explaining why these are important for safe and efficient operations. They outline the basic psychology, first in chapters 2 and 3, in relation to two key cognitive skills – situation awareness and decision-making. The next three chapters deal with social or interpersonal skills – team working, leadership and communication. Two further chapters cover the management of 'personal energy and emotional resources', that is the psychophysiology of stress and fatigue and their impact on performance. These last two topics, stress and fatigue, are included in CRM/NTS training courses but the component skills are less likely to be included in formal assessment schemes. This is because they are more difficult to reliably detect by observation unless the individual is extremely stressed or fatigued, when the failure to cope with the condition is very obvious. Each of these chapters has a similar structure:

- What is it? – definition and description of the skill category and some typical component elements.
- Why is it important for safety at work?
- What do we know about it?
- What can we do about it?
- What are the key points?

In the second part, the last three chapters deal with the more practical aspects of identifying, training and assessing non-technical skills and some of the legislative influences, mainly from the aviation regulators. Chapter 9 explains the basic procedures for identifying non-technical skills, either for training or in the selection of personnel. Chapter 10 describes how NTS are trained (typically on CRM programmes) and how these courses should be designed and evaluated. Chapter 11 considers how best to evaluate whether the skills being trained are transferring to the workplace, using behaviour rating systems for assessment, feedback or auditing. Several of these systems are described – in particular those we were involved in developing, namely the NOTECHS system designed as a prototype for the European airlines; ANTS, used by anaesthetists; and NOTSS for surgeons. Much of the material in this second part of the book is drawn from research and practice in civil aviation, as this industry has been at the forefront of non-technical skills training and assessment.

We hope that this book will provide a straightforward guide to the basic non-technical skills for maintaining safe and efficient operations.

Key points

- Human error in the workplace cannot be eliminated, but efforts can be made to minimise, trap and mitigate error.
- Non-technical skills (e.g. decision-making, team leadership, communication) can enhance safety and efficiency by reducing the likelihood of error and consequently the risk of adverse events.
- Errors and accidents in high risk work settings can be very expensive.

References

Barnett, M., Gatfield, D. and Pekcan, C. (2006) Non-technical skills: the vital ingredient in world maritime technology? In *Proceedings of the International Conference on World Maritime Technology.* London: Institute of Marine Engineering, Science and Technology.

Beaty, D. (1995) *The Naked Pilot: The Human Factor in Aircraft Accidents.* Marlborough, Wiltshire: Airlife.

Bogner, M. (1994) *Human Error in Medicine.* Hillsdale, NJ: Lawrence Erlbaum.

CAA (2006) *Crew Resource Management (CRM) Training. Guidance for Flight Crew, CRM Instructors (CRMIs) and CRM Instructor-Examiners (CRMIEs).* CAP 737. Version 2. Gatwick: Civil Aviation Authority. www.caa.co.uk

Campbell, R. and Bagshaw, M. (2002) *Human Performance and Limitations in Aviation. (3rd ed.)* Oxford: Blackwell.

Cullen, D. (1990) *The Public Inquiry into the Piper Alpha Disaster* (Cm 1310). London: HMSO.

Flin, R. (2003) 'Danger – Men at Work'. Management influence on safety. *Human Factors and Ergonomics in Manufacturing*, 13, 261–268.

Flin, R., Martin, L., Goeters, K., Hoermann, J., Amalberti, R., Valot, C. and Nijhuis, H. (2003) Development of the NOTECHS (Non-Technical Skills) system for assessing pilots' CRM skills. *Human Factors and Aerospace Safety*, 3, 95–117.

Flin, R., O'Connor, P and Mearns, K. (2002) Crew Resource Management: Improving teamwork in high reliability industries. *Team Performance Management*, 8, 68–78.

Flin, R., Yule, S., Paterson-Brown, S., Maran, N., Rowley, D. and Youngson, G. (2007) Teaching surgeons about non-technical skills. *The Surgeon*, 5, 107–110.

Hampson, S. and Coleman, A. (1995) *Individual Differences and Personality.* London: Longman.

Helmreich, R. (2000) On error management: lessons from aviation. *British Medical Journal*, 320, 781–785.

Helmreich, R., Klinect, J. and Wilhelm, J. (2003) Managing threat and error: Data from line operations. In G. Edkins and P. Pfister (eds.) *Innovation and Consolidation in Aviation.* Aldershot: Ashgate.

Hetherington, C., Flin, R. and Mearns, K. (2006) Safety at sea. Human factors in shipping. *Journal of Safety Research*, 37, 401–411.

Hopkins, A. (2000) *Lessons from Longford.* Sydney: CCH.

Jensen, R. (1995) *Pilot Judgment and Crew Resource Management.* Aldershot: Ashgate.

Klein, G. (1998) *Sources of Power. How People Make Decisions.* Cambridge: MIT Press.

Macleod, N. (2005) *Building Safe Systems in Aviation. A CRM Developers Handbook.* Aldershot: Ashgate.

McAllister, B. (1997) *Crew Resource Management: Awareness, Cockpit Efficiency and Safety.* Shrewsbury: Airlife.

Munsterberg, H. (1913) *Psychology and Industrial Efficiency.* Boston: Houghton Mifflin.

Musson, D. and Helmreich, R. (2004) Team training and resource management in health care: Current issues and future directions. *Harvard Health Policy Review*, 5, 25–35.

NRC. Three Mile Island Special Inquiry Group (1980). Human Factors Evaluation of Control Room Design and Operator Performance at Three Mile Island-2, Volume 1 (Final Report NUREG/CR-1270-V-1). Washington, DC: US Department of Commerce.

Okray, R. and Lubnau, T. (2004) *Crew Resource Management for the Fire Service.* Tulsa: PennWell.

Reason, J. (1987). The Chernobyl errors. *Bulletin of the British Psychological Society*, 40, 201–206.

Reason, J. (1990) *Human Error.* Cambridge: Cambridge University Press.

Reason, J. (1997) *Managing the Risks of Organizational Accidents.* Aldershot: Ashgate.

Salas, E., Edens, E. and Bowers, C. (2001) (eds.) *Improving Teamwork in Organizations.* Mahwah, NJ: LEA.

Salas, E., Wilson, K., Burke, C., Wightman, D. and Howse, W. (2006) Crew resource management training research and practice: A review, lessons learned and needs. In R. Williges (ed.) *Review of Human Factors and Ergonomics, Volume 2.* Santa Monica, CA: Human Factors and Ergonomics Society.

Vincent, C. (2006) *Patient Safety.* London: Churchill Livingstone.

Wagenaar, W. and Groeneweg, J. (1987) Accidents at sea: multiple causes and impossible consequences. *International Journal of Man-Machine Studies*, 27, 587–598.

Walters, A. (2002) *Crew Resource Management is No Accident.* Wallingford, UK: Aries.

Weick, K. (1991) The vulnerable system: an analysis of the Tenerife air disaster. In P. Frost, L. Moore, M. Louis and C. Lundberg (eds.) *Reframing Organizational Culture.* London: Sage.

Wiener, E., Kanki, B. and Helmreich, R. (1993) (eds.) *Cockpit Resource Management.* San Diego: Academic Press.

Sources for Table 1.1 in chronological order of event

Three Mile Island – NRC. Three Mile Island Special Inquiry Group (1980). Human Factors Evaluation of Control Room Design and Operator Performance at Three Mile Island-2, Volume 1 (Final Report NUREG/CR-1270-V-1). Washington, DC: US Department of Commerce.

Chernobyl – International Nuclear Safety Advisory Group (1986) *Summary Report on the Post-Accident Review Meeting on the Chernobyl Accident.* Vienna: International Atomic Energy Authority.

Herald of Free Enterprise – Sheen (1987) *M.V. Herald of Free Enterprise. Report of Court No 8074. Department of Transport.* London: HMSO.

Hillsborough – Taylor. P. (1989) *The Hillsborough Stadium Disaster. Interim Report. Home Office.* London: HMSO.

Piper Alpha – Cullen, D. (1990) *The Public Inquiry into the Piper Alpha Disaster.* (Cm 1310). London: HMSO.

USS Vincennes – Fogarty, J. (1988) *Formal Investigation into the Circumstances Surrounding the Downing of Iran Air Flight 655 on 3rd July 1988.* Washington: Department of Defence. Released 1993.

Kegworth – AAIB (1990) *Report on the Accident to Boeing 737-400 G-OBME near Kegworth, Leicestershire on 8 January 1989. Aircraft Accident Report 4/90.* London: HMSO.

Scandinavian Star – Norwegian Public Reports (1981) *The Scandinavian Star Disaster of 7 April 1990.* Oslo: Ministry of Justice and Police (English translation).

Betsy Lehman – Doctor's orders killed cancer patient: Dana-Farber admits drug overdose caused death of Globe columnist, damage to second woman. *The Boston Globe.* March 23, 1995.

Channel Tunnel – Channel Tunnel Safety Authority (1997) *Inquiry into the Fire of Heavy Goods Vehicle Shuttle 7539 on 18 November 1996.* London:CTSA.

Esso Longford – Esso Longford Gas Plant Accident: Report of the Longford Royal Commission: http://www.vgrs.vic.gov.au/public/longford.htm. Hopkins, A. (2000) *Lessons from Longford.* Sydney: CCH.

Graham Reeves – *The Times, January 13th 2004.* Royal College of Surgeons of England Report.

Wayne Jowett – Toft, B. (2001) *External Enquiry into the Adverse Incident that Occurred at the Queen's Medical Centre, Nottingham 4th January 2001.* London: Department of Health.

BP Texas City – Investigation Report Refinery Explosion and Fire. (March 2007) Washington: US Chemical Safety and Hazard Investigation Board.

Chapter 2

Situation Awareness

Introduction

Situation awareness can be explained simply as 'knowing what is going on around you'. It is the first of the cognitive skills that we will be considering, the other being decision-making.

The most common definition of situation awareness is that provided by Endsley (1995a: p36): 'the perception of the elements in the environment within a volume of time and space, the comprehension of their meaning and the projection of their status in the near future.' This concept has been used extensively in aviation training and research, and a more specific definition of situation awareness for flight crew illustrates the key components of this skill: 'situation awareness is a dynamic, multifaceted construct that involves the maintenance and anticipation of critical task performance events. Crew members must also have temporal awareness, anticipating future events based on knowledge of both the past and the present. It is crucial that individuals monitor the environment so that potential problems can be corrected before they escalate' (Shrestha et al., 1995: p52).

The terms situation awareness and situation assessment are often used synonymously. Situation assessment is sometimes defined as a process by which situation awareness, a state, is achieved (Sarter and Woods, 1991). In this chapter, we define situation awareness as the cognitive processes for building and maintaining awareness of a workplace situation or event. In Chapter 3, we use the term situation assessment to describe the first stage of decision-making in operational settings when a specific, focused diagnosis of the current situation is made in order to take action. This is based on the ongoing situation awareness (i.e. monitoring) and the need for a situation assessment is usually triggered by a significant change in the task environment.

Situation awareness is essentially what psychologists call perception or attention. It is a continuous monitoring of the environment, noticing what is going on and detecting any changes in the environment. It also functions as the first stage in the decision-making process (see Chapter 3). A generic set of elements for situation awareness is shown in Table 2.1, followed by a discussion of the importance of situation awareness in workplace safety. The next section describes how the brain processes information and presents a model for understanding the stages of situation awareness, followed by influencing factors, then discusses the training, maintenance and assessment of situation awareness. For background reading on situation awareness, see Endsley and Garland (2000), Banbury and Tremblay (2004) and Tenney and Pugh (2006).

Table 2.1 Components of situation awareness

Category	Elements
Situation awareness	Gathering information Interpreting information Anticipating future states

Situation awareness and safety

The term situation awareness comes originally from the military; Gilson (1995) says it was evident in the First World War when the concept of 'gaining an awareness of the enemy before the enemy gained a similar awareness' first appeared. Most of the recent research on situation awareness has been developed in aviation settings, both military and commercial, but the term is now being widely adopted in other occupations, such as anaesthesia (Gaba et al., 1995), nuclear plant process control (Patrick and Belton, 2003) and emergency response (Okray and Lubnau, 2004). Increasing interest in attention skills has been driven to some degree by the rapid developments of computer-based monitoring systems, automated control process, intelligent systems and other technological advances that serve to distance humans from the systems they are operating. Thus the importance has been recognised of the human operator, e.g. the control room supervisor, the fire officer, the anaesthetist, having a good 'mental model' (picture in their head) representing the status of the current task and the surrounding work environment.

One does not have to look very far to find accidents where problems in situation awareness were implicated. Reviews of the nuclear accidents at Chernobyl in 1986 and Three Mile Island in 1979 concluded that operators were working on the wrong mental model of the system. Other examples include the captain of the 'roll on – roll off' car ferry *Herald of Free Enterprise*, continuing to sail, unaware that his bosun was asleep and had failed to close the bow doors; the police officers at Hillsborough football ground, thinking initially they were dealing with a riot rather than a crowd crushing disaster; and surgeons who operate on the wrong site.

In 175 aviation accidents, poor situation awareness was found to be the leading causal factor (Hartel et al., 1991). Endsley (1995b) reviewed major air carrier accidents from 1989–1992 and found that situation awareness was a major causal factor in 88% of accidents associated with human error. These include 'controlled flights into terrain' accidents where a fully functioning plane has been flown into the ground (usually a mountain), as the pilots thought they were in a different location or at a safer altitude. In December 1995, a Boeing 757 crashed into a mountain near Cali, Columbia, killing 159 people. Analyses of the voice recording showed that while trying to resolve an error resulting from entering an incorrect navigation code, the pilots had not maintained awareness of the position of their aircraft in relation to the mountainous landscape. Although a cockpit alarm warned of ground proximity, they were unable to climb the aircraft in time to avoid the mountain.

Another situation awareness problem in aviation is when pilots land their aircraft at the wrong airport. One notable incident was in 1995, when a *Northwest Airlines* DC-10 from Detroit, USA, bound for Frankfurt, Germany, landed instead in Brussels, Belgium. This incident was apparently instigated when an air traffic controller erroneously changed the flight plan and destination in the en-route computers. The subsequent interchanges between ATC and the flight deck must have missed several opportunities to correct the pilots' mistaken situational awareness. (Some passengers watching the flight map realised that they were not en route to Frankfurt but (as in the Kegworth crash, see Box 3.1), this information was not conveyed to the pilots.) The pilots saw that the airport layout was not as expected as they emerged from cloud on their approach and sensibly continued the landing (*International Herald Tribune*, 2 October 2005, p1). See also Dismukes et al. (2007) and Dekker (2004) for analysis of similar problems attributed to failures in pilots' situation awareness.

When 200 managers of offshore oil production platforms and drilling rigs in the North Sea were asked what they thought were the main causes of offshore accidents, the leading causal factors they mentioned were attentional problems. While they did not use the term 'situation awareness', the most common answers were failures in situation awareness: 'not thinking the job through', 'carelessness', 'inadequate planning', 'inadequate risk assessment' (O'Dea and Flin, 1998). Sometimes they described this as being 'out of the loop'. Interviews with offshore industry drilling managers revealed a similar recognition of the problem and an awareness of the factors influencing situation awareness. A review of 332 drilling accidents showed that 135 were attributed to problems with workers' attention (Sneddon et al., 2006).

Retrospective accounts by system operators when failures in situation awareness have occurred include comments such as:

'I didn't realise that…'
'I didn't notice that…'
'I wasn't aware that…'
'We were very surprised when…'
'I was so busy attending to…'
'We were convinced that…'

As can be seen from the elements in Table 2.1, this cognitive skill is primarily about picking up and processing information from the work environment and using stored memories to make sense of it. Therefore, before considering situation awareness in more detail, a basic model of the human brain's information processing system is presented, as this helps to explain both the strengths and the limitations of human memory for both situation awareness and decision-making.

Information processing in the human brain

The human brain functions as a very sophisticated information processing machine. While it is often likened to a computer, there are some tasks that a computer can do much more efficiently, e.g. rapid calculations, and others where the brain is by far the superior device, e.g. recognising faces. We gather information from the world around us with our five sensory systems: vision, hearing, touch, taste and smell. As there is too much information available in the environment at any one time for our brains to process, we attend selectively to some things rather than others. The selection is driven partly by the environment – e.g. a sudden noise or change in light will attract our attention – but we are also guided by past experience. That is, information we have stored in memory – our knowledge of the world we live in – will guide us to attend to certain cues in the environment, as we know they are in some way meaningful or important to us.

This selective attention process essentially forms the basis of situation awareness. Information from the environment enters our cognitive system as physical or chemical signals via the sensory receptor cells in the eye, ear, nose, mouth and skin. These become chemical messages that are transported through the sensory nerves to the brain where they are interpreted and may be stored in the memory system.

The memory system has been studied extensively by psychologists over the last 40 years. Research into memory capacity and function has led to the development of a widely accepted model of the cognitive architecture – the structure of the brain's storage and processing system for information. A simplified view of memory proposes that there are three linked systems: sensory memory, short-term or working memory and long-term memory, each of which is briefly described below.

Sensory memory

The sensory memory holds incoming information for very brief periods of time – for vision, the iconic memory retains the image for about half a second and for acoustic signals, the echoic store lasts for about two seconds (Eysenck and Keane, 2005). We are rarely aware of these transient stores but when you notice an image persist very briefly after the visual stimulus has been removed, this is sensory memory. We appear to have little conscious control of these stores, however the persistence effect allows us extra time to process incoming information.

Working memory

Of more significance to situation awareness and decision-making is the second memory store, which is called short-term, or now more commonly, working memory – see Figure 2.1.

Working memory essentially contains our conscious awareness. It is a limited capacity store, holding on average about seven 'bits' or chunks of information – as expressed in a famous paper 'The Magical Number Seven, Plus or Minus Two' (Miller, 1956). One's personal working memory capacity is easy to measure – just see how many digits you can remember from a random list of numbers. If you can combine them into meaningful units or chunks (e.g. telephone area dialling codes or dates), then a longer list can be remembered. Working memory not only has a small storage capacity but it is also not very good at holding onto the information. If we do not devote mental attention to the numbers in the list we are trying to remember, for instance by repeating them over and over again, then we will just forget them – the memory trace of these numbers will decay. So we are very susceptible to losing the information being held in this store. Moreover, if we are distracted and switch our attention to something else, then again we will just lose the information. For instance, if you are given a long telephone number to remember while you go and find a quiet spot to use your phone, you will have to rehearse the numbers while you are walking. If someone comes up and interrupts you by saying hello, then if your mental repetition stops, the new information from the interruption displaces the numbers from your working memory and they are immediately forgotten.

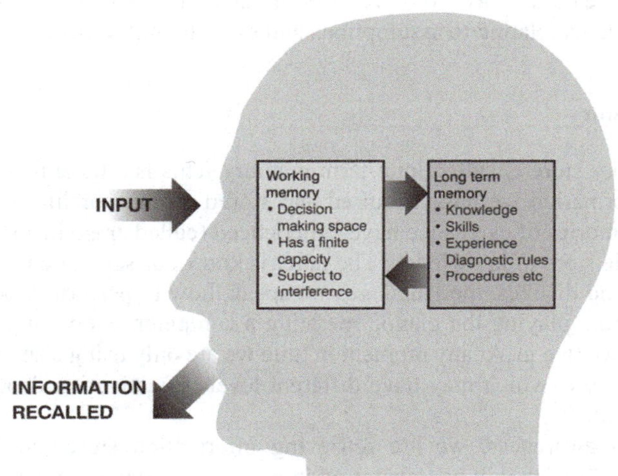

Figure 2.1 Model of working memory and long term memory

Our ability to maintain information in working memory is particularly important when engaged in safety-critical tasks. When a control room operator or an anaesthetist is working through a series of steps in a process or a pilot is remembering the clearance given on the radio by air traffic control, then they are using their working memory store to hold the information. They may well be using mental rehearsal to keep track of the procedures they have completed or the clearance level they have to remember. If they are distracted or interrupted while focusing their attention to hold this information, then they will probably forget which steps they have completed or the routing they were given by the control centre. In effect, the new information contained in the distraction erases the material that the working memory store was holding. Knowing when and how to preserve information in working memory is a key aspect of situation awareness. Distraction is discussed further below.

It is clear that prospective memory (remembering to do something in the future, i.e. to perform actions after a delay) is especially vulnerable to task interruption and distraction. An accident at Los Angeles International airport in 1991 occurred when an air traffic controller cleared an aircraft to hold in a take-off position and shortly afterwards directed another aircraft to land on the same runway, without clearing the first aircraft to take off beforehand. The controller forgot about the action required for the first plane because she had to switch her attention to other aircraft she was also managing (Loft et al., 2003). Dieckmann et al. (2006), in a simulator study with medical students, found that of 73 intentions to do something at a later time, 26% were missed in a busy, working situation.

With expertise, tasks (such as the physical skill of flying) become automatic – the procedures and actions become well known and are stored in the long-term memory. Therefore, this frees up our working memory to attend to other tasks, such as talk on the radio, address problems, as we do not have to attend to the well-practised tasks. However, when we do not have to think carefully about the steps in a task, this can cause other problems relating to assumptions and expectations, as discussed later.

Long-term memory

The main memory store is called long-term memory. This is a huge repository for all kinds of information we have acquired and stored during our life. It holds all our personal memories of events we have experienced (called episodic memory), as well as our whole store of knowledge. The latter is known as semantic memory and holds our likes and dislikes, the languages we speak, how to perform tasks such as driving, swimming, playing the piano, operating a computer, everything we know about the world we live in. At any moment in time we are only using a small fraction of this memory store, which may have different levels of activation depending on our current activities.

For situation awareness, we are retrieving information stored in long-term memory – some of which will be transferred into working memory, other associated knowledge may then move to a higher level of availability. Certain types of information are easier to retrieve from memory, for instance when the information is familiar, was accessed recently or is unusual, salient or of particular personal interest. For a more in-depth analysis of the role of memory in situation awareness, see Endsley and Garland (2000); Sarter and Woods (1991); Banbury and Tremblay (2004). More detailed explanations of the human perception and memory systems can be found in Baddeley (2004) or Eysenck and Keane (2005).

Model of situation awareness

There is an ongoing academic debate as to what 'situation awareness' actually means in terms of the mechanisms or models of the underpinning cognitive processes (Banbury and Tremblay, 2004; Dekker, 2004; Sarter and Woods, 1991). While this intellectual controversy fascinates the psychologists, practitioners are rather less intrigued. They appear to have embraced the term with enthusiasm, as it captures a cognitive skill that they immediately recognise and they appreciate its practical significance for their work.

The account presented here represents the best-known and most widely accepted model of situation awareness. It was developed by Mica Endsley, who was trained in both engineering and psychology. Her model of situation awareness that has three component stages or levels (see Figure 2.2) that correspond to the three skill elements shown earlier in Table 2.1.

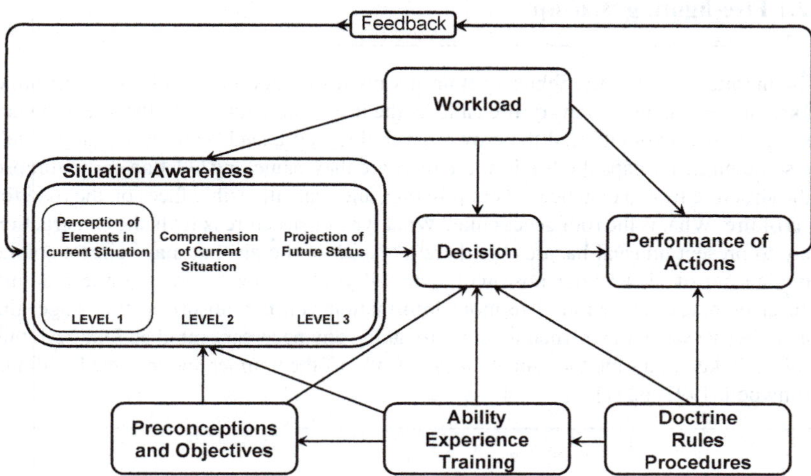

Figure 2.2 Model of situation awareness (adapted from Endsley, 1995, p35)

The model also illustrates that situation awareness influences the decisions regarding our actions in response to the situation. Several other factors can influence situation awareness, such as preconceptions and expectations, experience and task workload – these are discussed later in the chapter. Decisions and actions at work, as mentioned in Chapter 1, are influenced by organisational rules and cultural factors. Each of the three levels of situation awareness (sometimes described as 'What?', 'So what?', 'Now what?'), are discussed in the following sections.

Level 1: Gathering information

Endsley called this first stage 'perception of the elements in the current situation'. In a driving context, this would mean attending to the road conditions, the traffic, pedestrians, the speed of the car, etc. In the workplace this might be attending to visual information, such as instruments, computer screens, machinery state, weather conditions, co-workers' behaviour, as well as auditory information from alarms, conversation from co-workers, sounds of equipment, background noise. In some tasks there will also be attention to tactile information, a surgeon checking the condition of tissue or the tension of stitches, a mechanical technician checking heat, weight, pliability of material.

Essentially the worker is collecting information from the world around her to monitor the state of the work environment and progress on tasks she is engaged in. Officers in the emergency services rely heavily on these perceptual skills, as Box 2.1 illustrates.

Box 2.1 Fire-fighting 'size up'

> 'It is important for the firefighter to train his mind to "tune in and observe" essential features as he responds to every fire call. As the fire vehicle turns into the street to start looking for that hydrant, read the crowd psychology up ahead – are they trying to tell you something? Perhaps the fire is at the rear; are they panicking? Get an early glimpse of the structure from a distance, where possible, and scan all visible faces on the road for signs of fire. What is the roof access like? What type of structure is it? Is the construction likely to present unusual hazards? Is there a "haze" in the air that may suggest smoke is issuing out of view? Your nose will soon tell you!... your senses are finely tuning themselves ready for action. The more information you can absorb at this stage, the more effective you will be when it comes to taking any necessary rapid action... All this should be taken in during the time it takes to walk off the pumper and into the building.' (Grimwood, 1992: p241).

There are many reasons why we can fail to perceive the information that we need to correctly evaluate the situation. Endsley (1995a) lists the following errors relating to Level 1 situation awareness: gathering information:

- data were not available
- data were difficult to detect/perceive
- failure to scan or observe data
- misperception of data.

There are many examples of these types of problem – narrowing of attention is sometimes called tunnel vision or perceptual set or a fixation error. This refers to our attention becoming so focused on one element of the situation we fail to notice other cues, even if they are quite salient or important. Two oft-quoted examples are from aviation – these accidents were instrumental in the introduction of crew resource management training for pilots (see Kayten, 1993). In 1972 an Eastern Airlines L-1011 jet crashed in the Florida Everglades. The three flight-crew members became so fixated on trying to solve the puzzle of why a landing gear light had not illuminated, that they failed to notice that someone had inadvertently bumped the autopilot and turned off the altitude-hold. Six years later in 1978, a United Airlines DC-8 crashed at Portland, Oregon, after circling near the airport for an hour. While the crew tried to resolve a landing gear problem and prepare for an emergency landing, the aircraft ran out of fuel. In a medical setting anaesthetists can became so focused on trying to intubate a patient that they fail to notice that a critical amount of time has passed with the patient's brain being starved of oxygen. As shown above, our attention can be distracted from monitoring essential environmental cues because we are busy thinking about another task. The example below, Box 2.2, is from a police officer in command of a serious riot in London.

Box 2.2 Police officer's situation awareness

'We approached All Saints Road. This was a mistake because... we were suddenly confronted by a crowd of about 50 people who started attacking the car with missiles. I had been too preoccupied thinking how I was going to deal with the situation and what I was going to say to any of the "street leaders" to notice the route Sergeant Hole was taking' (Notting Hill riot, Moore, 2002: p74).

Psychologists have found it surprisingly easy to demonstrate the fallibility of human perceptual and memory systems. One attentional failure particularly relevant to dynamic environments is change blindness, where observers fail to notice that key elements have been altered between presentations of the same image. In a study of 'sustained attentional blindness' entitled 'Gorillas in our Midst', observers were asked to watch a basketball game (filmed in a laboratory) and to watch the team in white or the opposing team in black and to count the number of passes. During the game, a man in a gorilla suit walked across the game. Many of the observers failed to notice the gorilla at all, because they were so engrossed in watching the ball movements (Simons and Chabris, 1999). The same effect has been found with many professional groups including surgeons and pilots. The video has become so ingrained for some military pilots that they say to 'watch out for gorillas' during the pre-flight brief. The researchers who devised the gorilla study have also produced a range of video clips for demonstrating the frailties of visual perception (see websites at the end of this chapter).

We do seem to engage in some degree of self-monitoring of our current state of awareness or at least noticing when we have stopped gathering information relevant to the task in hand. If you have been reading this book and suddenly realised that you have not taken in any of the last three paragraphs, this reveals not only mind wandering but also some kind of monitoring function. Smallwood and Schooler (2006) have been studying mind wandering, sometimes called 'zoning out'. This is typically when our mind drifts away from the task in hand 'towards unrelated inner thoughts, fantasies, feelings and other musings' (p946). This could also be called daydreaming and is a very common experience. It is a form of distraction but attention has shifted to inner thoughts rather than a different external stimulus. To date mind wandering appears to have been mainly studied in the laboratory, but for jobs that can involve periods of vigilant monitoring such as control room operators or anaesthetists, then this would seem an area worthy of further investigation, given the safety implications.

Level 2: Interpreting the gathered information

At the second stage of situation awareness, 'comprehension of the current information', you have to process the incoming information to make sense of the current situation in order to understand what is going on and the significance of the cues that have been picked up. If you are driving and suddenly the cars in front brake, pedestrians are pausing to look at something, you hear a siren – these cues together will probably lead you to conclude that a traffic accident has occurred on the road ahead. The interpretation of the combination of cues is based on knowledge stored in long-term memory as to what patterns of information mean and signify in terms of response, e.g. brake lights coming on in the car in front means that you have to slow down or stop. Humans are very skilled at pattern matching, so this process can happen very quickly and with little conscious processing (i.e. limited use of working memory), so it feels automatic to the expert. But our interpretation of information can also be easily distorted by prior information, context and other factors. There are many examples of susceptibility to visual illusions (see Eysenck and Keane, 2005) and these are easy to demonstrate when teaching this aspect of perception.

In the work environment, experienced operators quickly learn to recognise and comprehend streams of information from control panels, equipment read-outs, radar plots, monitoring machines, anatomical conditions or financial spreadsheets. This process of categorisation and comprehension is facilitated by what are called 'mental models', that is, knowledge structures (sometimes called 'schema'), stored in memory, that represent particular combinations of cues and their meaning. This could be a mental map of a particular town or an airport runway layout, a plan of an incident command structure, or the structure of a device (e.g. component parts and related functions of a cooling system), a sequence of task activities or a set of behavioural responses (e.g. what normally happens at supermarket checkouts). These schema are generalised prototypes of classes of things or events rather than specific and detailed representations of every single previous encounter. The example below in Box 2.3 shows how various cues in combination are used by an anaesthetist to recognise a problem – i.e. make a diagnosis.

Box 2.3 Recognising a problem in anaesthesia

In a simulated case, a 19-year-old patient was having arthroscopy of the knee under general anaesthesia. On induction of the anaesthesia there was some difficulty opening his mouth (a subtle cue), but this caused no difficulty in placing a breathing tube in his trachea. Later the heart rate increased and did not respond to normal interventions. Expired carbon dioxide level increased slightly, as did body temperature. The anaesthetist made a presumptive diagnosis of malignant hyperthermia, a lethal metabolic abnormality that is triggered by anaesthetics. He obtained venous and arterial blood gas samples, which confirmed the diagnosis and then called for help to begin aggressive treatment of the crisis. Early treatment with a special drug resolved the acute emergency before it could become life threatening. (Gaba et al., 1995: p24)

Novices have fewer and less rich mental models and therefore have to spend more time and mental energy trying to comprehend patterns of cues by a process of systematic analysis and comparison with possible interpretations. Experienced practitioners faced with novel situations also have to interpret situations by using significant mental effort (a process that places a high load on working memory).

The resulting mental model produces expectations about the characteristics of a given situation. When the individual has a store of mental models for the domain they work in, a situation does not need to be an exact match to a previous situation to be recognised – but needs to present enough features for a categorisation into type to be made. Endsley (1997) explains that these mental models provide: 'a) for the dynamic direction of attention to critical cues, b) expectations regarding future states of the environment (including what to expect as well as what not to expect) based on the projection mechanisms of the model [see below] and c) a direct single step link between recognized situation classifications and typical actions' (p275).

The friendly fire incident in northern Iraq in 1994, when two American fighter planes accidentally shot down two US Black Hawk helicopters carrying 26 peacekeepers, illustrates the effects of expectations on perception. According to Snook (2000), 'There is little doubt that what the F-15 pilots expected to see during their visual pass [i.e. enemy Hind aircraft] influenced what they actually saw.' In the words of one of the pilots, 'I had no doubt when I looked at him that he was a Hind. The Black Hawk did not even cross my mind when I made that visual identification' (p80).

The mental model for a given task or situation is formed from not only experience but also from any briefing information that has been given. In many workplaces pre-job briefings are given prior to any tasks being started, so that those involved have a clear idea of what is to be done, who is to undertake each task, what the risks are and how they are to be minimised or managed. Of course, while this is normally a good idea, if the briefing is not accurate, then the wrong mental model may be activated, as a fire commander from London explained, in Box 2.4.

Box 2.4 Firefighters' expectations of the task ahead

> We were travelling to the scene of the call and our control had informed us that we were heading to a 'car alight' (a car that had been deliberately set on fire). As we approached the location discussing how we were going to respond, we began to notice some cues we had not been expecting – there were more people watching the event, there was much more smoke than normal, the scene appeared to be noisier than it should be for a vandalised car. We were considering what kind of 'car alight' this was and how we should tackle it, when we began to realise this was actually a totally different kind of incident – it was in fact a bomb explosion we were going to have to deal with.

The incorrect information had caused them to activate and start to operate on the wrong mental model; they then attempted to fit the incoming information from the scene into it. Emergency services control centres can only work on the information they are given by first responders or the public. If inaccurate, the resulting briefings to responding officers could produce erroneous expectations driven from the wrong mental model (Box 2.5).

Box 2.5 Police firearms incidents

Police firearms officers in London on 22 September 1999 were sent to deal with a report from a member of the public concerning an Irish man armed with a gun. When the police officers challenged this man, he responded in what they interpreted to be a life-threatening manner and they shot him dead. In fact the man, Harry Stanley, was Scottish and he was carrying a table leg (Independent Police Complaints Commission, 2006).

On 22 July 2005, a week after four suicide-bomb attacks in London by Muslim terrorists, police officers mistakenly shot to death an innocent Brazilian, Charles de Menezes, on an underground train, having been briefed that he was another suicide bomber (Crown Prosecution Service, 2006).

So while mental models are an efficient way of directing the selection and interpretation of new information, there can be problems if the wrong model is activated or created. One of the principal risks, as shown above, is confirmation bias, when the model is actually wrong for the present situation but incoming information is interpreted to match this model ('bending the facts' to fit). Disconfirming cues are ignored, rejected or argued away. Roger Green, an aviation psychologist, described a case where pilots flying a Boeing 747 into Nairobi in cloudy conditions did not catch the first number when the air traffic controller told them to descend to 7,500 feet. They knew 500 must be wrong for the altitude they were at but rather than check with the controller, they decided their clearance must have been to 5,000 feet. Nairobi is at 5,300 feet above sea level, so as Green says, 'at this point they were motoring in for a subterranean approach'. Even though an alarm sounded, only when the aircraft broke through the clouds and the pilots had new visual cues, did they suddenly realise their mistake and take corrective action (*BBC Horizon*, 1984).

It is not only pilots who can be somewhere other than where they think they are. Many other professionals can make similar mistakes; engineers and technicians can have accidents believing that they are working on a particular structure when in fact they are operating on an adjacent one. Again, in these cases, there may be bias to attend to information that appears to confirm the initial interpretation and discounting cues that do not match the assumed mental model. In medicine, gall bladder surgery is prone to mistakes in situation awareness, where the bile duct can be cut by mistake, potentially leading to liver failure (see Box 1.2). Way et al. (2003) analysed 252 laparoscopic bile duct injuries according to the principles of 'visual perception, judgment and human error' and found that the errors stemmed mainly from surgeons' misperception (i.e. poor situation awareness) rather than problems in skill or judgement.

Endsley (1995a) suggested that there could be several reasons why we can fail to comprehend the situation we are experiencing:

- lack of/poor mental model
- use of incorrect mental model
- over-reliance on default values in model
- memory failure.

With experience, the individual has more and richer mental models of situations and tasks that they have acquired over years of work. Part of this level 2, comprehension, stage also involves an assessment of the level of risk in the situation. In studies of pilots' risk perception, it was found that captains use different cues to make risk judgements compared with less experienced co-pilots (Orasanu and Fischer, 1997). Thomson et al. (2004) also found differences in 'accuracy' between situational risk assessments made by more and less experienced military pilots. As the experts have richer mental models and consequently know what cues are informative, this finding is hardly surprising but it does suggest that identifying the cues that experts use to make these judgements might help novices to make more rapid and more accurate risk assessments.

Level 3: Anticipating future states

The third level of situation awareness, called 'projection of future status', builds on the second level and means considering what might happen next. Having comprehension of the situation, understanding what it means, then using your stored knowledge from past experience, you are able to think ahead about how the situation is likely to develop in the immediate future. Sarter and Woods (1991) described this as 'mental simulation of future system state and behaviour to eliminate surprises. On the one hand, the resulting expectancies may facilitate perception because they can make it easier to remain vigilant and to adequately allocate attention. On the other hand, they involve the potential for ignoring or misinterpreting the unexpected' (p51). For example, if there is an accident ahead when you are driving, then you know that this is going to stop the traffic, so unless you can turn off quickly, you are probably going to be stuck in this queue, perhaps for an hour or longer. Defensive driving courses emphasise that you have to continually be alert for cues indicating threatening situations developing ahead of where you are now.

In dynamic work environments, where conditions are continually changing, this projection component of situation awareness is extremely critical, sometimes called 'being ahead of the curve'. Pilots talk of 'being ahead of the plane'. Apparently military pilots who fly the fast jets are taught, 'If you know where you are now, it's too late because you were there five miles ago.' For safe execution of hazardous tasks, workers must be able to think ahead of the current state.

Box 2.6 Offshore oil drilling: predicting the well

In offshore oil and gas exploration, especially when working in a new location, drilling teams are constantly reviewing the current conditions and trying to anticipate future conditions. 'When we were within reservoir, we came upon higher pressures than expected. The solution to our current mud weight calculation might be tied to source of pressure, and if we can understand that, then might be able to come up with a solution. Plus, we were also trying to predict ahead and see what this might mean, e.g. if this is the situation at 2,000m then what might be the implications when we get to 2,500m? Is it back to original plan or a completely new one?'

Surgeons place great store on situation awareness, especially what they call 'anticipation'. Table 2.2 gives sample statements related to each of the three levels of situation awareness, from interviews with consultant surgeons describing challenging cases (Flin et al., 2007).

Table 2.2 Examples of situation awareness in surgeons

Situation assessment	Consultant surgeons' statements
Perception (SA level 1)	'The bone was extremely soft.' 'The spinal fluid was still leaking out.' 'We took the swabs off to find that it was pouring blood.'
Comprehension (SA level 2)	'I thought that the most likely thing was that the patient was becoming coagulopathic' [blood coagulating]. 'It is two hours into the operation and what we do is review the x-rays and stop the operation.' 'Having thought about the situation, we decided…'
Projection (SA level 3)	'What we're trying to avoid is…' 'We decided that it would be technically very difficult to repair the hole in the lining of the spinal canal.'

Consultant surgeons teach their trainees that they must be able to predict the effects of particular actions and interventions. They must consider the consequences and risks with each step of a procedure and continually have back-up plans in place. Failures in situation awareness for trainees in any profession are likely to be due to the lack of or an inadequate mental model, or simply not thinking ahead of the present situation.

Jones and Endsley (1996) studied 143 aviation accidents to determine what level of situation awareness failure was implicated for pilots and air traffic controllers. They found that 78% of the accidents were problems relating to level 1, not having the information that was needed to make sense of the situation. Further, the most common level 1 error is a failure of scan (35% of all SA errors; Jones and Endsley, 1996). Far fewer problems (17%) occurred when all the information had been gathered but was then misunderstood and only 5% were related to failures to think ahead when the situation had been correctly interpreted. Sneddon et al. (2006) studied reports from 135 accidents on offshore drilling rigs and also found that 67% were attributed to level 1 SA. (It could be argued that this failure is easier to identify in retrospect than failures in comprehension and projection.)

There are common 'clues' that can indicate that you or your co-worker or the whole team are possibly 'losing' the correct situation awareness (CAA, 2006; Okray and Lubnau, 2004):

- ambiguity – information from two or more sources does not agree
- fixation – focusing on one thing to the exclusion of everything else
- confusion – uncertainty or bafflement about a situation (often accompanied by anxiety or psychological discomfort)
- lack of required information
- failure to maintain critical tasks (e.g. flying the aircraft, monitoring fuel)
- failure to meet expected checkpoint or target
- failure to resolve discrepancies – contradictory data, personal conflicts
- a bad gut feeling that things are not quite right.

Factors affecting situation awareness

Situation awareness is essentially about level of concentration or attention, and there is good evidence about factors that can affect this cognitive skill, some of which are shown in Figure 2.2. Each person has a certain amount of capacity for picking up new information and maintaining mental awareness of it (the amount can change depending on conditions). This may be likened to the capacity of a vessel, such as a jug. The present information load is represented by the liquid; when the jug is not full the person can still manage to attend to incoming information (more liquid). But when the jug is full, no new liquid can be added, unless some of the existing liquid is removed or displaced. The ideal mental state in risky environments is for workers to have some spare capacity in case the information load they have to cope with suddenly rises. The type of image shown in Figure 2.3 is sometimes used in situation awareness training to represent this metaphor.

It is well known that both fatigue and stress can reduce the quality of situation awareness. When we are tired, our cognitive capacity and processing of new information can be reduced (i.e. smaller jug). Fatigue (Chapter 8) appears to reduce the capacity for attention, both in terms of attending to new cues in the environment and for holding information in conscious awareness. Accidents where fatigue has been implicated (such as ship groundings and collisions) are probably due mainly to the influence of fatigue on situation awareness. Stress (Chapter 7), often manifested as anxiety, has a similar detrimental effect, probably because the person is preoccupied with other problems or worries that are taking up attentional resources. Sneddon et al. (under review) found for offshore drill crews that levels of work stress influence workers' situation awareness and this in turn was related to unsafe behaviours. Certain chemicals may influence attentional capacity. Stimulants, typically caffeine, are often used to counteract the detrimental effects of fatigue. Levels of blood glucose may make the individual feel more alert or more sleepy, and thus can influence situation awareness.

As situation awareness is very dependent on working memory (as explained above), it is affected by distraction, interruption, stimulus overload. These can be highly risky for workers engaged in complex tasks. Distractions and interruptions occur surprisingly often in safety-critical domains, such as operating theatres (Healey et al., 2006) and when teams are resuscitating patients (Marsch et al. 2005). Therefore knowing not to distract or interrupt fellow workers who are trying to retain information in working memory is an important element of teamworking – as part of the element 'considering others'. In some hospitals, the nurses dispensing drugs are now wearing tabards printed with 'DISPENSING DRUGS. DO NOT INTERRUPT'. Distraction has been recognised as a serious threat to safety in aviation, and Dismukes et al. (1998) at NASA have been studying ways of coping with this for pilots. Novices are particularly likely to have their task performance disrupted by distraction. As a worker becomes more skilled, then he or she will be able to perform tasks in a much more automatic fashion, now less concentration is required to perform the task, i.e. the load on working memory has been reduced.

Figure 2.3 Jug representing total situation awareness capacity, liquid indicating information and current mental load

Training situation awareness

Training in situation awareness, for example as part of crew resource management courses (see Chapter 10), tends to relate mainly in demonstrating its importance in maintaining safe operations, to providing information on how the brain processes and stores information, explaining models of situation awareness and the factors influencing this. Tips may be provided on how to maintain higher attention in particular work situations or tasks. Several years ago, British Airways introduced an additional situation awareness training package for all their pilots.

There are training guidelines available (e.g. Prince, 2000; Endsley and Robertson, 2000), as well as other sources of training material (see CAA, 2006). There have been few robust studies conducted to test whether there are long-term differences in performance relating to situation awareness training. A recent European project, ESSAI (Enhanced Safety through Situation Awareness Integration in training), has developed instructional materials for airline pilots and has shown that these can help to enhance SA skills (Banbury et al., 2007). Gaba et al. (1995) proposed that the following aspects of situation awareness could be taught to anaesthetists:

- practice in scanning instruments and the operating theatre environment
- more use of checklists to ensure relevant data are not missed
- knowing how to allocate attention more effectively
- practice in multi-tasking
- training in pattern recognition and matching of cues to disease and fault conditions.

Endsley and Robertson (2000) advise training in task management, recognition of critical cues, development of comprehension, projection, planning, information seeking and self-checking activities. Practical training in situation awareness can also be conducted using simulation facilities where specific cues and events can be manipulated, along with workload and distracting conditions. The scenario may be stopped at any time to allow an assessment of the trainees' situation awareness followed by review and coaching. Saus et al. (2006) gave scenario-based training with a freeze technique and reflection, linked to the stages of situation awareness, to student police officers using a shooting simulator. They found that this enhanced both subjective and observer ratings of SA as well as shooting performance. Some of the situation awareness measurement tools, described below, can be used in such training sessions to provide formative evaluations and structured feedback.

Maintaining situation awareness

Given the importance of situation awareness in many safety-critical tasks, advice can be offered on how to minimise the risks of reduced SA:

a. Good briefing – so that the operator understands the nature and risks of a given task before starting to work on it. In some occupations (e.g. flight crews, power plant crews, offshore oil installation crews), pre-job briefing, sometimes with formal risk assessment, is an accepted safety procedure.
b. Fitness for work – workers whose physical or mental fitness for duty is diminished (fatigue, drug addiction, infections, stress) are likely to have poorer situation awareness skills.
c. Minimising distraction and interruption during critical tasks – this can be controlled to some degree by an individual but generally requires co-operation from other team members. When distraction has occurred during a task sequence, then it is suggested to 'back up' a few steps to ensure the task sequence has been maintained.
d. 'Sterile cockpit' – In the USA the Federal Aviation Administration introduced what is known as 'the sterile cockpit' rule. This prohibits crew members from performing non-essential duties or other activities while the aircraft is at a critical stage of the flight. (These are taxiing, taking off, initial climb, final approach and landing.) So the crew must not have conversations during this time that are not related to the flight operation. The sterile cockpit helps to reduce distraction and to maintain situation awareness. It is a rule that could be usefully transposed as guidance into other workplaces for critical phases

of operations. Walters (2002: p43), in Box 2.7, gives a good example of why sterile cockpit is so important.

Box 2.7 Non-sterile cockpit

> Six minutes before touchdown, an Eastern Airlines DC-9 with 82 on board was descending towards runway 36 at Charlotte in conditions of patchy fog. During the approach the crew discussed politics, used cars and the US economy. Two minutes before touchdown, the conversation switched to identification of a local amusement park which they had just passed and shortly after receiving their final clearance, the Captain remarked 'all we got to do is find the airport…'. Three seconds later, the aircraft crashed 3.3 miles short of the runway with 74 fatalities. The National Transportation Safety Board cited poor cockpit discipline as a cause.

e. Updating – Regularly comparing a mental model of the situation with real-world cues.
f. Monitoring – Be sensitive to self-monitoring cues that 'zoning out' is occurring or at least be aware of the conditions when this is likely to happen.
g. Speaking up – Encouraging staff to speak up and admit when they are not sure of the procedure/goal/next step. See Chapter 4 on assertiveness.
h. Time management – Avoiding the 'hurry-up syndrome' by early planning, preparation and resisting imposed time pressure. In a study at NASA (McElhatton and Drew, 1993) examined 31 aviation accidents and found that in 55% of them, the pilots were behind schedule. Rushing to complete tasks in a tight time-frame is not conducive to good situation awareness.

In a study based in nuclear power plant control room (Patrick and Belton, 2003), six areas were identified where improved skills would improve situation awareness:

- planning (e.g. check plant status before use)
- problem-solving (e.g. do not discount symptoms of a problem by hypothesising an indication problem)
- attention (e.g. overfocused on procedure at the expense of monitoring the plant)
- team co-ordination (e.g. review team activities regularly)
- knowledge (e.g. misinterpretation of displayed information)
- communication (e.g. verbalise future actions).

Situation awareness is also regarded as an essential skill for emergency services personnel, such as police officers and firefighters. Okray and Lubnau (2004: p68) say that periodically during an operation, the firefighter should ask questions such as: Am I aware of what is going on around me? Are things happening as they are supposed to be happening? If not, why not? If things should go wrong, what is the plan? Does the leader know all the answers to these questions? One of the UK fire brigades uses a similar prompt set (from Klein, 1999):

- What is the immediate goal of your team?
- What are you doing to support that goal?
- What are you worried about?
- What is the current problem, size and intentions?
- What do you think this situation will look like in __ minutes, and why?

As most workplaces are staffed by teams rather than individuals working in isolation, then it is important that teams working on co-operative tasks have some degree of shared situation awareness and shared mental models for the task. In essence this refers to all the team members 'singing from the same hymn sheet' or 'being on the same page'. That is, they have a shared understanding of the tasks in hand, they know who is responsible for what actions and they are aware of their shared goal. Reader et al. (2007) have found that teams in intensive care units do not always have a completely shared impression of the patients' condition in terms of future risks. It has become increasingly apparent that a lack of a shared mental model can contribute to adverse events (Wilson et al., 2007). Team mental models and communication are discussed in Chapter 4, along with suggestions on how to recognise when others are showing good situation awareness or when this is diminished – or they are working on a different mental model from the rest of the team.

There are also a number of issues relating to workplace conditions, instrument and equipment design and whether these can be enhanced to provide better environments for enhancing situation awareness. This is beyond the scope of this book; see Endsley (2003) for more information.

Assessing situation awareness

Situation awareness skills for a given profession can be assessed using workplace or simulator observations, often with the use of behavioural rating scales. The Situation Awareness Rating Scales (SARS) are ratings completed by an observer watching an operator perform a task (Bell and Lyon, 2000). As this is a cognitive skill, it is impossible to observe directly, so the judges have to attend to specific task actions and communications that indicate the individual is gathering information or developing an understanding of the situation or thinking ahead.

Most of the non-technical skills behavioural rating systems contain a category on situation awareness; again, observed behaviours are used to infer aspects of this skill. For example, in the Anaesthetists' Non-Technical Skills rating system (ANTS, Fletcher et al., 2004) some behaviours indicating skill in situation awareness are:

- conducts frequent scan of the environment
- increases frequency of monitoring in response to patient condition
- keeps ahead of the situation by giving fluids/drugs.

Behaviours indicative of poor situation awareness skill are:

- does not respond to changes in patient state
- responds to individual cues without confirmation
- does not consider potential problems associated with case.

These types of instruments are designed not to interfere with the task and particular examples from aviation and medicine are discussed further in chapter 11. A number of other techniques have been used to measure situation awareness by tracking eye movements or by assessing the individual's mental model of the situation at a given point in time (Stanton et al., 2005). These tend to be applied in studies of task performance in simulators where the proceedings can be stopped without any risk. In the Situation Awareness Global Assessment Technique (SAGAT: Endsley, 1988), the task is stopped and the operators are asked questions to determine their current knowledge of the task and related circumstances. For example, in simulated medical emergency management training, the scenario can be halted and team members individually interviewed as to the state of the patient and the current treatment plan (Flin and Maran, 2004). On occasion, these have been found to be so discrepant, that the trainer may ask, 'Are you all treating the same patient?'

In the Situation Awareness Rating Technique (SART: Taylor, 1990), a self-rating scale is used during a task to determine current level of attentional capacity and understanding of the present task and situation. There are also some newer measures that have been designed to assess workers' general levels of situation awareness by self-report questionnaires (Sneddon et al., under review; Wallace and Chen, 2005). Details of measurement techniques can be found in Stanton et al. (2005), Endsley and Garland (2000) or Tenney and Pew (2006). For specific domains, Wright et al. (2004) discuss measures for simulated medical tasks, Salmon et al. (2006) review measures suited to command and control environments, and a Eurocontrol report (2003) describes the development of measures for air traffic management.

Conclusion

Situation awareness is a term used in industry to describe the cognitive skills that involve selection and comprehension of information from the world around us to make sense of our work environment. Problems relating to situation awareness are commonly attributed as causes of accidents in dynamic task settings such as flying aircraft, piloting ships, operating control rooms, warfare, fire-fighting, policing and acute medicine. Research into this skill has enhanced our understanding of the factors influencing situation awareness and more emphasis is now placed on developing and assessing competence in situation awareness in high-risk occupations.

Key points

- A three-level model of situation awareness encompasses the basic elements of this skill category: gathering information, comprehension and anticipation.
- The limitations in situation awareness relate principally to the structure of the human memory system, especially the working memory component.

- Factors that affect situation awareness include stress, fatigue, expertise, workload and distraction.
- Both classroom and simulator training are used to develop knowledge and practice of situation awareness.
- A range of techniques are available to assess levels of individual situation awareness.

Key texts

Banbury, S. and Tremblay, S. (2004) (eds.) *Situation Awareness: A Cognitive Approach*. Aldershot: Ashgate.

CAA (2006) *Crew Resource Management (CRM) Training*. CAP 737. (2nd ed). Appendix 6 Situation Awareness. Appendix 4 Information Processing. Gatwick: Civil Aviation Authority. www.caa.org

Endsley, M. and Garland, D. (2000) (eds.) *Situation Awareness. Analysis and Measurement*. Mahwah, NJ: LEA.

Websites

Endsley's company: www.satechnologies.com

Papers from Royal Aeronautical Society conference on situation awareness: www.raes-hfg.com/xsitawar.htm, www.raes-hfg.com/xsitass.htm

ESSAI project – European project on situation awareness in aviation: www.nlr.nl/public/hosted-sites/essai/pages/reports.html

Visual Cognition video clips for training: www.viscog.com

References

Baddeley, A. (2004) *Your Memory: A User's Guide*. London: Carlton.

Banbury, S. and Tremblay, S. (2004) (eds.) *Situation Awareness: A Cognitive Approach*. Aldershot: Ashgate.

Banbury, S, Dudfield, H., Hormann, J. and Soll, H. (2007) FASA: Development and validation of a novel measure to assess the effectiveness of commercial airline pilot situation awareness training. *International Journal of Aviation Psychology*, 17, 131–152.

Bell, H. and Lyon, D. (2000) Using observer ratings to assess situation awareness. In M. Endsley and D. Garland (eds.) *Situation Awareness. Analysis and Measurement*. Mahwah, NJ: LEA.

CAA (2006) *Crew Resource Management (CRM) Training*. CAP 737. (2nd ed). Appendix 6 Situation Awareness. Appendix 4 Information Processing. Gatwick: Civil Aviation Authority. www.caa.org

Crown Prosecution Service (2006) *Charging decision on the fatal shooting of Jean Charles de Menezes*. Retrieved on 4 September 2006 from www.cps.gov.uk/news/pressreleases/146_06.html

Dekker, S. (2004) *Ten Questions about Human Error*. Mahwah, NJ: Lawrence Erlbaum.

Dieckmann, P., Reddersen, S., Wehner, T. and Rall, M. (2006) Prospective memory failures as an unexplored threat to patient safety. *Ergonomics*, 49, 526–543.

Dismukes, K., Berman, B. and Loukopoulos, L. (2007) *The Limits of Expertise. Rethinking Pilot Error and the Causes of Airline Accidents*. Aldershot: Ashgate.

Dismukes, K., Young, G. and Sumwalt, R. (1998) Cockpit interruptions and distractions. *ASRS Frontline*, 10, 1–8. www.asrs.src.nasa.gov/directline

Endsley, M. (1988) Design and evaluation for situation awareness enhancement. In *Proceedings of the Human Factors and Ergonomics Society 32nd Annual Meeting*. (Vol 1, pp97–101). Santa Monica: HFES.

Endsley, M. (1995a) Toward a theory of situation awareness in dynamic systems. *Human Factors*, 37, 32–64.

Endsley, M. (1995b) A taxonomy of situation awareness errors. In R. Fuller, N. Johnson and N. McDonald (eds.) *Human Factors in Aviation Operations*. Aldershot: Avebury.

Endsley, M. (1997) The role of situation awareness in naturalistic decision making. In C. Zsambok and G. Klein (eds.) *Naturalistic Decision Making*. Mahwah, NJ: Lawrence Erlbaum Associates.

Endsley, M. (2003) *Designing for Situation Awareness*. London: Taylor & Francis.

Endsley, M. and Garland, D. (2000) (eds.) *Situation Awareness. Analysis and Measurement*. Mahwah, NJ: LEA.

Endsley, M. and Robertson, M. (2000) Training for situation awareness in individuals and teams. In M. Endsley and D. Garland (eds.) *Situation Awareness. Analysis and Measurement*. Mahwah, NJ: LEA.

Eurocontrol (2003) *The Development of Situation Awareness Measures in ATM Systems*. Brussels: Eurocontrol. HSR/HSP-005-REP-01.

Eysenck, M. and Keane, M. (2005) *Cognitive Psychology: A Student's Handbook* (5th ed). Hove: Psychology Press.

Flin, R. and Maran, N. (2004) Identifying and training non-technical skills for teams in acute medicine. *Quality and Safety in Health Care*, 13 (Suppl 1), i80–i84.

Flin, R., Youngson, G. and Yule, S. (2007) How do surgeons make intra-operative decisions? *Quality and Safety in Health Care*, 16, 235–239.

Gaba, D., Howard, S. and Small, S. (1995) Situation awareness in anesthesiology. *Human Factors*, 37, 20–31.

Gilson, R. (1995) Situation awareness. *Human Factors*, 37, 3–4.

Grimwood, P. (1992) *Fog Attack. Firefighting Strategies and Tactics*. Redhill, Surrey: FMJ Publications.

Hartel, C., Smith, K. and Prince, C. (1991) Defining aircrew situation awareness: Searching for mishaps with meaning. In D. Jensen (ed.) *Proceedings of the 6th International Symposium on Aviation Psychology*. Columbus Ohio: OSU.

Healey, A., Sevdalis, N. and Vincent, C. (2006) Measuring intra-operative interference from distraction and interruption observed in the operating theatre. *Ergonomics*, 49, 589–604.

Independent Police Complaints Commission (2006). *Harry Stanley – IPCC Publishes Decision and Report.* Retrieved 4 September 2006 from www.ipcc.gov.uk/pr090206_stanley.htm

Jones, D. and Endsley, M. (1996) Sources of situation awareness errors in aviation. *Aviation, Space and Environmental Medicine*, 67, 507–512.

Kayten, P. (1993) The accident investigator's perspective. In E. Wiener, B. Kanki and R. Helmreich (eds.) *Cockpit Resource Management*. San Diego: Academic Press.

Klein, G. (1999) *Situation Calibration Card.* Decision Skills Training. Instructor Guide (NOO178-97D-1043). Fairborn, OH: Marine Corps Warfighting Laboratory.

Loft, S., Humphreys, M. and Neal, A. (2003) Prospective memory in air traffic controllers. In G. Edkins and P. Pfister (eds.) *Innovation and Consolidation in Aviation*. Aldershot: Ashgate.

McElhatton, J. and Drew, C. (1993) Time pressure as a causal factor in aviation accidents. In R. Jensen (ed.) *Proceedings of the Seventh International Symposium on Aviation Psychology*. Columbus, Ohio: Ohio State University Press.

Marsch, S., Tschan, F., Semmer, N., Spychiger, M., Breuer, M. and Hunziker, P. (2005) Unnecessary interruptions of cardiac massage during simulated cardiac arrests. *European Journal of Anaesthesiology*, 22, 831–833.

Miller, G. (1956) The magical number seven, plus or minus two: Some limits on our capacity for processing information. *Psychological Review*, 63, 81–97.

Moore, T. (2002) Police commander. In R. Flin and K. Arbuthnot (eds.) *Incident Command: Tales from the Hot Seat*. Aldershot: Ashgate.

O'Dea, A. and Flin, R. (1998) Site management and safety leadership in the offshore oil and gas industry. *Safety Science*, 37, 39–57.

Okray, R. and Lubnau, T. (2004) *Crew Resource Management for the Fire Service*. Tulsa, Oklahoma: PennWell.

Orasanu, J. and Fischer, U. (1997) Finding decisions in natural environments: The view from the cockpit. In C. Zsambok and G. Klein (eds.) *Naturalistic Decision Making*. Mahwah, New Jersey: LEA.

Patrick, J. and Belton, S. (2003) What's going on? *Nuclear Engineering International*, January, 36–40.

Prince, C. (2000) *Guidelines for Situation Awareness Training*. Report to FAA. Available on www.crm-devel.com

Reader, T., Flin, R., Mearns, K. and Cuthbertson, B. (2007) Team cognition in the intensive care unit: an investigation of team situation awareness. Paper presented at the Society for Industrial and Organizational Psychology, New York, April.

Salmon, P., Stanton, N., Walker, G. and Green, D. (2006) Situation awareness measurement: A review of applicability for C4i environments. *Applied Ergonomics*, 37, 225–238.

Sarter, N. and Woods, D. (1991) Situation awareness: a critical but ill-defined phenomenon. *International Journal of Aviation Psychology*, 1, 45–57.

Saus, E., Johnson, B., Eid, J., Riisem, P., Andersen, R. and Thayer, J. (2006) The effect of brief situational awareness training in a police shooting simulator: an experimental study. *Military Psychology*, 18, S3–S21.

Shrestha, L., Prince, C., Baker, D. and Salas E. (1995) Understanding situation awareness: concepts, methods, training. *Human Technology Interactions in Complex Systems*, 7, 45–83.

Simons, D. and Chabris, C. (1999) Gorillas in our midst: sustained inattentional blindness for dynamic events. *Perception*, 28, 1059–1074.

Smallwood, J. and Schooler, J. (2006) The restless mind. *Psychological Bulletin*, 132, 946–958.

Sneddon, A., Mearns, K. and Flin, R. (2006) Situation awareness in offshore drill crews. *Cognition, Technology and Work*, 8, 255–267.

Sneddon, A., Mearns, K. and Flin, R. (under review) Workplace situation awareness.

Snook, S. (2000) *Friendly Fire: The Accidental Shootdown of US Black Hawks Over Northern Iraq*. New Jersey: Princeton University Press.

Stanton, N., Chambers, P. and Piggott, J. (2001) Situation awareness and safety. *Safety Science*, 39, 189–204.

Stanton, N., Salmon, P., Walker, G., Baber, C. and Jenkins, D. (2005) *Human Factors Methods*. Aldershot: Ashgate.

Taylor, R. (1990) Situation awareness rating technique (SART): The development of a tool for aircrew systems design. In *Situational Awareness in Aerospace Operations*. Neuilly sur Seine: NATO-AGARD.

Tenney, Y. and Pew, R. (2006) Situation awareness catches on: What? So what? Now what? In R. Williges (ed.) *Review of Human Factors and Ergonomics Vol. 2*. San Diego: Human Factors and Ergonomics Society.

Thomson, M., Onkal, D., Avcioglu, A. and Goodwin, P. (2004) Aviation risk perception: A comparison between experts and novices. *Risk Analysis*, 24, 1585–1595.

Wallace, J. and Chen, G. (2005) Development and validation of a work-specific measure of cognitive failure: implications for occupational safety. *Journal of Occupational and Organizational Psychology*, 78, 615–632.

Walters, J. (2002) *Crew Resource Management is No Accident*. Wallingford, UK: Aries.

Way, L., Stewart, L., Gantert, W., Kingsway, L., Lee, C., Whang, K. and Hunter, J. (2003) Causes and prevention of laparoscopic bile duct injuries. *Annals of Surgery*, 237, 460–469.

Wilson, K., Salas, E., Priest, H. and Andrews, D. (2007) Errors in the heat of battle: Taking a closer look at shared cognition breakdowns through teamwork. *Human Factors*, 49, 243–256.

Wright, M., Taekman, J. and Endsley, M. (2004) Objective measures of situation awareness in a simulated medical environment. *Quality and Safety in Health Care, 13 (Suppl1)*, i65–i71.

Chapter 3

Decision-Making

Introduction

Decision-making can be defined as the process of reaching a judgement or choosing an option, sometimes called a course of action, to meet the needs of a given situation. In most operational work settings, there is a continuous cycle of monitoring and re-evaluating the task environment, then taking appropriate action. Decision-making does not just involve one method – different decision-making techniques may be used at different times, depending on circumstances. Conditions for decision-making can vary in relation to time pressure, task demands, feasibility of options and what level of constraint, support and resource exists for the decision-maker. Four main components of decision-making are shown in Table 3.1 below.

Table 3.1 Elements of decision-making

Category	Elements
Decision-making	Situation assessment/Defining problem Generating and considering one or more response options Selecting and implementing an option Outcome review

There is an extensive literature on decision-making from business management, economic, philosophical and military sources. This chapter draws almost exclusively on psychological research, especially relating to decision-making skills in higher-risk work settings. It focuses principally on decisions made at an operational, rather than at a tactical or strategic level, that is real-time decisions made by front-line staff.[1] We draw on a relatively new area of study, called naturalistic decision-making, where researchers examine how experienced practitioners make decisions in real-world settings. Using a basic model of decision-making, several different methods that people use to make operational decisions are outlined, followed by suggestions for measurement and training.

1 There are very useful accounts of flawed decision-making at a tactical level, such as the decisions to launch the space shuttles *Challenger* (Vaughan, 1996) and *Columbia* (Starbuck and Farjoun, 2005).

Decision-making and safety

Decision-making skills are important in most work domains, but are especially critical in high-risk settings when the individuals involved may be functioning under time pressure and stress, as Box 3.1 illustrates.

Box 3.1 Pilots shut down the wrong engine

On the evening of Sunday 8 January 1989, the flight deck crew of a British Midland Boeing 737-400 felt a strong vibration while on a flight from Heathrow to Belfast. A burning smell and fumes were present in the passenger cabin, which led the pilots to conclude that there was a problem with the right engine. (There was a connection between the cabin air conditioning system and the right engine on the earlier types of B737 that they had flown but this had been modified on the 737-400 to the left engine, and the pilots were probably unaware of this change.)

As the aircraft was levelled out while the crew assessed the problem, the auto-throttle reduced power on both engines and the vibration stopped. However, this happened to be coincidental, as the problem was actually located in the left engine, which had suffered a catastrophic turbine failure.

Believing that the problem had been dealt with, the captain instructed the co-pilot to shut down the right engine and they made preparations for an emergency landing at their home base, East Midlands airport. The pilots tried to review which engine had the problem but were interrupted several times by air traffic control and by having to reprogram the flight computer for the diverted landing. Their attempt to verify the location of the problem remained unfinished. The severely damaged left engine failed completely as they increased thrust on final approach into East Midlands airport and with neither engine running, they crashed onto the M1 motorway near Kegworth. From 8 crew and 118 passengers, there were 47 fatalities and 79 people injured.
(UK Air Accident Investigation Board, 1990).

An analysis of aircraft accidents in the USA between 1983 and 1987 revealed that poor crew judgement and decision-making were contributory causes in 47% of cases (NTSB, 1991). The availability of data from cockpit voice recordings and simulator observations has resulted in pilots' decision errors and the consequences being particularly well documented and analysed (FAA, 1991; Jensen, 1995). But problems with decision-making are not unique to pilots. Across all high-risk occupations, there are abundant examples of the importance of decision-making skills for safe task performance (Flin et al., 1997; Montgomery et al., 2005; Zsambok and Klein, 1997).

Notable failures in decision-making leading to multiple fatalities include:

- indecisiveness by police incident commanders at Hillsborough football ground in a crowd emergency (Taylor, 1990)

- delayed decision by the installation manager to evacuate the burning Piper Alpha oil platform, compounded by delayed decisions to shut off oil production by managers on two linked platforms (Cullen, 1990)
- the decision by the USS Vincennes commander to shoot down what he thought was an incoming missile (but it was an Iranian passenger airliner) (Fogarty, 1988; Rogers and Rogers, 1992)
- the decision by escaping firefighters in a wildfire at South Canyon not to drop their tools (Weick, 1996)
- the decision of the control room operators on the Chernobyl nuclear plant to continue testing a voltage generator when there was a low power level (Reason, 1997)
- the decision of two US pilots to shoot down what they thought were enemy helicopters (but were in fact US helicopters, carrying UN peacekeepers) (Snook, 2000).

Clearly it is very easy for others to be wise after the event and these actions would have made sense to the decision-makers when they were taken. Moreover, as explained in the first chapter, the likelihood of worker errors or poor decisions is influenced by technical, system and safety culture factors at the worksite.

The need to understand decision-making in safety-critical work settings has resulted in a new field of decision research, known as naturalistic decision-making.

Naturalistic decision-making (NDM)

Decision-making in time-pressured, risky, dynamic work environments has attracted the attention of psychologists specialising in the study of human performance. They discovered that classical (i.e. rational or normative) decision theory was of limited application to uncertain, time-pressured settings, where reaching a satisfactory solution to gain control of a problem tends to be the norm – as opposed to trying to reach an optimal or perfect solution. Classical decision theory, as espoused by economists and business analysts, offers sophisticated mathematical techniques for choosing between options by identifying key features of each and then weighing them against the desired goal state, e.g. multi-attribute utility theory (see Connolly et al., 2000). Much of the underpinning research is laboratory based, uses static, well-defined problems with undergraduates undertaking the experiments, rather than expert decision-makers studied in their own work environments. While a range of sophisticated evaluation techniques have emerged from this research, many of the derived methods are too complex and time-consuming for operational staff to use in risky environments. However, the basic approach of systematically comparing options is still valuable and is discussed below under choice decision-making.

In the late 1980s, following a sequence of major incidents where poor decision-making was implicated, psychologists began to develop techniques to study expert decision-makers in their normal environments, such as flight decks or military command posts (Flin et al., 1997; Klein et al., 1993; Zsambok and Klein, 1997). This new approach to decision-making research was called naturalistic decision-making

(NDM). The aim of NDM researchers is to describe how experts make decisions under conditions of high uncertainty, inadequate information, shifting goals, high time pressure and risk, usually working with a team and subject to organisational constraints (Hoffman, 2006; Lipshitz et al., 2001; Montgomery et al., 2005; Salas and Klein, 2001). This naturalistic approach is of value in time-pressured workplaces other than aviation, emergency services and the military. It has also been applied to the study of decision-making in acute medicine (e.g. surgery, anaesthesia, casualty, intensive care) where uncertainty, sub-optimality and value-based judgements are common (Falzer, 2004; Flin et al., 2007a; Gaba, 1992). In hazardous industries, such as nuclear power generation, NDM has been used to analyse decisions of control room supervisors (Carvalho et al., 2005) and emergency managers (Crichton et al., 2005).

Model of decision-making

In operational work environments, a continuous cycle of monitoring the situation, assessing the state of events, then taking appropriate actions and re-evaluating the results is required. This is sometimes called dynamic decision-making and there are many models that have been produced to illustrate this process in different occupations, such as fire-fighting or anaesthesia (see Flin, 1996, for examples). The various models tend to portray the same underlying stages, and our general model of decision-making, based on the NDM approach, is given in Figure 3.1. This has been adapted from research with airline pilots (Orasanu, 1995, see Figure 3.2) and is applicable to a range of operational work settings. The model portrays a two-stage process: i) carry out a situation assessment and ii) use a decision method for choosing a course of action. In Orasanu's research, pilots were found to rely heavily on situation assessment, and then, depending on their estimate of available time and risk, they would use one or more methods of decision-making, such as intuition, rules and analysis of options. The idea is that the decision-making method is chosen to meet the demands of the situation (see Hammond, 1988).

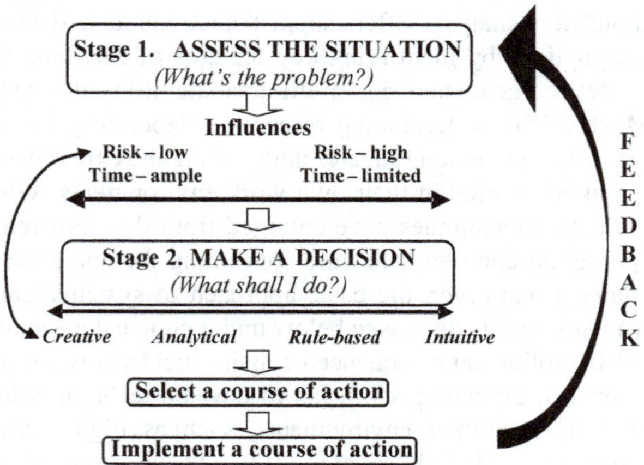

Figure 3.1 Simplified model of decision-making

So there are two stages in the cognitive process to reach a decision: i) What is the problem? ii) What shall I do? These are now discussed in turn.

i) Situation assessment – what is the problem?

Situation assessment is defined here as the process by which a focused survey and evaluation of the work environment takes place. Fire ground commanders call this 'size-up'. It is an attempt to make sense of the present situation, usually when there is recognition of a problem or change of state requiring a decision as to an appropriate response action. In some circumstances, the individuals will have been on task for some time, but in other events, they will arrive on-scene – such as a police officer attending an incident or a doctor called to help a colleague.

This first situation assessment stage is critical for decision-making. It differs from the continuous cognitive monitoring known as situation awareness (Chapter 2) where we attend to the environment and events around us. In the situation assessment phase of decision-making, a much more conscious effort is directed at identifying and understanding a new or altered situation. This step involves very similar perceptual and attentional processes to situation awareness and it is also influenced by factors such as expertise, workload and expectations.

As explained in Chapter 2, when workers become familiar with their job surroundings and the tasks to be undertaken, they begin to retain memories of typical events and their circumstances. As expertise develops, these become stored mental models that allow rapid pattern matching from memory to facilitate the situation awareness process. When continuous monitoring detects a significant change of state, then a more focused situation assessment takes place. Anaesthetists Rall and Gaba (2005) call this 'problem recognition'. It involves attending to a selection of available cues, assembling them into a pattern and searching the long-term memory store to achieve recognition of the problem. It may be that this assessment only reaches the conclusion that the situation is dangerous and that remedial actions need to be taken, without full diagnosis of the problem being possible. In the words of a police commander, 'There's 1,000 things happening, you're aware of 100, and you can only do something about 10. Thus the ability to perceive, understand and focus on a few key aspects of an unfolding situation is key to performance as an incident commander.' (Sarna, 2002: p40).

As Figure 3.1 illustrates, the first step of the decision-making process is diagnosing the current situation. At this point, the decision-maker, often with a team involved, builds a mental model to explain the situation encountered. In a study of surgical situation assessments, Dominguez et al. (2004) interviewed general surgeons who were shown a videotape of a patient undergoing a laparoscopic (key-hole surgery) gall bladder removal. The tape was stopped at three points and the surgeon was asked questions relating to the procedure, e.g. should it be converted to open surgery? They were also asked to rate their 'comfort level' with the operation at this point. While the tape was playing, they were asked to think aloud and to provide a continuous commentary on the events they were watching (as if they were doing the case or had been called to help). The results showed the importance of situation assessment in surgeons' decision-making – judging comfort level, assessing risks, predicting outcomes and also the role of self monitoring (e.g. 'I'll need to be careful here').

If the situation assessment is incorrect, then it is likely that the resulting decision and selected course of action that is taken in response will not be suitable. In the example in Box 3.1 of the Kegworth air crash, the pilots' situation assessment was that the right-hand engine was vibrating and generating smoke on the flight deck, when in fact the fire was in the engine on the left side. Their decision (chosen course of action) was correct for the problem they thought they had but was fatal for the actual situation. Similarly, as mentioned above, the commander of the US warship *USS Vincennes* ordered an incoming aircraft to be shot down after his team had mislabelled the tracks on their radar as being a hostile military aircraft. It was actually an Iranian passenger airliner carrying 290 passengers, none of whom survived. Again the response selected was wrong because the commander and his warfare team had not formed the correct situation assessment (Fogarty, 1988; Klein, 1998). Certainly for pilots, more decision errors occur because they misunderstand the situation, than cases where they correctly diagnose the situation and then make the wrong responses (Orasanu et al., 1993).

Faulty situation assessment arises from a number of factors, described in Chapter 2. Cues in the situation may be misinterpreted, misdiagnosed or ignored, resulting in an incorrect mental picture being formed of the problem. Alternatively, risk levels may be miscalculated or the amount of available time may be misjudged (Orasanu et al., 2001). In operational policing, officers are making key decisions in the opening minutes of an encounter and they have to estimate personal risk as well as situational risk at this time (Flin et al., 2007b). Decision failures can also occur when conditions change so insidiously that the operators do not update their situation assessments often enough. Anaesthetists Rall and Gaba (2005: p3027) discuss how a case in the operating theatre can 'go sour': 'A slow but steady and sustained blood loss in a child during surgery may result in few or subtle changes in haemodynamics for some time until a rapid decompensation occurs. If the weak signs of the developing problem were not detected, the ensuing catastrophe may seem to have occurred "suddenly".'

Likewise, in the offshore oil and gas drilling industry, the term 'the slippery slope' is used to describe this realisation that the current situation has altered to some extent from the expected situation and that remedial actions are required to return to the planned path. In the words of a drilling superintendent: 'Things start going down the slippery slope when you lose control and start doing things without thinking them through – do things or take actions which you then regret.' Similarly, from a drilling engineer: 'We realised too late that we had gone off the rails and that we had, no contingency plans for this situation – we were down the slippery slope.'

A team of psychologists at NASA lead by Orasanu has been studying airline pilots' decision-making. They have observed pilots flying in the simulator and have also examined reports of problem situations causing accidents and near-misses. Orasanu and Fischer (1997) have shown that the estimation of available time and level of risk during this situation assessment phase is critical, as this determines the type of decision method the pilot will then adopt. As the model in Figure 3.2 illustrates, where there is very little time and high risk, pilots use faster strategies, such as applying a known rule. When there is more time (even with variable risk), they may opt for a slower but more rigorous choice method to compare and evaluate alternative courses of action.

In terms of time estimation, studies from aviation (as well as accounts from emergency medicine and policing) indicate that experienced practitioners tend to be more accurate with estimates of available time than less experienced colleagues, the latter tending to underestimate this. Experts tend to also be aware of more strategies that they can use to 'buy time' in a problem situation.

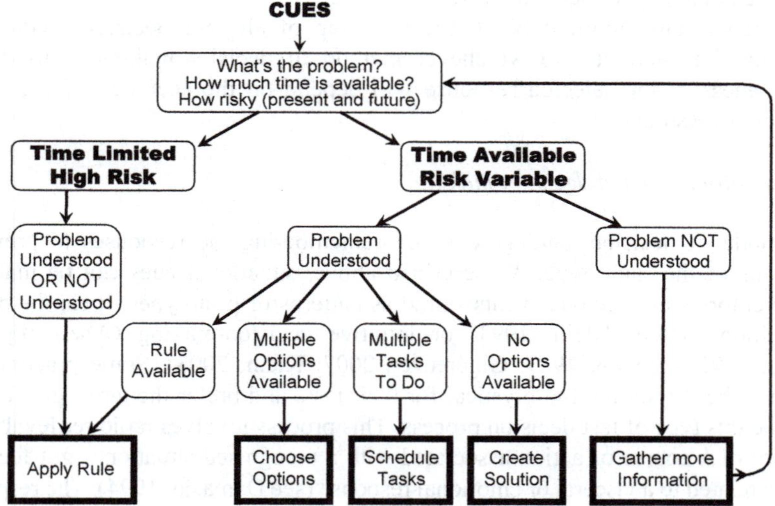

Figure 3.2 Airline pilots' decision-making (Orasanu, 1995 p1260)
Reproduced with permission from Proceedings of the Human Factors and Ergonomics Society.

The framework above underlies the earlier decision-making model shown in Figure 3.1, and the different decision methods, such as choose an option or apply a rule, are discussed in the next section.

ii) Decision-making methods – what shall I do?

The second stage of decision-making, as shown in Figures 3.1 and 3.2, is the process of choosing a course of action to meet the needs of the situation assessment. There appear to be four principal methods involved. These are:

a. recognition-primed (intuitive)
b. rule-based
c. choice through comparison of options
d. creative.

In the recognition-primed and rule-based methods, only one response option is considered at a time. In choice decision-making, several possible courses of action are generated then compared simultaneously. In the creative option, the situation is judged to be so unfamiliar that it requires a novel response. These four methods are considered in turn below. In some situations, doing nothing or waiting to see what happens may be the optimal course of action. As a trainee surgeon was advised, 'Don't just do something, stand there.' However, novices typically experience more stress and, as this appears to be relieved by taking action, they are less likely to wait and watch than experienced practitioners.

To repeat an important point, the first step of all four decision methods is assessing the situation. So whichever method of decision-making is used, the effectiveness of the selected response will depend on the accuracy of the initial situation assessment.

a) Recognition-primed decision-making

This mode of decision-making relies on remembering the responses to previous situations of the same type. As described above, situational cues can be matched with memories of previous events stored as patterns or prototypes. It can be called recognition-primed (Klein, 1993) or intuitive decision-making (Abernathy and Hamm, 1993; Claxton, 1997; Gigerenzer, 2007; Klein, 2003). Some practitioners find the label 'intuition' too mystical for their taste and prefer the term 'gut feel' to describe this type of fast decision process. This process involves rapid retrieval from memory of a course of action associated with a recognised situation – gut feelings may be related to a visceral or emotional response (see Damasio, 1994). The response is recalled so quickly that the appropriate course of action response for this situation has probably been stored along with the episodic memory of previous similar events. It could be a recollection of a rule or a personal/observed technique used in a previous encounter with this situation. In this case, choosing a course of action is likely to be experienced almost as an automatic process, with little conscious deliberation.

Klein (1993) developed a model of the recognition-primed decision-making (RPD) process. He was studying fire commanders to discover how they compared optional courses of actions in risky, time-pressured situations. The fire commanders 'argued that they were not "making choices", "considering alternatives" or "assessing probabilities". They saw themselves as acting and reacting on the basis of previous experience; they were generating, monitoring and modifying plans to meet the needs of situations. Rarely did the fire commanders contrast even two options. … Moreover, it appeared that a search for an optimal choice could stall the fire ground commanders long enough to lose the control of the situation' (p139). Klein realised that they were using a decision process that was founded on situation assessment and classification, and then rapid retrieval of a suitable course of action. Their aim was to find a satisfactory solution to control the situation rather than an optimal course of action. Once the commanders recognised they were dealing with a particular type of event, they usually knew the typical response to deal with it. They would quickly evaluate the feasibility of that course of action, imagining how they would implement it, to check whether anything important might go wrong. If they envisaged any problems, then the plan might be modified, but only if they rejected it would they consider another method.

The key features of recognition-primed decisions are:

- The emphasis is on reading the situation, rather than on generating different options for possible actions.
- The recalled response to the situation is based on past experience.
- The response generated, while it may not be the optimum, should result in a course of action that is workable and satisfactory.

The three versions of the RPD model are shown below in Figure 3.3.

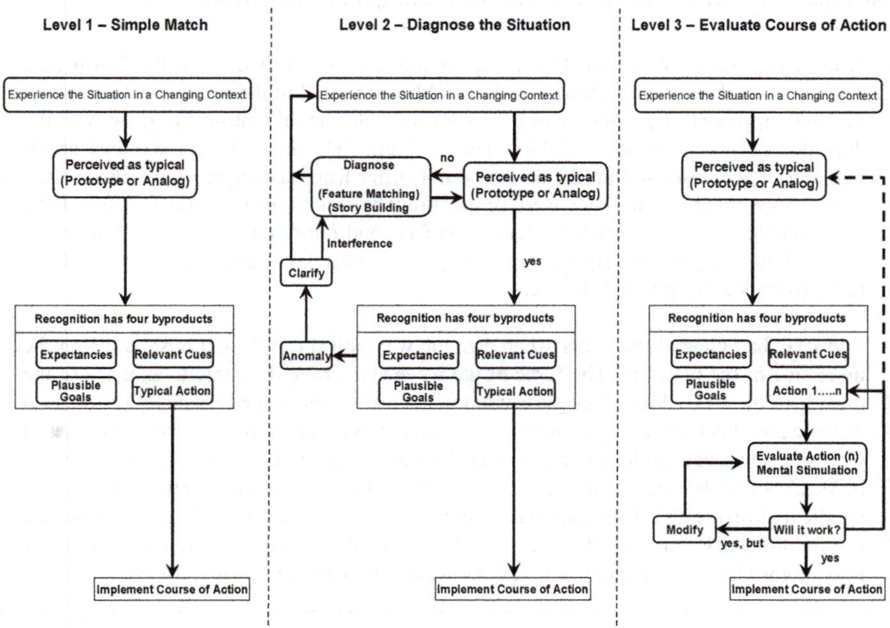

Figure 3.3 Recognition-primed decision-making (Klein, 1995)
Reprinted with permission of Klein Associates.

Klein (1998) explains that the RPD model 'fuses two processes: the way decision makers size up the situation to recognize which course of action makes sense, and the way they evaluate that course of action by imagining it' (p24).

The model has a basic format level 1 (simple match), where the decision-maker recognises the type of situation, knows the appropriate response and implements it. When faced with an unfamiliar or more complex situation (level 2 diagnose), then there will be a more pronounced situation assessment (diagnostic) phase. This may involve a simple feature match, where the decision-maker thinks of several possible interpretations of the situation and searches for cues in the environment that match one of them. In some cases, the decision-maker may find it necessary to make sense of the available cues by constructing a plausible explanation or story that would account for them. In the level 3 version (evaluation), the decision-maker is less confident of the course of action and so conducts a brief mental simulation to check for any possible problems that could arise from implementing it. Klein calls this 'preplaying a course of action', and if this does raise concerns, then an attempt will be made to modify or adapt this response before it is rejected. Only if it is rejected would a second course of action be considered. Hence the decision-maker is never comparing more than one option at a time and the process may feel very automatic or intuitive.

Box 3.2 Recognition-primed decision-making on a saturation dive

A team of three commercial divers was saturated at 300 feet deep in the North Sea. They had finished their work for the day and had returned to the pressurised personnel transfer capsule (PTC), which was then winched onto the oil platform to mate with the deck decompression chamber (DDC). The PTC and DDC are both kept at a pressure of 300 feet of sea water so that the divers do not suffer from decompression sickness (the bends), as would happen if they were simply brought back to sea level. Once in the DDC, the divers rest and sleep before entering the PTC and returning to work for another 12 hours. When the job is finished, the divers spend several days being decompressed back to a sea-level atmospheric pressure.

The PTC had cleared the water when a rogue wave hit the PTC and pushed it into the underside of the platform. The force of the wave also drove the camera, attached to the outside of the PTC, through its glass window. The air in the divers' specially pressurised capsule now began to escape through the broken glass. The diving supervisor on the oil platform saw what had happened, heard the escaping gas, realised that the divers would be dead before he could mate the PTC and the DDC, and immediately hit the winch handle to drop the PTC back to 300 feet below sea level. Fortunately, the divers managed to scramble back into their dive gear and breathing apparatus and were able to transfer into another PTC provided by a neighbouring oil platform (O'Connor, 2005).

There is now a considerable body of evidence that many professionals use RPD, as examples are found from firefighters, intensive care nurses, soldiers, pilots and offshore installation managers (see Klein, 1998). For anaesthetists, many problems require decisions to be made under uncertainty, 'with quick action to prevent a rapid cascade to a catastrophic adverse outcome. For these problems deriving a solution through formal deductive reasoning from "first principles" is too slow. In complex, dynamic domains the initial responses of experts to the majority of events stem from pre-compiled "rules" or "response plans" for dealing with a recognized event... once the event is identified, the response is well known.' (Gaba, 1994: p2654). While this type of decision-making can also be used where there is no time pressure, it is particularly appropriate for extreme situations requiring rapid, irreversible decisions in seconds or even milliseconds (Orasanu, 1997). For instance, when pilots have to decide whether to reject a take-off and stay on the runway (Harris and Khan, 2003) or police firearms officers have to decide to shoot a suspect or withhold their fire (Mitchell and Flin, in press). Gladwell (2005), in his book *Blink*, gives a number of examples of this type of rapid decision process in policing and other situations. In the case from saturation diving (Box 3.2), the recognition-primed decision (i.e. rapid situation assessment and immediate response) of a diving supervisor saved three men.

Klein also proposes two incremental versions of the basic RPD model. When the situation is unclear, the decision-maker relies on a story-building strategy to make sense of the available cues in the situation. When unsure if the selected course of action is appropriate, this can be mentally simulated – to imagine if it will work or if there are likely to be any problems (see Klein, 1998, for details). Decision errors, however, can occur with RPD if the situation is wrongly assessed and the matched rule for that incorrect situation is retrieved from memory and applied.

The advantages and disadvantages of RPD are summarised below.

Recognition-primed decisions	
Positive	Negative
• very fast • requires little conscious thought • can provide a satisfactory, workable option • is useful in routine situations • reasonably resistant to stress	• requires that the user be experienced • may be difficult to justify • can encourage looking only for evidence to support one's model, rather than considering evidence that may not support that model (confirmation bias)

In summary, the RPD method is useful for experienced practitioners working with relatively familiar situations, especially under high time pressure, and it may be more resistant to the effects of acute stress than other strategies. This is because it is making very limited use of working memory and is drawing mainly on long-term memory (see Chapter 2). Recognition-primed decision-making is unlikely to be used by novices, as by definition they have limited domain experience and thus possess fewer memories of relevant events. For them, almost every situation is unfamiliar and requires conscious deliberation and application of taught rules.

(b) Rule-based decision-making

Rule-based decision-making is often referred to as procedure-based decision-making, but here the term rule-based is adopted. The rule-based method involves identifying the situation encountered and remembering or looking up in a manual the rule or procedure that applies. This process can involve more conscious effort than recognition-primed decision-making. The individual is actively searching their memory store to recall the matching rule or physically consulting a procedures manual or a checklist to find the given response. High-risk industries are 'governed' by standard procedures and operating personnel are often required to consult the manual before taking an action. So for instance on an aircraft flight deck, the pilots would normally be expected to have a set of emergency procedures memorised, or consult the flight manual when a problem arises. To illustrate, here is the emergency procedure for what to do if there is a brake failure in a T-34 Mentor (the first propeller-driven training aircraft flown by US Navy student pilots). The student pilots must be able to carry out this procedure from memory.

1. Maintain directional control with rudder and the remaining brake. Use beta [changing the orientation of the propeller to provide reverse thrust] as necessary to aid deceleration.
If going into unprepared terrain is anticipated:
2. Condition Lever ----------------- FUEL OFF
3. T-Handle -------------------------- PULL
4. Canopy ---------------------------- EMERGENCY OPEN
5. Battery Switch ------------------- OFF
When aircraft comes to rest:
6. Harness ---------------------------- RELEASE
7. Parachute -------------------------- UNFASTENED
8. Evacuate Aircraft (www.bryanweatherup.com/gouge/Primary/C4001.html)

Rule-based decision-making is used extensively by novices who learn standard procedures for frequent or high-risk situations. Procedures are also useful for experts. For example, if pilots have memorised the procedure for dealing with an engine fire, then they do not need to think about what they have to do to shut down the engine and extinguish the fire. This permits them to react quickly to the situation, and allows them to think about other important factors, such as where to land. With practice, this becomes automatic and the rule can be retrieved from memory with little conscious effort, where it effectively becomes intuitive decision-making. This is the reason that trainee pilots devote a lot of time to learning and practising both standard and emergency procedures. Figure 3.4 shows a sequence of procedures, in the form of 'if, then' rules in a simple decisions tree, which were developed to guide offshore installation managers in the appropriate actions to take in a platform emergency.

The use of procedures is also helpful when justifying a particular decision after the event, as the protocol can be blamed if the response was incorrect. Over-reliance on rule-based decision-making may cause a degree of skill decay (Skriver and Flin, 1996). If an unexpected, unfamiliar situation arises and no rule exists, will the decision-maker be able to formulate a novel course of action? In industry, there is such a strong emphasis on procedures that organisations attempt to write a procedure for every eventuality. This can result in such voluminous collections of procedures manuals that few of the rules can be recalled or easily located in the event of an emergency. Larken, a retired Royal Navy Rear Admiral, comments that while procedures are very useful, 'I have doubts on the value of procedures in the spectrum of true crisis and emergency decision-making. In my experience serious decision-making starts when procedures run out, and they will, in my experience, run out very soon in an incident with "fire in its belly"' (2002: p118). Benefits and risks of rule-based decisions are shown below.

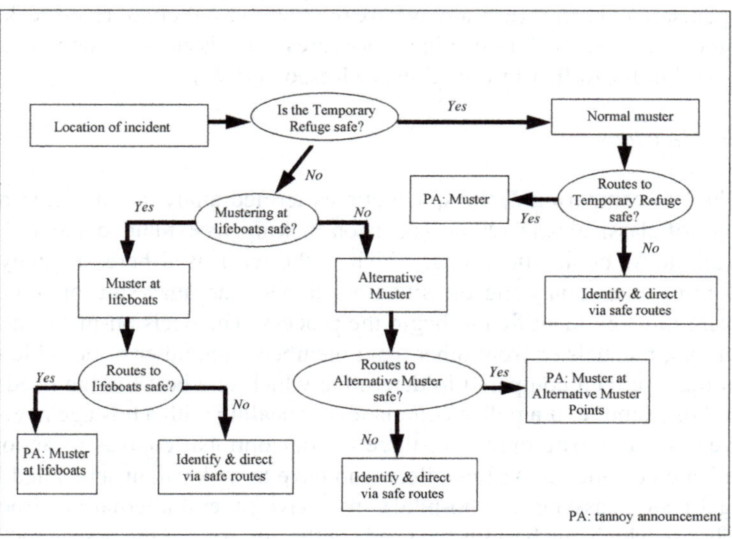

Figure 3.4 Steps for making decisions in an offshore emergency (Skriver, 1998)

Rule-based decisions	
Positive	Negative
• good for novices • can be rapid, if rule has been learned • gives a course of action that has been determined by experts • easy to justify as 'following prescribed procedures' • not necessary to understand the reason for each step	• can be time-consuming, if manual has to be consulted • cannot find written rule or procedure • if interrupted, it is easy to miss a step • rule may be out of date or inaccurate • may cause skill decay • may not understand the reason for each step • the wrong procedure may be selected

Decision errors can occur with this method if the wrong rule is selected (there is a particular risk of selecting a familiar one) to match the situation that the decision-maker finds themselves in. The rule may have been forgotten or the procedure difficult to locate in the manuals or on the computer. One of our researchers watched an exercise for a major accident scenario where a highly trained control room team spent 20 minutes systematically enacting a set of procedures that were completely wrong for the emergency they were trying to manage.

Accident analyses can reveal that an individual chose to violate a particular rule. Often organisational factors are at play here relating to a culture of rule-breaking and whether there are systems for updating procedures – this issue is beyond the scope of this chapter, but for further information see Reason (1997).

(c) Choice decisions

The method of comparing options, sometimes termed analytical decision-making, is the focus of classical research into decision-making, as explained above. Orasanu (1997) calls these choice decisions, which is the term used here, as analytical is often taken to mean analysing the situation, rather than analysing options. Again the problem must be identified to begin the process. The decision-maker generates from memory, manuals or from other team members, a number of possible courses of action that are then compared to determine which one best fits the needs of the situation. For example, if a police commander is dealing with a hostage negotiation, for a given situation, she may have three or four optional courses of action to be compared and considered. Airline pilots who have to make an unscheduled landing (e.g. to get a sick passenger to hospital), may have several alternative airports and have to choose which one best fits the needs of the aircraft and the passenger. Ideally, for this choice method, all the relevant features of the options should be identified and then weighed in terms of their match for the requirements of the situation. This is the kind of decision-making we often engage in when trying to select a holiday destination or choose a new laptop computer. There are a multitude of mathematical and statistical techniques to aid in selection of the optimal choice, such as decision trees, multi-attribute decision theory, Bayesian modelling. But these often require considerable time, mental effort and, for most people, the use of paper and pencil to work them out. The advantage of the choice decision method is that the chances of reaching an optimal solution are greatly enhanced when all possible alternatives have been carefully evaluated.

In practice, most decision-makers use a number of heuristics or shortcuts, such as simple calculations, to make comparisons between options and these are subject to a range of decision biases: picking the most familiar, choosing the most recently considered, selecting the first one to come to mind. Experienced decision-makers sometimes use what are called 'fast and frugal heuristics' (Gigerenzer and Goldstein, 1996) – rules of thumb for making rapid decisions.

Where there has been a pre-task analysis of options and possible contingencies, these can be considered and compared fairly rapidly in a demanding situation. With sufficient preparation, this might also allow the faster recognition-primed decision (RPD) method, described above. For example, Larken (2002) – describing naval commanders' decision-making in battle – says 'much can be gained by maintaining the strategy and the plans derived analytically and thinking through action contingencies ("what-if-ing") to be applied on an RPD basis' (p114).

In the oil industry, rules for choosing between options in the form of decision trees are typically generated prior to the start of an operation, which guide how the work should be carried out. As a drilling engineer comments, 'Decision trees help to deal with uncertainty in the operations during the planning stage – but when it comes to a situation during operations, we never actually use a decision tree. We know what we need to do based on our experiences but we do discuss as a team what these are. Thinking about issues beforehand helps, without needing to use a decision tree.'

Decision errors can occur with choice decisions when not all the relevant options are considered or the wrong choice is made as to the best course of action. Some other costs and benefits are shown below.

Choice decisions	
Positive	Negative
• fully compares alternative courses of action • can be justified • more likely to produce an optimal solution • techniques available	• requires time • not suited to noisy, distracting environments • can be affected by stress • may produce cognitive overload and 'stall' decision-maker

(d) Creative decision-making

While this is the basis of innovation, it is a method infrequently used in high time-pressure environments, as it requires devising a novel course of action for an unfamiliar situation. According to Orasanu and Fischer (1997), pilots rarely appear to use this method successfully during an in-flight problem. A famous example is the DC-10 (United Airlines flight 232) 'one chance-in-a-billion' centre engine failure that caused severing of the hydraulic pipes and consequent loss of all flight controls. The crew worked out a novel solution using differential engine thrust on the remaining two engines to regain some pitch and roll control and managed to crash land the aircraft on the runway at Sioux City airport, saving many lives (Haynes, 1992). Another occasion where an original course of action had to be generated in a high-risk situation is the Apollo 13 moon mission. An explosion shortly after take-off damaged the spacecraft's oxygen tanks. The mission control team in Houston, in conjunction with the astronauts on board, had to devise a series of solutions to a novel and escalating set of problems.[2] In a different in-flight emergency, a surgeon with the assistance of another doctor (both passengers on the flight), operated on a woman passenger whose lung had collapsed in transit from Hong Kong to London. The surgery to make an incision in her chest wall involved some creative decision-making, as it was conducted without anaesthetic, using a metal coat-hanger, an *Evian* water bottle, *Sellotape* [sticky tape], a bottle of brandy and the aircraft's first aid kit. The woman was in a life-threatening condition but thanks to their intervention made a good recovery (*The Independent*, 24 March 1995).

While this appears to be a decision method used infrequently and is not normally recommended for high-risk domains unless there is no alternative, some military

2 Well illustrated in the film *Apollo 13* starring Tom Hanks as Commander Jim Lovell.

commanders have argued that they have to use this type of decision-making during warfare. 'Creative decision-making would appear to be an approach needed when the exponent finds him/herself plus team completely outside the envelope of previous experience or anticipated out-turns. It is the operational journey without maps' (Larken, 2002: p115). Creative, like choice, decision-making should be undertaken during the planning and procedure development phases of operations management, when there is adequate time to design and evaluate novel courses of action. The positive and negative features of creative decisions are shown below.

Creative decisions	
Positive	Negative
• produces solution for unfamiliar problem • may invent new solution	• time-consuming • untested solution • difficult in noise and distraction • difficult under stress • may be difficult to justify

Table 3.2 Examples of surgeons' decision-making

Decision-making method	Consultant surgeons' statements
(i) Recognition-primed (intuitive)	'I am under extreme time pressure – there is no time to make decisions – the bleeding must be controlled rapidly and I have 20 minutes before the kidney dies. I tell the anaesthetist immediately as I find the source of bleeding and arrange for it to be clamped. I need to keep the good kidney alive so get some cold saline into the kidney.'
(ii) Rule-based	'If damage is occurring then you want to stop, especially according to clinical governance guidelines. Part of the expertise lies in doing but the other part is recognising when you are struggling and knowing that "first do no harm", so I decided to stop and get a second opinion.'
(iii) Choice	'There were three options to consider and at this stage we had to balance the potential risk of problems in the post-op phase with the risks of doing something intra-operatively.'
(iv) Creative	'None of the usual joints would work so we had to adapt a different one in order to make it fit.'

In dynamic work settings, decision-makers will typically switch between recognition-primed, rules and choice methods depending on the available time, familiarity and task demands. As Crosskerry (2005) notes, when comparing intuitive and analytical (choice) methods in relation to clinical decision-making, 'the trick is in matching the appropriate cognitive activity to the particular task' (pR1). To illustrate these different methods, some examples are given in Table 3.2 from interviews with consultant surgeons who were describing their decision-making during challenging operations (Flin et al., 2007a).

Summary of decision-making strategies

In summary, as Figure 3.1 illustrates, operational decision-making can be conceptualised as a two-stage cognitive process. The first step is situation assessment, that is, working out the nature of the problem. The second step is deciding what to do in response to this particular situation. The available evidence from professionals working in high-risk domains is that they make decisions in a number of different ways: recognition-primed, rule-based, choice and creative. The method, or methods, of decision-making that are ultimately employed will depend on a number of factors: Is the decision-maker an expert or a novice? Are there rules that should be applied? Is the situation familiar? How much time is available? What is the level of risk?

Having chosen a course of action, it should then be implemented. While this may sound rather obvious and seem to be hardly worth stating, it is a step that can be missed, especially under demanding conditions. A simulator tutor told us that she has watched trainees who decide what to do but then fail to take the necessary action, presumably because they become distracted, they forget or they are beginning to 'freeze' with stress (see Chapter 8).

The final part of good decision-making is to review the result of implementing the chosen course of action: Has there been an improvement in the situation? Is the response that was taken now solving the problem? When pilots are trained in decision-making, this is emphasised as a critical step in the process, checking that their decision is producing the desired result. In relation to the need to repeatedly review whether the chosen option or plan is working, Orasanu (2005) has found that airline pilots have a tendency to make what she calls plan-continuation errors. These are characterised by the pilots' reluctance to abandon their original course of action, even when the risks have significantly changed – for example unwillingness to divert from the chosen plan increases as they approach their destination.

Factors influencing decision-making

Competence in decision-making is significantly influenced by technical expertise, level of experience, familiarity with the situation and practice in responding to problem situations. As decision-making is a cognitive skill, it is affected by many of the same factors as situation awareness, namely stress, fatigue, noise, distraction and interruption. In stressful situations, decision-making may be particularly vulnerable, especially choice decisions where time and mental effort are required to evaluate and compare optional courses of action. The negative effects of acute stress on cognitive processes can be: overselective attention (tunnel vision), loss of working memory capacity, restrictions in retrieval from long-term memory, with simple retrieval strategies being favoured over more complex ones. Shifts in strategy, such as speed/accuracy trade-offs, can be observed with people under stress behaving as if they were working under time pressure, when in fact there is none (Orasanu, 1997).

Some psychologists argue that acute stress does not necessarily have a detrimental effect on decision-making, rather stress may affect the way information is processed, as outlined above. So of the four modes of decision-making described, stress has most impact on choice and creative methods, as these require heavy use of cognitive resources, such as the working memory space, and we know working memory capacity is reduced under stress. In contrast decision methods, such as recognition-primed decision-making, which are relatively light on cognitive processing, seem to be less affected by stress (Stokes et al., 1997). Similarly, if rules can be easily recalled, or checklists easily located, then this method will also generally function well in stressful conditions (Figure 3.5).

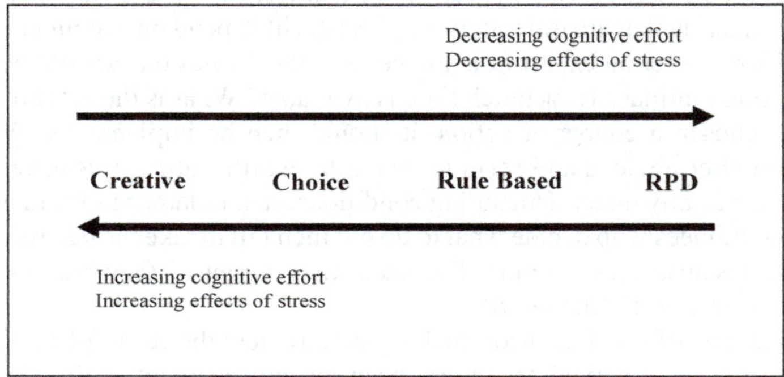

Figure 3.5 Relative effects of stress on decision-making method

Klein (1998) says that some of the changes to cognition that happen under stress are in fact adaptive. For instance, simple retrieval strategies from memory make less demand on already depleted cognitive resources. He also points out that decision-making in stressful conditions may also be due to inadequacy of available information and the distracting effects of starting to experience a stress reaction (see Chapter 8 for more information on stress).

Fatigue is another common condition that can influence the quality of decision-making. Even one night of sleep loss can impair flexibility, increase perseveration errors and ability to appreciate an updated situation (Harrison and Horne, 1999). In a recent Australian study, Petrilli et al. (2006) tested the decision-making skills of Boeing 747-400 flight crews who had just flown an aircraft into Sydney following an international pattern (non-rested) compared with crews who had just had four days' rest. The crews were required to fly an hour-long flight in the simulator during which a critical decision event occurred. They analysed performance in relation to situation awareness, options and planning, decision implementation and evaluation. Their results showed that the more fatigued crews took longer to implement decisions and were more risk averse. Interestingly the tired crews engaged in a number of protective behaviours, such as more cross-checking. This study shows that workers who are aware of conditions that can influence the quality of their decision-making can engage in compensatory behaviours.

Other factors that can influence the quality of decision-making relate to differences in personal style and social interactions with co-workers, discussed in the following chapters. Walters (2002: p15) describes a number of what he calls decision traps for pilots, and these are applicable in other domains:

- jumping to solutions
- not communicating
- being unwilling to challenge the experts
- complacency ('you worry too much')
- assuming you don't have the time
- failure to consult
- failure to review.

Awareness of the 'traps' for a given situation can be increased by analysis of incidents, post-event discussion and structured workplace or simulator observations with feedback. The various factors that can affect the quality of decision-making should be included within a decision-making training module.

Training decision-making

Traditionally the training of decision-making focused primarily on learning complex techniques to make optimal choices between options – focusing frequently on choice methods (Klein, 1998). As discussed above, while these methods can be very effective, they tend to be ill-suited to dynamic or high-risk work environments where practitioners have to take decisions quickly, in distracting conditions and often hampered by inadequate information and changing goals.

In some operational environments, practitioners are taught a process or sequence for decision-making. For instance, there are several acronyms used to teach pilots the basic steps for decision-making in flight. The best known of these is DODAR, according to Walters (2002), which has been used by *British Airways*:

D – Diagnosis	What is the problem?
O – Options	What are the options?
D – Decision	What are we going to do?
A – Assign the tasks	Who does what?
R – Review	What happened/what are we doing about it?

Another version, with the acronym FOR-DEC (Hörmann, 1995), is used by Lufthansa:

F – Facts
O – Options

R – Risks and benefits
D – Decision
E – Execution
C – Check

A feature of the FOR-DEC method is an inbuilt 'cycling' back to the start of the process as each step progresses to check that the situation has not changed.

Both of these step models emphasise the key phases of gathering information and reviewing the decision. They suggest generating options, while under severe time pressure this could be very difficult, and neither says anything about different methods of making a decision, although this may be taught along with the basic model. Klein (1998) has argued that the problem with all of these step models is that they imply decision-making is a linear process, when in fact it is very cyclical and interactive (see Figure 3.6).

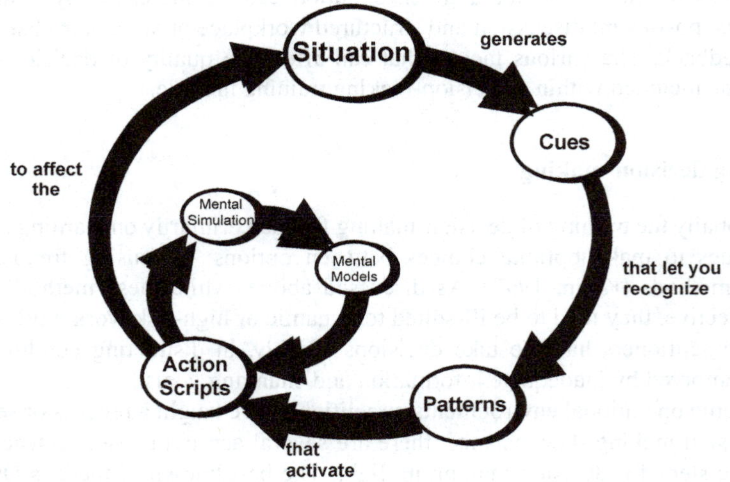

Figure 3.6 Cyclical nature of decision-making (Klein, 2003)
Source: 'Intuition At Work: Why Developing Your Gut Instincts Will Make You better At What You Do', by Gary Klein, Ph.D., copyright © 2003 by Gary Klein, Ph.D. Used by permission of Doubleday, a division of Random House, Inc.

Since the advent of the naturalistic decision-making (NDM) approach, there has been more interest in how to train decisions-makers who work in demanding, real-world environments. Psychologists working with the military have pointed out that the emphasis for cognitive skills training is shifting from 'what to think' to 'how to think'. Fallensen and Pounds (2001: p55) comment that 'prescriptive, formal models provide limited guidance on how to think in complex, ill defined situations' and that what is required is an understanding of natural reasoning processes and how to reinforce and extend natural thinking skills. One concept that practitioners find useful is 'metacognition', that is, thinking about thinking (Cohen et al., 1998) – awareness of your own thinking processes and strategies. For instance, operators can think if they are using the best decision-making strategy or if they should be switching to another method. Decision skills training under the NDM approach aims to train people to make better decisions by facilitating the development of expertise, rather than teaching generic decision-making strategies, and support, rather than attempting to replace the strategies people use (Cohen et al., 1998; Klein and Wolf, 1995; Pliske et al., 2001). The focus has been on developing cue and pattern recognition skills for improved situation assessment, guided decision-making practice in simulators and low-fidelity settings, and feedback on performance (Cannon-Bowers and Bell, 1997). Leaders who may have to make decisions in high-pressure situations should also be given structured practice in stressful situations (Useem et al., 2005).

A novel method of recording decision-makers at work was devised by Omodie et al. (1997), who had fire commanders wear head-mounted cameras during real operations and then used the videotape afterwards for discussion and debriefing. The tape recorded what the commander had been looking at, as well as an audio recording of decisions in commands and conversation. Such recordings can also provide valuable training materials for instruction using case studies. Two low-fidelity methods of training decision-making – tactical decision games and story-telling – are described next.

Tactical decision games

Simulators are not available for all work environments and other decision training techniques have been developed for computer-based or paper-based delivery. One particular method suitable for many professional groups is tactical decision games (TDG), initially devised by the US Marines (Schmitt, 1994). A TDG is a facilitated simulation using brief written scenarios, ranging in complexity, designed to exercise non-technical skills, especially decision-making. Participants in a TDG, usually four to ten individuals, are presented with the scenario and may be asked to take on assigned roles. The scenario (written by a subject-matter expert) describes a problem situation, but this may contain misleading or ambiguous information, or critical information may be missing. It culminates in a dilemma that must be resolved. Participants are asked to decide upon actions to be taken to contain or resolve the situation. Only a limited amount of time (e.g. 2–5 minutes) is allowed for each decision – so there can be a considerable degree of time pressure on the decision-makers. Each participant presents their own solution to the dilemma, then an open discussion is facilitated where differences and similarities between suggested courses of action are discussed. The facilitator can extend the exercise by introducing new information, requiring additional decisions, while participants are presenting their solutions. The TDG debrief focuses on why particular decisions were made, and the whole exercise provides a structured technique for sharing expertise across participants. TDGs have been used to train decision-making for high-risk domains, such as nuclear power plants, prisons and anaesthesia (see Crichton et al., 2002, for further details).

Story-telling

It could be argued that all that tactical decision games do is formalise the technique of story-telling, a practice that is ubiquitous across high-risk workplaces. In our experience, pilots, police officers, air traffic controllers, military officers, firefighters, scrub nurses, oilfield drillers, surgeons, anaesthetists, and so on, like nothing better than exchanging 'war stories' with each other. This happens in corridors, coffee rooms, during work and after work, and significant learning in these occupations is achieved by sharing experiences in this way. Professionals store these anecdotal accounts and use them to enrich their own memorised patterns of significant events for their work setting. Okray and Lubnau (2004: p240), discussing decision-making by firefighters, say, 'One of the most effective training programs that exists in every fire department is the story teller... These boring old firehouse stories that get told over and over again are a way of sharing our experiences with other firefighters... Through the accurate use of storytelling, we can increase our ability to respond to fire scenes.' Klein (1998), a master story-teller himself, has long argued for the power of stories to build expertise in decision-making. He advocates the use of stories as a useful form of vicarious experience for people who have not actually had these experiences; they help people to understand situations and make sense of events and to pick up lessons. Story-telling has been shown to be a valuable method of training firefighters (Joung et al., 2006).

Assessing decision-making skills

Assessing competence in professional decision skills can be carried out by written or oral examination, although these methods tend to tap stored knowledge and typically the test decisions are taken in non-operational conditions. Situational judgement questionnaires and interview methods (Motowildo et al., 1990) present candidates with a workplace problem situation and a number of alternative responses. They are designed to test whether candidates will decide to opt for the same actions as experts from their profession and are sometimes described as low-fidelity simulations.

The preferred method of assessing decision-making in many safety-critical domains is to use simulators, as found in aviation and an increasing number of other work settings (Lee, 2005; Riley, in press). In this case, specific events requiring decisions under different conditions can be written into the training or test scenarios. Observations of performance in the workplace can also be used, although this is not always practical. The cognitive process of decision-making is clearly impossible to observe directly, but where the decision-maker is required to communicate with others or work with a team, then there should be a number of behaviours that are indicative of the decisions being made and sometimes also the process of decision-making. In Chapter 11, methods for rating observed behaviours to assess non-technical skills, including decision-making, are described. To give one illustration, the category of *Decision-making* within the NOTECHS non-technical skills framework (Appendix 1) was based on information processing frameworks from aviation psychology (e.g.

Stokes et al., 1997), pilot decision-making models (e.g. Hörmann, 1995; Orasanu, 1993; see Figure 3.2), as well as input from experienced pilots.

Examples of good decision-making behaviours in NOTECHS are:

- gathers information and identifies problem
- reviews causal factors with other crew members
- considers and shares risks of alternative courses of action.

Examples of poor decision-making behaviours are:

- fails to diagnose problem
- does not search for information
- fails to inform crew of decision path being taken.

Finally, there are some psychometric questionnaires available that assess preferred styles of decision-making (e.g. Scott and Bruce, 1995; or *Myers-Briggs Type Indicator*, Myers et al., 1998); however, they reveal nothing of decision-making competence. Such instruments ask questions such as:

- Do you prefer to think carefully about all the options when making a decision?
- Are you comfortable if you have to make a decision quickly?
- Do you rely on gut feel?
- Do you like to consult others before you reach a decision?

These can be used in training to provide some feedback to individuals on preferred decision-making approaches and to discuss differences across a team.

Conclusion

Decision-making is a critical skill in most workplaces but is especially important in higher-risk settings. It is normally triggered by some change of state in the work environment that is detected by ongoing situation awareness (discussed in Chapter 2). The basic steps of decision-making – situation assessment, then selecting a course of action and reviewing it – were outlined. Four methods of reaching a decision were described: recognition-primed, rule-based, choice and creative. The appropriateness of a given method depends on the level of expertise of the decision-maker and situational constraints. Decision-makers rarely work alone in safety-critical workplaces and the issue of team decision-making is considered in Chapter 5.

Key points

- Decision-making in operational environments is usually a two-stage process: situation assessment and selecting a course of action.
- There are different methods of selecting a course of action: recognition-primed, rule-based, choice or creative.

- Decision-making can be influenced by a number of factors, such as stress and fatigue.
- Training for decision-makers should focus on strengthening situation assessment skills, such as cue recognition, as well as practice in deciding what action to take.
- Decision training can be based on a number of methods, with increasing use of both low- and high-fidelity simulation.

Websites

Klein Associates: www.decisionmaking.com

Cognitive Engineering and Decision-making Technical Group of Human Factors and Ergonomics Society (naturalistic decision-making): http://cedm.hfes.org

Society for Judgment and Decision-making (study of normative, descriptive and prescriptive theories of decision-making): www.sjdm.org

Key texts

Cannon-Bowers, J. and Salas, E. (1998) (eds.) *Making Decisions Under Stress.* Washington: APA Books.

Flin, R. (1996) *Sitting in the Hot Seat: Leaders and Teams for Critical Incident management.* Chichester: Wiley.

Flin, R., Salas, E., Strub, M. and Martin, L. (1997) (eds.) *Decision-making under Stress.* Aldershot: Ashgate.

Klein, G. (1998) *Sources of Power. How People Make Decisions.* Cambridge: MIT Press.

Klein, G. (2003) *Intuition at Work.* New York: Doubleday.

Montgomery, H., Lipshitz, R. and Brehmer, B. (2005) (eds.) *How Professionals Make Decisions.* Mahwah, NJ: Lawrence Erlbaum.

References

AAIB (1990) *Report on the Accident to Boeing 737-400 G-OBME near Kegworth, Leicestershire on 8th January 1989.* Aircraft Accident Report 4/90. London: HMSO.

Abernathy, C. and Hamm, R. (1993) *Surgical Intuition.* Philadelphia: Belfus.

Cannon-Bowers, J. and Bell, H. (1997) Training decision makers for complex environments: Implications of the naturalistic decision making perspective. In C. Zsambok and G. Klein (1997) (eds.) *Naturalistic Decision Making.* Mahwah, NJ: Lawrence Erlbaum.

Carvalho, P., Santos, I. and Vidal, M. (2005) Nuclear power plant shift supervisor's decision-making during microincidents. *Industrial Ergonomics*, 35, 619–644.

Claxton, G. (1997) *Hare Brain, Tortoise Mind. Why Intelligence Increases when you Think Less.* London: Fourth Estate.

Cohen, M., Freeman, J. and Thompson, B. (1998) Critical thinking skills in tactical decision making: A model and a training strategy. In J. Cannon-Bowers and E. Salas (eds.) *Making Decisions Under Stress.* Washington: APA Books.

Connolly, T., Arkes, H. and Hammond, K. (2000) (eds.) *Judgment and Decision Making.* Cambridge: Cambridge University Press.

Crichton, M., Flin, R. and McGeorge, P. (2005) Decision making by on-scene incident commanders in nuclear emergencies. *Cognition, Technology and Work,* 7, 156–166.

Crichton, M., Flin, R. and Rattray, W. (2002) Training decision makers – tactical decision games. *Journal of Contingencies and Crisis Management*, 8, 208–217.

Crosskerry, P. (2005) The theory and practice of clinical decision-making. *Canadian Journal of Anaesthesia*, 52, R1–R8.

Cullen, D. (1990) *The Public Inquiry into the Piper Alpha Disaster.* London: HMSO.

Damasio, A. (1994) *Descartes' Error*. New York: Putnam.

Dominguez, C., Flach, J., McDermott, P., McKellar, D., Dunn, M. (2004) The conversion decision in laparoscopic surgery: Knowing your limits and limiting your risks. In K. Smith and P. Johnston (eds.) *Psychological Investigations of Competence in Decision Making.* Cambridge: Cambridge University Press.

FAA (1991) *Aeronautical Decision Making. Advisory Circular 60–22.* Washington: Federal Aviation Administration.

Fallensen, J. and Pounds, J. (2001) Identifying and testing a naturalistic approach for cognitive skill training. In E. Salas and G. Klein (eds.) *Linking Expertise and Naturalistic Decision Making.* Mahwah, NJ: LEA.

Falzer, P. (2004) Cognitive schema and naturalistic decision making in evidence-based practices. *Journal of Biomedical Informatics*, 37, 86–98.

Flin, R. (1996) *Sitting in the Hot Seat: Leaders and Teams for Critical Incident Management.* Chichester: Wiley.

Flin, R., Pender, Z., Wujec, L., Grant, V. and Stewart, E. (2007b, in press) Police officers' situation assessments. *Policing*, 30, 310–323.

Flin, R., Salas, E., Strub, M. and Martin, L. (1997) (eds.) *Decision Making under Stress.* Aldershot: Ashgate.

Flin, R., Youngson, G. and Yule, S. (2007a) How do surgeons make intra-operative decisions? *Quality and Safety in Health Care*, 16, 235–239.

Fogarty, W. M. (1988). Formal investigation into the circumstances surrounding the downing of a commercial airliner by the USS Vincennes (Tech Rep). Washington, DC: Department of Defense.

Gaba, D. (1992) Dynamic decision making in anaesthesia: Cognitive models and training approaches. In D. Evans and V. Patel (eds.) *Advanced Models of Cognition for Medical Training and Practice.* Berlin: Springer-Verlag.

Gaba, D. (1994) Human work environment and simulators. In R. Miller (ed.) *Anesthesia.* (4th ed.) New York: Churchill Livingstone.

Gigerenzer, G. (2007) *Gut Feelings. The Intelligence of the Unconscious*. New York: Viking.

Gigerenzer, G. and Goldstein, D. (1996) Reasoning the fast and frugal way: Models of bounded rationality. *Psychological Review*, 103, 650–669.

Gladwell, M. (2005) *Blink. The Power of Thinking without Thinking.* London: Penguin.

Harris, D. and Khan, H. (2003) Response time to reject a takeoff. *Human Factors and Aerospace Safety*, 3, 165–175.

Haynes, A. (1992) United 232: Coping with the 'one chance-in-a-billion' loss of all flight controls. *Flight Deck*, 3, Spring, 5–21.

Hammond, K. (1988) Judgment and decision making in dynamic tasks. *Information and Decision Technologies*, 14, 3–14.

Harrison, Y. and Horne, J. (1999) One night of sleep loss impairs innovative thinking and flexible decision making. *Organizational Behavior and Human Decision Processes*, 78, 128–145.

Hoffman, R. (2006) (ed.) *Expertise out of Context.* Mahwah, NJ: Lawrence Erlbaum.

Hörmann, H.-J. (1995). FOR-DEC: A prescriptive model for aeronautical decision making. In R. Fuller, N. Johnston and N. McDonald (eds.) *Human Factors in Aviation Operations.* Aldershot: Avebury.

Jensen, R. (1995) *Pilot Judgment and Crew Resource Management.* Aldershot: Ashgate.

Joung, W., Hesketh, V. and Neal, A. (2006) Using 'war stories' to train for adaptive performance: Is it better to learn from error or success. *Applied Psychology: An International Review*, 55, 282–302.

Klein, G. (1993) A recognition-primed decision (RPD) model of rapid decision making. In G. Klein, J. Orasanu, R. Calderwood and C. Zsambok (eds.) *Decision-making in Action.* New York: Ablex.

Klein, G. (1995) Naturalistic decision-making. Individual and team training. Seminar presented at the Robert Gordon University, Aberdeen.

Klein, G. (1998) *Sources of Power. How People Make Decisions.* Cambridge: MIT Press.

Klein, G. (2003) *Intuition at Work.* New York: Doubleday.

Klein, G., and Wolf, S. (1995). Decision-centred training. Paper presented at the Human Factors and Ergonomics Society, 39th Annual Meeting, San Diego.

Klein, G., Orasanu, J., Calderwood, R. and Zsambok, C. (1993) (eds.) *Decision Making in Action.* New York: Ablex.

Larken, J. (2002) Military commander – Royal Navy. In R. Flin and K. Arbuthnot (eds.) *Incident Command: Tales from the Hot Seat.* Aldershot: Ashgate.

Lee, A. (2005) *Flight Simulation.* Aldershot: Ashgate.

Lipshitz, R., Klein, G., Orasanu, J. and Salas, E. (2001) Taking stock of naturalistic decision making. *Journal of Behavioral Decision Making*, 14, 331–352.

Mitchell, L. and Flin, R. (in press) Police officers' decisions to shoot. *Journal of Cognitive Engineering and Decision Making.*

Montgomery, H., Lipshitz, R. and Brehmer, B. (2005) (eds.) *How Professionals Make Decisions.* Mahwah, NJ: Lawrence Erlbaum.

Motowildo, S., Dunette, M. and Carter, G. (1990) An alternative selection procedure: The low fidelity simulation. *Journal of Applied Psychology*, 75, 640–647.

Myers, I. et al. (1998) *Manual for Myers-Briggs Type Indicator.* Mountain View, CA: Consulting Psychologists Press.

NTSB (1991) *Annual Review of Aircraft Accident Data: US Air Carrier Operations Calender Year 1988.* NTSB/ARC-91/01. Washington: National Transportation Safety Board.

O'Connor, P. (2005). *A Navy diving supervisor's guide to the nontechnical skills required for safe and productive diving operations.* Research Report 05–09. Panama City, FL: Navy Experimental Diving Unit.

Okray, R. and Lubnau, T. (2004) *Crew Resource Management for the Fire Service.* Tulsa: PennWell.

Omodie, M., Wearing, A. and McLennan, J. (1997) Head mounted video recording. A methodology for studying naturalistic decision-making. In R. Flin, E. Salas, M. Strub and L. Martin (1997) (eds.) *Decision Making under Stress.* Aldershot: Ashgate.

Orasanu, J. (1995) Training for aviation decision making: The naturalistic decision making perspective. In *Proceedings of the Human Factors and Ergonomic Society, 39th Annual Meeting, San Diego.* Santa Monica, CA: The Human Factors and Ergonomics Society.

Orasanu, J. (1997) Stress and naturalistic decision making: Strengthening the weak links. In R. Flin and K. Arbuthnot (Eds.) *Incident Command: Tales from the Hot Seat.* Aldershot: Ashgate.

Orasanu, J. (2005) Risk taking behaviours in pilots. Invited address. Quincentennial Conference, Royal College of Surgeons of Edinburgh. Available on Royal Society of Edinburgh website. www.rse.org

Orasanu, J., Dismukes, K. and Fischer, U. (1993) Decision errors in the cockpit. In *Proceedings of the Human Factors and Ergonomics Society 37th Annual Conference.* San Diego: HFES.

Orasanu, J. and Fischer, U. (1997) Finding decisions in natural environments: The view from the cockpit. In C. Zsambok and G. Klein (1997) (eds.) *Naturalistic Decision Making.* Mahwah, NJ: Lawrence Erlbaum.

Orasanu, J., Martin, L. and Davison, J. (2001) Cognitive and contextual factors in aviation accidents: Decision errors. In E. Salas and G. Klein (eds.) *Linking Expertise and Naturalistic Decision Making.* Mahwah NJ: Lawrence Erlbaum.

Petrilli, R., Thomas, M., Dawson, D. and Roach, G. (2006) The decision making of commercial airline crews following an international pattern. In *Proceedings of the Australian Aviation Society Conference.* Sydney, November.

Pliske, R., McCloskey, M. and Klein, G. (2001). Decision skills training: Facilitating learning from experience. In E. Salas and G. Klein (eds.) *Linking Expertise and Naturalistic Decision Making.* Mahwah, NJ: LEA.

Rall, M. and Gaba, D. (2005) Human performance and patient safety. In R. Miller (ed.) *Miller's Anesthesia.* Philadelphia: Elsevier.

Reason, J. (1997) *Managing the Risks of Organizational Accidents.* Aldershot: Ashgate.

Riley, R. (in press) *Handbook of Medical Simulation.* Oxford: Oxford University Press.

Rogers, W. and Rogers, S. (1992) *Storm Center: USS Vincennes and Iran Air Flight 655.* Annapolis, MD: Naval Institute Press.

Salas, E. and Klein, G. (2001) (eds.) *Linking Expertise and Naturalistic Decision Making.* Mahwah, NJ: Lawrence Erlbaum.

Sarna, P. (2002) Managing the spike: The command perspective in critical incidents. In R. Flin and K. Arbuthnot (eds.) *Incident Command: Tales from the Hot Seat.* Aldershot: Ashgate.

Schmitt, J. (1994) *Mastering Tactics. Tactical Decision Game Workbook.* Quantico, Virginia: Marine Corps Association.

Scott, S. and Bruce, R. (1995). Decision making style: The development and assessment of a new measure. *Educational and Psychological Measurement*, 55, 818–831.

Skriver, J. (1998) Decision making in offshore emergencies. Unpublished PhD thesis. University of Aberdeen.

Skriver, J. and Flin, R. (1996) Decision making in offshore emergencies: Are standard operating procedures the solution? In *Proceedings of Third International Conference on Health and Safety in Oil Exploration and Production, New Orleans, 9–12 June.* Paper SPE 35940. Richardson, Texas: Society of Petroleum Engineers.

Snook, S. (2000) *Friendly Fire: The Accidental Shootdown of US Black Hawks over Northern Iraq.* New Jersey: Princeton University Press.

Starbuck, W. and Farjoun, M. (2005) (eds.) *Organization at the Limit: Lessons from the Columbia Disaster.* Malden, MA: Blackwell.

Stokes, A., Kemper, K. and Kite, K. (1997) Aeronautical decision making, cue recognition and expertise under pressure. In C. Zsambok and G. Klein (eds.) *Naturalistic Decision Making.* Mahwah, NJ: LEA.

Taylor, P. (1990) *The Hillborough Stadium Disaster. Final Report.* London: Home Office, HMSO.

Useem, M., Cook, J. and Sutton, L. (2005) Developing leaders for decision making under stress: Wildland firefighters in the south Canyon fire and its aftermath. *Academy of Management Learning and Education*, 4, 461–485.

Vaughan, D. (1996) *The Challenger Launch Decision.* Chicago: University of Chicago Press.

Walters, A. (2002) *Crew Resource Management is No Accident.* Wallingford: Aries.

Weick, K. (1990) The vulnerable system: An analysis of the Tenerife air disaster. *Journal of Management*, 16, 571–593.

Weick, K. (1993) The collapse of sensemaking in organizations: The Mann Gulch disaster. *Administrative Science Quarterly*, 38, 628–652.

Weick, K. (1996) Drop your tools: An allegory for organizational studies. *Administrative Science Quarterly*, 41, 301–313.

Zsambok, C. and Klein, G. (1997) (eds.) *Naturalistic Decision Making.* Mahwah, NJ: Lawrence Erlbaum.

Chapter 4

Communication

Communication is a major part of good teamwork (Nieva et al., 1978) and is fundamental to workplace efficiency and safety. Communication is the exchange of information, feedback or response, ideas and feelings. It provides knowledge, institutes relationships, establishes predictable behaviour patterns, maintains attention to the task, and is a management tool (Kanki and Palmer, 1993). It can be subdivided into four components:

- *What* – the information to be communicated
- *How* – the means to communicate the information
- *Why* – the reason for the communication, and
- *Who* – the person(s) to whom the information is being communicated.

This chapter will examine general communication skills in the workplace. Problems with communication are described, and examples provided of the impact of poor communication on safety. Models of communication, and different types of communication, are presented, as well as suggestions to improve communication. The main components of communication are shown below in Table 4.1.

Table 4.1 Elements of communication

Category	Element
Communication	• Send information clearly and concisely. • Include context and intent during information exchange. • Receive information, especially by listening. • Identify and address barriers to communication.

Communication behaviour is a skill that can be structured by organisational policy (e.g. standard operating procedures) and can be shaped by training. Communication also refers to qualities or styles of interaction while messages are communicated. In either a team or group setting, communication is a means by which tasks are accomplished. Actions are co-ordinated by issuing instructions, stating intentions, and sending and receiving information (Wiener et al., 1993). Communication is a key activity in co-ordination between humans and plays a vital role in ensuring the successful completion of tasks (Leplat, 1991).

The failure to exchange information and co-ordinate actions is one factor that differentiates between good and poor team performance (Driskell and Salas, 1992). Errors in communication can occur as individuals fail to pass on information, communicate incorrect information and delay in making decisions. Several shipping accidents have revealed difficulties in communicating mental models between teams on the same vessel and/or between separate agencies involved in a crisis situation. One communication problem observed in the simulator at the Warsash Maritime Academy is related to the sharing of situational awareness between members in a team and also between distributed teams. Video observations from their simulator exercises suggest that team leaders can find it difficult to articulate their understanding of the situation to other team members. This difficulty is not limited to intra-team communication but can work at an inter-team level too. It is apparent that one team can easily become oblivious to the information needs of a separate team when under stress, for example, bridge and engine room teams habitually fail to update each other as a training scenario unfolds (Barnett et al., 2006).

Poor communication has frequently been cited as a contributor to accidents in other workplaces: examples include the handover failure between the outgoing and incoming shifts in the *Piper Alpha* disaster in 1988 (Box 4.1), the lack of communication between the rescue co-ordination centre and the passenger ship *Scandinavian Star* during the fire in 1990 and, famously, the communication between the air traffic control tower and the pilots that contributed to the Tenerife air crash in 1977 (see Box 1.1).

Box 4.1 *Piper Alpha* disaster

Situated 110 miles northeast of Aberdeen, Scotland, in the North Sea, Occidental's *Piper Alpha* platform had an explosion on its production deck on 6 July 1988. The resulting fire spread rapidly and was followed by a series of smaller explosions. Furthermore, two other platforms (Texaco's *Tartan* and Occidental's *Claymore*) were feeding into the same oil export line as *Piper Alpha*, and did not shut down until one hour after the initial mayday call. This resulted in oil from these platforms flowing back towards *Piper Alpha*, considerably exacerbating the intensity of the fire. Of the 266 persons who were on board the platform, only 61 survived.

Of the many factors contributing to the *Piper Alpha* disaster was the failure of transmission of relevant information at shift handover. Knowledge that a pressure safety valve had been removed and replaced by a blind flange was not communicated between shifts. Lack of this knowledge led to the incoming night shift taking actions that are presumed to have initiated the disaster. As a result the UK offshore oil industry had to revise its permit to work procedures (see later for a discussion of shift handover).

These communication problems do not just occur in industry. In medicine, a study of 47 cases of perinatal death (depending on the definition, this is the time from the 20th to 28th week of gestation and ends 1 to 4 weeks after birth) or permanent disability indicated that communication issues topped the list of identified root causes (72%; JCAHO, 2004). Reader et al. (2006), examining adverse events in intensive care units, found that communication featured as a prime cause in many reported incidents.

The exchange of information is a core activity for decision-making, situation awareness, team co-ordination and leadership, as discussed in other chapters. The ability to communicate with others is a skill that can be developed and trained. Effective communication enhances information-sharing, perspective-taking and genuine understanding. The importance of communication for effective performance, reducing errors and improving safety cannot be overemphasised.

Models of communication

Communication can typically be described as either one-way or two-way. Each of these models of communication can be experienced in different situations.

One-way communication, shown in Figure 4.1, appears simple. The information or message that the sender wants to convey is encoded into words or other signals by the sender, which are then transmitted to one or more receivers, who then decode the information to identify the meaning.

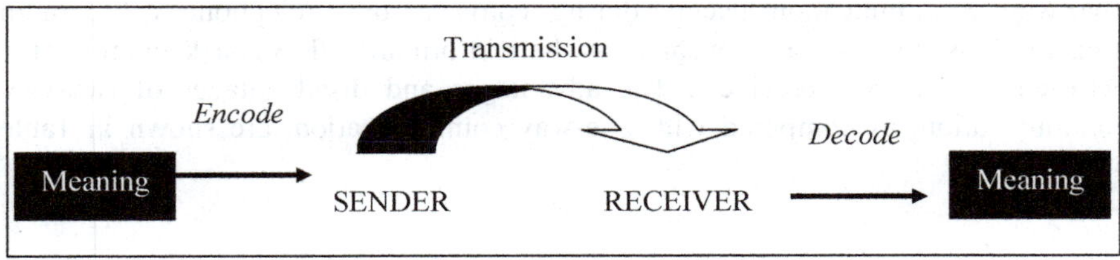

Figure 4.1 Simplified model of one-way communication

Examples of this form of communication include spoken or written instructions, email, voicemail, tannoy messages or television. There are certain advantages and disadvantages to one-way communication, as listed in Table 4.2.

Table 4.2 Advantages and disadvantages of one-way communication

Advantages	Disadvantages
• rapid • looks and sounds 'neat' • the sender feels in control	• generally requires planning • the responsibility lies with the sender • no feedback • the receiver may not pay adequate attention

Two-way communication involves the sender transmitting information to the receiver, who has the opportunity to respond and so in turn becomes the sender and transmits information back to the receiver, forming a closed feedback loop (see Figure 4.2).

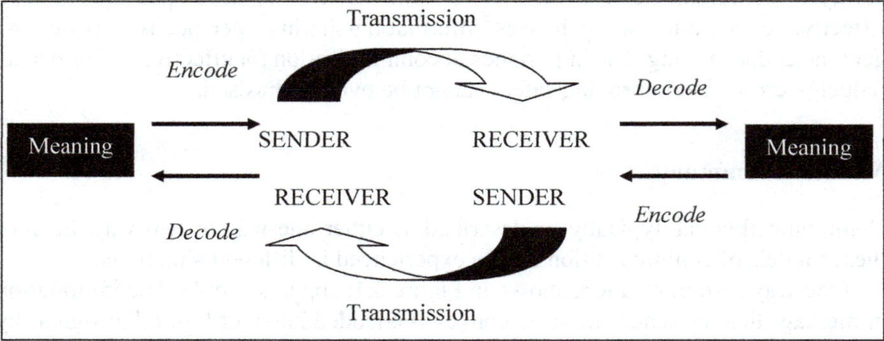

Figure 4.2 Simplified model of two-way communication

Two-way communication occurs during conversations, telephone calls, radio transmissions, email or other exchanges where information flows back and forwards between senders and receivers. The advantages and disadvantages of two-way communication, as compared with one-way communication, are shown in Table 4.3.

Table 4.3 Advantages and disadvantages of two-way communication

Advantages	Disadvantages
• potentially more accurate, reliable and effective • permits checking and correction of details • requires less planning • receivers have more confidence in themselves and make more correct judgements of accuracy • sender and receiver have responsibility • sender and receiver work together to achieve mutual understanding	• generally takes longer • receiver also has to communicate in return

Although one-way communication is faster and hence more efficient, two-way communication is more accurate because it relies on both the sender and receiver to work together to ensure that information is understood. As can be seen in the Garuda Airlines crash outlined below (Box 4.2), even with two-way communication it is possible for there to be confusion between the sender and receiver about simple direction commands.

Box 4.2 A fatal failure in two-way communication – Garuda Airlines Flight 152

On 26 September 1996, Garuda Airlines Flight 152 flew into a mountain 15 minutes before it was due to land at Medan, Indonesia, on a flight from Jakarta, Indonesia. The aircraft crashed 20 miles from the airport. An air traffic control (ATC) error resulted in the plane being routed into mountainous terrain that was obscured by smoke and haze due to forest fires in the area. All 234 passengers and crew on board were killed. The following is the ATC tower's conversation with the Garuda Indonesia Airlines (GIA) Airbus A300.
ATC: GIA 152, turn right heading 046 report established localiser.
GIA 152: Turn right heading 040 GIA 152 check established.
ATC: Turning right sir.
GIA 152: Roger 152.
ATC: 152 Confirm you're making turning left now?
GIA 152: We are turning right now.
ATC: 152 OK you continue turning left now.
GIA 152: A (pause) confirm turning left? We are starting turning right now.
ATC: OK (pause) OK.
(Aviation Safety Network, 2004).

The key difference between one-way and two-way communication is the role of feedback. Feedback enables the sender and receiver to ensure that the meaning within the information has been understood. It closes the communication loop and is the simplest way of preventing any misunderstanding in the receiver's interpretation of the original meaning of the message.

Feedback can range from a simple nod during a conversation to a written comment on a paper communication. Three types of feedback exist:

1. *Informational* – the receiver provides a non-evaluative response, e.g. provides an objective statement in response to the initial statement.
2. *Corrective* – the receiver challenges or corrects the sender's message, e.g. questions or queries the initial statement to gain clarification.
3. *Reinforcing* – the receiver acknowledges clear receipt of the message, e.g. checks understanding of the message.

Feedback should include quality, quantity, relevance and manner. Without feedback, it is difficult to know if a message was received.

Types of communication

Language is a form of communication consisting of signs and symbols. Language can be verbal (spoken or written) or non-verbal (i.e. without words, such as sign language, gestures or tone of voice; see Box 4.3 for an example of the importance of tone).

Box 4.3 The importance of tone to US naval aviators

> Landing planes on aircraft carriers, especially at night, is one of the most complex and dangerous tasks carried out by naval aviators. One of the systems in place to help pilots achieve this objective is the landing signal officer (LSO). LSOs are naval aviators who have received special training to control the final approach and landing of aircraft on aircraft carriers. The LSOs watch the aircraft from the flight deck and help to guide the aviators onto the carrier through voice communication with the pilots over the radio.
>
> The LSOs have been trained to use not only words, but also the tone and volume of their voice to provide information to the pilots to help them to land. To illustrate, if a pilot is a little below the glide slope (the prescribed descent of an aircraft coming in to land) the LSO will softly say 'a little power'. However, if the pilot is well below the glide slope the LSO will loudly say 'POWER'. The pilots have been trained to add power in accordance with the level of volume with which the LSO gives the command.

An often quoted (and misquoted) series of studies of communication were carried out by Mehrabian in the late 1960s (e.g. Mehrabian and Ferris, 1967; Mehrabian and Wiener, 1967). These experiments were based upon situations in which there were ambiguities between the spoken word and the tone or facial expressions of the individual receiving the message. The researchers concluded that the amount of attention the receiver pays to the components of the sender's message can be broken down into:

Words:	7%
Tone:	38%
Other non-verbal clues:	55%
(e.g. gesture, posture, facial expression)	

Although the generalisability of these findings can be challenged, the implications are that non-verbal signals are important. Communication in the absence of non-verbal communication can result in the words being misunderstood. Further, when we are unsure about the meaning of words and we do not know the individual, we pay more attention to non-verbal cues. Therefore, people should think about what is the most effective communication medium for delivering a message.

West (2004) comments that the richness of transfer is determined by the medium of information exchange:

- The least rich in information is written.
- Slightly richer information is transferred by telephone conversations or videoconferencing.
- The most rich information transfer occurs during face-to-face conversations. This allows both verbal and non-verbal communication to take place, i.e. the words used combined with voice inflection, facial expression, body posture and gestures.

Email and letters are generally useful for exchanging information, communicating news and asking questions. On the other hand, face-to-face meetings are best used for expressing concerns, complaints and finding resolutions to conflicts. Each of these types of communication will be discussed below.

Spoken communication

Spoken communication is both social and functional. At a social level, it helps to build relationships and is instrumental in helping a team to carry out its task. The actual information that is exchanged and the language and timing are also important. However, the content of speech on its own is often insufficient for effective communication. It may be not *what* is said that is important but *how* it is said.

Non-verbal communication

Non-verbal communication is the way people communicate, either deliberately or unintentionally, without using words. It is a major signal of emotional state, as well as conveying other information. This is often referred to as body language and there are many books available with good examples (e.g. Morris, 2002; Pease and Pease, 2005). Malandro et al. (1989) identified four classes of non-verbal communication that are relevant to the workplace: (1) facial expression and eye behaviour; (2) body movement and gestures; (3) touching behaviour; and (4) voice characteristics and qualities.

In addition to conveying emotion, researchers have identified five ways in which non-verbal communication complements verbal communication:

1. repeats what is being said (e.g. saying 'no' and shaking your head)
2. adds to or reinforces verbal communication (e.g. saying 'I'm angry' and frowning)
3. highlights verbal communication by emphasising certain words or phrases (e.g. Are you *sure* that's right?)
4. contradicts the verbal message (e.g. saying 'I'm angry with you' while smiling)
5. substitutes for verbal behaviour (e.g. shrugging your shoulders to indicate you don't care).

However, just as with verbal communication, non-verbal communication can be ambiguous and open to misinterpretation, especially when communicating with someone from a different culture. As workplaces become much more multicultural, recognising that gestures that have one meaning in some cultures can have a completely different meaning in others becomes particularly important. For example, a thumbs up in most Western countries means good luck or approval, whereas in Middle Eastern cultures this is an obscene gesture. The 'OK' sign frequently used in scuba diving (forming a circle between forefinger and thumb with the other fingers raised) can be seen as an obscene gesture in Germany. The simple head nod is accepted as meaning 'Yes' in western Europe, China and North America, whereas in Sri Lanka and many eastern European countries, a nod actually means 'No' (see Morris, 2002, for other examples).

Non-verbal cues can also be subtle. A random gesture may be given a certain meaning when none was intended. Similarly, people often use non-verbal cues to discern if someone is lying (Zuckerman et al., 1981) but even professional interviewers are surprisingly inaccurate in detecting deceit from communication (Vrij et al., 2004).

Written communication

Written communication, frequently electronic, is regularly used in the workplace. This can also be open to misunderstanding and misinterpretation and care must be taken to ensure that it is clear, precise and informative (see Box 4.4).

Box 4.4 Medication error

On 11 April 2001 at a children's hospital in Washington, D.C., a surgeon prescribed two '.5 mg' intravenous doses of morphine for the management of post-operative pain for a 5-month-old baby girl. However, a unit secretary did not see the decimal point and transcribed the order by hand onto a medication administration record as '5 mg'. An experienced nurse followed the directions on the medication administration record and intravenously gave the baby 5 mg of morphine initially and another 5 mg dose two hours later. About four hours after the second dose, the baby stopped breathing and suffered a cardiac arrest (Goldstein, 2001).

One of the major causes of medication errors is the use of abbreviations, dose expressions and handwriting that is difficult to read. The JCAHO (2001) identifies a number of potentially dangerous abbreviations. For example, when 'U' (short-hand for units) is handwritten, it can often look like a zero, or using trailing zeros (e.g., 2.0 vs. 2) or use of a leading decimal point without a leading zero (e.g., .2 instead of 0.2), as in the case described above, can lead to ten-fold dosing errors. JCAHO (2001) makes a number of recommendations to prevent these types of errors:

- Develop a list of unacceptable abbreviations and symbols that is shared with all prescribers.
- Develop a policy to ensure that medical staff refer to the list and take steps to ensure compliance.
- Establish a policy that if an unacceptable abbreviation is used, the prescription order is verified with the prescriber prior to its being filled.

Remote communication

Co-located team members work in the same place and mainly communicate through face-to-face interactions. Teams or team members that are not co-located predominantly use information technology to communicate, collaborate, share information and co-ordinate their efforts (Brown et al., 2006; Townsend et al., 1998). Thus physically distant team members lack the capacity for direct communication and, by implication, the ability to engage in informal and spontaneous interaction. Physical distance between team members has been shown to have detrimental effects on many aspects of teamwork: it causes a lack of shared identity, lack of mutual awareness of contextual aspects of members' work environments and lack of awareness of members' roles and responsibilities (Mortensen and Hinds, 2001). However, many teams have to work with members that are not co-located. Examples of these types of teams include strike aircraft being talked onto a target by personnel on the ground, offshore oil platforms communicating with personnel back on the 'beach' and dive teams in which the divers communicate with top-side personnel over the radio.

In theory, computer-based systems such as groupware technologies and company intranets should support information exchange by providing a mechanism through which communication and co-ordination can be achieved and by making explicit the distribution of knowledge within the group. In practice, however, it has been found that in groups using computer-mediated communication technologies, information exchange was less complete and discussion more biased; they also took more time to exchange less information than co-located groups (Hightower and Sayeed, 1996; Hollingshead, 1996). Similarly, Driskell et al. (2003) found that computer-based technologies that are supposed to aid communication can have a significant negative impact on how teams perform by making it more difficult to communicate and interpret information, resulting in breakdowns in mutual knowledge and shared situation awareness.

Barriers to communication

According to Reason (1997), communication problems that contribute to organisational accidents can be categorised as:

- system failures in which the necessary channels of communication do not exist, or are not functioning, or are not regularly used;
- message failures in which the channels exist but the necessary information is not transmitted; and
- reception failures in which the channels exist, the right message is sent, but it is either misinterpreted by the recipient or arrives too late (p135).

Each of these communication problems can lead to human error and affect safety and performance. A communicator may neglect to advise other team members of critical or updated information. Further, poor communication arises when misunderstandings or misinterpretations occur. This can be caused by team members making different assessments of the situation (see Chapters 2 and 6), i.e. 'reading' the situation differently or working on assumptions rather than checking facts or soliciting feedback.

Hazards that can reduce the quality of communications include (CAA, 2006):

- failures during the transmitting process (e.g. sending unclear or ambiguous messages, language problems)
- difficulties due to the transmission medium (e.g. background noise, distortion)
- failures during receiving (e.g. expectation of another message, misinterpretation of the message, disregarding the message)
- failures due to interference between the rational and emotional levels of communication (e.g. arguments), and
- physical problems in listening or speaking (e.g. impaired hearing, wearing equipment such as oxygen masks or personal protective equipment).

Barriers to communication can emerge due to both internal and external factors, as summarised in Table 4.4. Internal barriers can be attributed to the individual, whereas external barriers can be attributed to the environment.

Table 4.4 Internal and external barriers to communication

Internal	External
• language difference	• noise
• culture	• interference or distraction
• motivation	• separation in location, time
• expectations	• lack of visual cues (e.g. body language, gestures, facial expressions)
• past experience	
• prejudice	
• status	
• emotions/moods	
• deafness	
• voice level	

Recommendations for improving communication in teams

Effective communication is crucial for most of the non-technical skills discussed in this book. Drawing from research carried out in a range of high-reliability industries, it is possible to make a number of recommendations for improving communication, especially between team members. Four aspects of communication are described: explicitness, timing, assertiveness and active listening.

Explicitness

Explicitness means clearly stating the desired action and who should do it. Explicitness in communication is required for the avoidance of ambiguity. Being explicit means that the receiver does not assume what the sender knows or thinks. It ensures the inclusion of an adequate amount of information to reduce the chance of errors. However, the message should be as brief as possible, giving the required information, but not overloading the receiver. It is important that only the most relevant information is given due to the cost in attention and cognitive resources for both the sender and the receiver (particularly during periods of high workload). US naval aviation refers to this as 'comm-brevity". Further, it will be helpful for the receiver if standardised communication patterns are used. Kanki et al. (1989) found that cockpit crews that communicate in standard (and therefore predictable) ways were better able to co-ordinate their tasks, whereas teams that did not use standard communication phraseology were less effective in their performance. It would appear that only in uncertain tasks is a flexible communication pattern beneficial (Tushman, 1977). The phonetic alphabet (A – alpha, B – bravo, etc) is often used as a standard method of communicating letters of the alphabet. Although this method is used in aviation, it is suggested that other organisations should also adopt the phonetic alphabet when spelling out words that cannot be understood, or describing pieces of a plant that are identified with alphabetical and numerical codes.

Timing

The sender should be sensitive to other activities with which the receiver may be engaged. The sender should provide the information at the most relevant time, that is, not too early and not too late. They should determine the current task demands of the person to whom they are speaking. They should not talk to them when they are clearly dealing with something that is urgent, or are very busy, unless the information is important. The problems of distraction and task interruption were outlined in Chapter 2. Orasanu (1993) found that the captains of high-performing airline crews talked less during high-workload conditions than during normal conditions. The number of commands was lower and communications concerned with planning and strategy increased. Further, team effectiveness can be enhanced when team members provide information before it is requested (Volpe et al., 1996).

Assertiveness

The need for assertive behaviour in more junior team members has been sharply highlighted in aviation research (Jentsch and Smith-Jentsch, 2001). Flight deck voice recordings revealed that co-pilots could be reluctant to question the captain's decisions, and there have been aircraft crashes in which this lack of assertiveness has been a contributory factor (e.g. Tenerife, 1977; Washington, 1982; see Box 4.5 below). Further evidence comes from simulator studies. For example, Foushee and Helmreich (1988) found that when aircraft captains feigned incapacity during a final landing approach, 25% of the planes 'crashed' because the co-pilot failed to take over control.

Box 4.5 Lack of assertiveness of a first officer: Air Florida Flight 90 incident

Air Florida Flight 90, a Boeing 737, was a scheduled flight to Fort Lauderdale, Florida, from Washington National Airport, Washington, D.C., on 13 January 1982. There were 77 passengers and 5 crew members on board. The flight's scheduled departure time was delayed about 1 hour 45 minutes due to a moderate to heavy snowfall that necessitated the temporary closing of the airport (NTSB, 1982).

Following take-off from runway 36, which was made with snow and/or ice adhering to the aircraft, the aircraft failed to gain sufficient height, crashed into the barrier wall of the northbound span of the 14th Street Bridge and plunged into the ice-covered Potomac River. Four passengers and one crew member survived the crash. Further, when the aircraft hit the bridge, it struck seven occupied vehicles and then tore away a section of the bridge barrier wall and bridge railing. Four people in the vehicles were killed.

The NTSB determined that the probable causes of this accident were the flight crew's failure to use engine anti-ice during ground operation and take-off, their decision to take off with snow/ice on the airfoil surfaces of the aircraft, and the captain's failure to reject the take-off during the early stage when his attention was called to anomalous engine instrument readings by the first officer (the engines were set at substantially less than normal take-off thrust due to a partially ice-blocked probe causing false, high thrust readings on the engine gauges). The first officer was aware that there was something wrong, but the language he used was not sufficiently explicit and direct to get the captain's attention. 'The first officer commented "that don't seem right, does it?" "Ah, that's not right." The captain's only response was, "yes it is, there's eighty (knots)." The first officer again expressed concern: "Naw, I don't think that's right." Again there was no response from the captain… The first officer continued to show concern as the aircraft accelerated through a "hundred and twenty (knots)"' (NTSB, 1982: p64).

Contributing to the accident were the prolonged ground delay between de-icing and the receipt of air traffic control take-off clearance, during which the aeroplane was exposed to continual precipitation, and the limited experience of the flight crew in jet transport winter operations.

More recent studies in other domains, such as healthcare, have shown that status differences in groups affect communication behaviours, such as speaking up or challenging another's behaviour (Edmondson, 2003; Reader et al., in press). Assertiveness can be understood as a disposition situated between one that is too passive and one that is too aggressive:

- *Passive* – failing to stand up for yourself, or standing up for yourself in such a way that others can easily disregard your words or actions. This communication mode typically involves a failure to honestly express feelings, which may be with the purpose of avoiding conflict. A passive mode can be appropriate when the issues being discussed are minor.

- *Assertive* – standing up for yourself in such a way as not to disregard the other person's opinion. This communication mode respects the boundaries of all parties, results in fewer emotional outbursts, but takes practice. Assertiveness requires persistence, objectivity and validation, to stay focused on the issue and avoid becoming defensive or emotional. Assertiveness may also involve inquiry and advocacy. *Inquiry* is defined as asking questions to acquire additional information. This requires the communicators to be able to ask questions to acquire additional information that may be required to establish a position or to resolve a question. *Advocacy* is defined as the assertiveness of the individual in stating and defending a position.
- *Aggressive* – standing up for yourself, but in such a way as to disregard the other person's opinion. Aggressive behaviour is demonstrated by the use of verbal and non-verbal cues, such as defensive, superior statements, insincere or overblown descriptions of importance and a dominating posture. However, more aggressive behaviour may be required in a time-pressured situation, such as an emergency.

Non-verbal cues, such as eye contact, body posture and gestures, as well as verbal cues, such as the content of speech, together constitute assertive behaviour (Ross and Altmaier, 1994). Situations requiring an assertive mode include refusing a request, saying no and handling criticism (see Table 4.5 for suggestions for adopting an assertive stance, and Jentsch and Smith-Jentsch, 2001, or Patterson, 2000, for more details).

Table 4.5 Suggestions for adopting an assertive stance

Verbal	Non-verbal

• Content:
  • Decide what you want to say and state it specifically and directly.
  • Be honest: 'I'm annoyed at what you did!'
  • Stick to the statement; repeat it, if necessary.
• Use 'I' statements.
• Assertively deflect any responses from the other person that might undermine you.
• 'Broken record technique':
  • 'I hear what you are saying, but...'
• Offer a solution.
• Obtain feedback.
	• eye contact
	• body posture
	• gestures
	• facial expression
	• voice tone, inflection, and volume
	• timing

Active listening

Listening is an active process and requires effort on behalf of the listener. There is little to be gained from teaching junior team members to be assertive if senior team members are not taught to listen. Even in ideal circumstances only a small part of what is heard is actually listened to (usually only about one-third) if the listener is interested, even less if the listener is not. The earlier part of a communication also tends to be listened to more than the later – when the listener may be trying to find a place to interrupt. This is sometimes called 'gap searching'. Below in Table 4.6 is a list of dos and don'ts that will aid in effective listening.

Table 4.6 Dos and don'ts for effective listening

Do:	Don't:
• be patient • ask questions • be supportive • paraphrase • make eye contact • use positive body language	• debate what is being said in your mind • detour (i.e. look for a key word to change the subject) • interrupt • finish the other person's sentence • pre-plan (work out what the person will say next) • tune out • fixate on other tasks

For team members to perform effectively, it is necessary for them to have all of the communication skills described above. The use of these skills is particularly important during briefings.

Briefing

Briefing is crucial to allow team members to develop a shared mental model (see Chapter 2 for a discussion of this concept). It is possible to divide briefing into three types: shift handover, pre-mission/pre-task brief and debriefing. Each type of briefing will be discussed below.

Shift handover

In industries that operate continuous processes, continuity is maintained across shift changes via shift changeover. In many organisations, such as healthcare, new working time legislation means that more shift handovers are now taking place. The major barrier to such continuity is failure of effective handover from the outgoing to the incoming shift due to poor communication. The goal of shift handover is the accurate, reliable communication of task-relevant information across shift changes, thereby ensuring continuity of safe and effective working (Lardner, 1996).

Poor communication, especially during shift handovers, has lead to incidents occurring in high-hazard industries. Miscommunication, or failures of communication, have been cited as being among the contributory causes in analyses of accidents, and the rate of accidents at or near shift changeover increases (see Box 4.6).

Box 4.6 Examples of the contribution to accidents made by failures in shift handover

1. *Sellafield beach incident* (HSE, 1983). Sellafield is a site in northern England on the coast of the Irish Sea consisting of a number of nuclear reprocessing facilities. Due to failure of communication between shifts, a tank that was assumed to contain liquid suitable for discharge to sea – but that actually contained highly radioactive material – was discharged to sea, creating an environmental hazard. A written description of the tank contents was carried forward from one shift log to the next, across several consecutive shifts. This resulted in the written description of the tank contents being changed. There was a failure to describe the tank's contents in unambiguous terms and transcription errors were made as the written log book contents were copied from page to page.

2. *BP Texas City refinery explosion.* A series of failures by personnel before and during the start-up of the isomerisation process unit in the Texas City (Texas, USA) refinery in March 2005 led to an explosion and fire that killed 15 workers and injured more than 170 people. The isomerisation unit is used to convert raffinate, a low-octane blending feed, into higher-octane components for unleaded petrol. Part of the unit consists of a 164-foot tower. The fluid level in the tower at the time of the explosion was nearly 20 times higher than it should have been. The nightshift started to fill the tower, and this was continued by the following dayshift. The dayshift supervisor arrived late and did not receive any handover from the operator who had been filling the tower (he left one hour before shift change), or from the nightshift supervisor. The handover between two crucial operators involved in the process occurred at the refinery gate, and its duration and quality was less than adequate (US Chemical Safety and Hazard Investigation Board, 2005).

Due to the numbers of communication problems that can occur in healthcare related to shift handovers and other transfers of responsibility, the British Medical Association (BMA, 2004) has published an advice booklet, *Safe Handover, Safe Patients*. It outlines five questions that should be asked when trying to improve shift handovers:

- *WHO should be involved?* The handover should be a face-to-face meeting between the off-going and oncoming personnel. Further, management (e.g. ward managers) should also be present at the handover to ensure that they are aware of any issues or ongoing concerns, and demonstrate that they recognise the importance of an effective shift handover.

- *WHEN should handover take place?* Handovers should be held at a fixed time, with sufficient time allowed to exchange all of the pertinent information. Lardner (1996: p14) identified a number of particular instances in which high levels of care should be taken to conduct a thorough shift handover. These situations include:
  - 'during plant maintenance, particularly when this work continues over a shift change;
  - when safety systems have been over-ridden;
  - during deviations from normal working;
  - following a lengthy absence from work; and
  - when handovers are between experienced and inexperienced staff.'

- *WHERE should handover take place?* The handover should occur close to the most used worksite sites, and be in a space that is sufficiently large, so that on-going and off-going staff can attend. The refinery gate is not the appropriate place for a handover (see Box 4.6). Attempts should be made to reduce distractions from telephones, alarms, beepers, etc.

- *HOW should handover happen?* The handover should be formal, predictable and follow a consistent format. Leonard et al. (2004) suggest that a situational briefing tool, known by the acronym SBAR, is an effective guide for conducting a handover of a patient in a medical setting. SBAR stands for: (1) situation – what is wrong with the patient?; (2) background – what is the clinical background or context?; (3) assessment – what do I think the problem is?; and (4) recommendations – what would I do to correct the problem?

  As much as possible, staff should give their full attention to the handover, rather than monitoring equipment. The BMA (2004) recommends that shift handover should be designated as 'bleep-free', except for emergencies.

- *WHAT should be handed over?* Written notes, or an electronic log book, should be completed by members of the off-going shift and handed over to the oncoming shift. Two-way verbal communication should complement a written log. Oncoming shift personnel should not be passive receivers of the information. Every log should follow a standard format, with verbal communication used to clarify the written information (see Table 4.7 for a summary of recommendations for conducting an effective briefing).

Pre-mission/task brief

The five questions posed by the BMA (2004) for shift handovers are equally applicable to a pre-mission or task brief (see Table 4.7). All of the team members, both junior and senior, should attend the brief; obviously, the pre-mission/task brief should be carried out before the task; the briefing should be carried out near the worksite, with sufficient room for all of the team members to attend; and as with a shift handover, the brief should follow a formal and predictable structure.

Table 4.7 Recommendations for conducting an effective briefing

	Shift handover	Pre-mission/task brief	Debrief
Who?	All members of the shift, including senior management	All team members	
When?	At a fixed time	Prior to the mission/ task	After the mission/ task
Where?	Close to the worksite, where all team members can be present		
How?	Two-way communication, following a formal and predictable structure		May be formal or informal
What?	Written notes and two-way communication		All the information pertaining to team performance

US military aviation mandates a strict structure for a pre-flight brief. To illustrate, a typical pre-flight brief that is given by military aviators will include the following information:

1. Time hack – ensure that everybody in the crew has the same time.
2. General – call signs, gross weight, take-off time, etc.
3. Target or destination.
4. Navigation/flight planning – e.g. mission plan, obstacles.
5. Communications – frequencies, procedures/discipline, etc.
6. Weapons – loading, arming, etc.
7. Weather – local, en route, etc.
8. Emergencies – aborts, crew co-ordination, etc.
9. Special instructions – intelligence, safety, etc.

Generally, the team leader should lead the brief. However, it is important that input is encouraged, and accepted, from other team members. All of the pertinent information relating to the task should be discussed, including both technical and non-technical issues. Two-way verbal communication should be used. However, whether spoken information is supplemented with written communication will depend on the task. See Makary et al. (2006) for a pre-task briefing checklist for operating theatre teams.

Debrief

It is easy for a team to fail to do a debrief after a task. Often the last thing people may want to do after completing a task is to spend time discussing their performance. Nevertheless, taking the time to carry out a debriefing following an event is something that should be emphasised to foster learning (see Table 4.7). Therefore it is important that senior management and team leaders encourage debriefing. During a debrief the team leader is crucial in ensuring that an effective debriefing is completed. However, it is recognised that the team leader may not always be the best person to give the debrief. Tannenbaum et al. (1998) identify eight team behaviours that should be exhibited during a debrief:

1. Provide a self-critique early in a post-incident review. An effective briefing environment should be one in which team members feel comfortable admitting when they were confused or made a mistake. Therefore, to encourage this behaviour, the briefer should model this behaviour during the briefing.
2. Accept feedback and ideas from others. To optimise learning in a debriefing, team members must be willing to accept feedback on their behaviour from others. To encourage this, the briefer must model this behaviour.
3. Avoid person-focused feedback and concentrate on task-focused feedback. Feedback regarding evaluative judgements about particular team members should be avoided. Rather, performance feedback in terms of behaviours that should be changed instead of personal attacks on individual team members should be employed.
4. Provide specific, constructive suggestions when providing feedback. The provision of specific, constructive feedback (as opposed to vague, negative feedback) is an effective way for team leaders to help identify and summarise areas of improvement for both individuals and the team.
5. Encourage active team member participation during debriefings and do not simply state one's own views of the team's performance. Briefers should guide and facilitate team debriefs rather than simply lecturing the team members.
6. Guide briefings to include a discussion of not only the technical aspects of the task, but also non-technical aspects, e.g. teamworking. Team members do not always understand the specific teamwork behaviours related to their performance of a task.
7. Refer to prior briefings and team performance when conducting subsequent debriefs. This provides a sense of continuity, reliability and consistency.
8. Tell the individual team members or team as a whole when they demonstrate an improvement in performance.

Training and assessing communication

Training in communication is typically included as a module of crew resource management (CRM) training. Typically, this covers aspects of communication such as listening and clarity of expression, non-verbal communication, assertiveness and the other features of communication described above. However, Kanki and Smith (2001) propose that when communication is taught as a standalone module of CRM, this is missing the overarching importance of communication to both the technical and non-technical aspects of performance in high-reliability organisations such as aviation. They propose that communication training should be separated into three objectives. These objectives are described below:

- *Communication to achieve technical objectives.* Communication training should be combined with technical training. The use of standard terminology and phraseology should be reinforced during technical training.
- *Procedural communication.* Standard procedural communication should be memorised, rehearsed and used. For example, from when they first enter a cockpit, aviation students are taught to perform a positive three-way communication when handing the controls from one pilot to another (i.e. Pilot 1: 'I have the controls.', Pilot 2: 'You have the controls.' Pilot 1: 'I have the controls.').
- *Communication to achieve CRM objectives.* As outlined earlier in this chapter, communication underpins all of the other non-technical skills. Therefore, communication training should be blended into each module of CRM training.

Communication can be assessed for effectiveness using techniques such as behavioural observation scales (see Chapter 11). Observed communication can then be assessed against these exemplars. Beaubien et al. (2004) present a behavioural checklist for communication (see Table 4.8):

Table 4.8 Behavioural observation scale for communication (Beaubien et al., 2004)

Communication example behaviours:
Team leader: • establishes a positive work environment by soliciting team members' input • listens without evaluating • identifies bottom-line safety conditions, and • establishes contingency plans.
Team members: • verbally indicate their understanding • verbally indicate their understanding of contingency plans • provide consistent verbal and non-verbal signals, and • respond to queries in a timely manner.

Conclusion

The skill of communication is integral to many technical skills, as well as to other non-technical skills, such as decision-making, situation awareness, teamwork, leadership and stress management, for safe and effective performance. Across organisations, the importance of effective communication should be recognised at all levels. The content of the communication (e.g. the words used) must be combined with the type of communication and non-verbal cues to encourage effective transmission and comprehension. Especially in teams, this is more effective where standard communication structures are introduced, for example log books and formalised briefings and debriefings. Different modes of communication, such as face-to-face conversations, email, telephone, should be employed to suit communication requirements related to the situation.

Key points

- Effective communication is a skill that can be learned, developed and improved.
- The use of two-way communication that involves feedback helps to avoid misunderstandings in the communication process and provides more effective communication than one-way.
- Lack of communication, poor communication, use of assumptions and lack of feedback all contribute to problems with communication.
- Non-verbal communication and active listening are as important as spoken communication.
- Workplace incidents where poor communication has been implicated should be reviewed to capture lessons that may be relevant.
- Communication is vital for effective teamwork, especially for teams that may be distributed geographically and that rely on remote communication.
- Standard protocols will enhance communication between incoming and outgoing shifts. A good shift or task handover protocol should be established to reduce communication difficulties.

Suggestions for further reading

British Medical Association (2004) *Safe Handover: Safe Patients.* London: British Medical Association. (download from www.bma.org.uk)

CAA (2006) *Crew Resource Management (CRM) Training. CAP 737*. Hounslow, Middlesex: Civil Aviation Authority.

Kanki, B.G. and Palmer, M.T. (1993) Communication and crew resource management. In E.L. Wiener, B.G. Kanki and R.L. Helmreich (eds.) *Cockpit Resource Management.* New York: Academic Press.

Lardner, R. (1996). *Effective Shift Handover: A Literature Review*. Offshore technology report *OTO 96 003*. London: HMSO.

Morris, D. (2002) *Peoplewatching. The Desmond Morris Guide to Body Language.* London: Vintage.

Patterson, R. (2000). *The Assertiveness Workbook: How to Express your Ideas and Stand up for Yourself at Work and in Relationships*. Oakland, CA: New Harbinger Publications.

Searles, G.J. (2005). *Workplace Communication: The Basics*. London: Longman.

Websites

UK Health and Safety Executive. Human factors: Safety critical communication: www.hse.gov.uk/humanfactors/comah/safetycritical.htm

References

Aviation Safety Network (2004) ATC transcript Garuda Flight 152 - 26 September 1997 [online]. Available at: http://aviationsafety.net/investigation/cvr/transcripts/atc_ga152.php

Barnett, M., Gatfield, D. and Pekcan, C. (2006) Non-technical skills: the vital ingredient in world maritime technology? In *Proceedings of the International Conference on World Maritime Technology*. London: Institute of Marine Engineering, Science and Technology.

Beaubien, J.M., Goodwin, G.F., Costar, D.M. and Baker, D. (2004). Behavioural Observation Scales (BOS). In N.A. Stanton, A. Hedge, K. Brookhuis, E. Salas and H. Hendrick (eds.) *Handbook of Human Factors and Ergonomic Methods*. Boca Raton, FL: CRC Press.

Brown, K., Huettner, B. and James-Tanny, C. (2006) *Managing Virtual Teams: Getting the Most from Wikis, Blogs and other Collaborative Tools.* London: Wordware.

BMA (2004) *Safe Handover: Safe Patients.* London: British Medical Association.

CAA (2006) *Crew Resource Management (CRM) training. CAP 737*. Hounslow, Middlesex: Civil Aviation Authority.

CAA (2004) *CAP716 Aviation Maintenance Human Factors* (EASA/JAR145 Approved Organisations). Sussex: CAA Safety Regulation Group.

Cullen (1990) *The Public Inquiry into the Piper Alpha Disaster* (Cm 1310)). London: HMSO.

Driskell, J.E. and Salas, E. (1992) Collective behaviour and team performance. *Human Factors*, 34, 277–288.

Driskell, J., Radtke, P. and Salas, E. (2003) Virtual teams: effects of technological mediation on team performance. *Group Dynamics: Theory, Research, and Practice*, 7, 297–323.

Edmondson, A. (2003) Speaking up in the operating room. How team leaders promote learning in interdisciplinary action teams. *Journal of Management Studies*, 40, 1419–1452.

Foushee, H.C. and Helmreich, R. (1988) Group interaction and flightcrew performance. In E. Wiener and D. Nagel (eds.) *Human Factors in Aviation.* San Diego, CA: Academic Press.

Goldstein, A. (2001). Overdose kills girl at children's hospital. *Washington Post*, 20 April.

Hightower, R. and Sayeed, L. (1996) Effects of communication mode and prediscussion information distribution characteristics on information exchange in-groups. *Information Systems Research*, 7, 451–465.

Hollingshead, A.B. (1996) The rank order effect in group decision making. *Organizational Behavior and Human Decision Processes*, 68, 181–193.

HSE (1983) *The Contamination of the Beach Incident at British Nuclear Fuels Limited, Sellafield, November 1983*. London: HMSO.

JCAHO (2004) *Sentinel Event Alert, Issue No. 30*. Oak Brook, Illinois: Joint commission of accreditation of healthcare organizations.

JCAHO (2001) Medication errors related to potentially dangerous abbreviations. *Sentinel event alert, Issue No. 23*. Oak Brook, Illinois: Joint commission of accreditation of healthcare organizations.

Jentsch, F. and Smith-Jentsch, K. (2001) Assertiveness and team performance: More than "Just Say No". In E. Salas, C. Bowers and E. Edens (eds.) *Improving Teamwork in Organisations.* Mahwah, NJ: Lawrence Erlbaum.

Kanki, B.G., Lozito, S. and Foushee, H.C. (1989) Communication indices of crew coordination. *Aviation, Space and Environmental Medicine*, January, 56–60.

Kanki, B.G. and Palmer, M.T. (1993) Communication and crew resource management. In E.L. Wiener, B.G. Kanki and R.L. Helmreich (eds.) *Cockpit Resource Management.* San Diego: Academic Press.

Kanki, B.G. and Smith, G.M. (2001) Training aviation communication skills. In E. Salas, C.A. Bowers and E. Edens (eds.) *Improving Teamwork in Organizations.* Mahwah, NJ: Lawrence Erlbaum.

Lardner, R. (1996) *Effective Shift Handover: A Literature Review*. Offshore technology report OTO 96 003. London: HMSO.

Leonard, M., Graham, S. and Bonacum, D. (2004) The human factor: the critical importance of teamworking and communication in providing self care. *Quality and Safety in Health Care*, 13, i85–i90.

Leplat, J. (1991). Organisation of activity in collective tasks. In J. Rasmussen, B. Brehmer and J. Leplat (eds.) *Distributed Decision Making: Cognitive Models for Co-operative work.* London: John Wiley.

Makary, M. et al. (2006) Operating room briefings: Working on the same page. *Joint Commission Journal on Quality and Patient Safety*, 32, 351–356.

Malandro, L, Barker, L. and Barker, D. (1989) *Nonverbal Communication*, 2nd ed. Reading, MA: Addison-Wesley.

Mehrabian, A. and Ferris (1967) Inference of attitudes from nonverbal communication in two channels. *Journal of Consulting Psychology*, 31, 248–252.

Mehrabian, A. and Wiener, M. (1967) Decoding of inconsistent communications. *Journal of Personality and Social Psychology*, 6, 109–114.

Morris, D. (2002) *Peoplewatching. The Desmond Morris Guide to Body Language.* London: Vintage.

Mortensen, M. and Hinds, P. (2001) Conflict and shared identity in geographically distributed teams. *International Journal of Conflict Management*, 12, 212–238.

Nieva, V. F., Fleishman, E. A. and Rieck, A. (1978) *Team dimensions: Their identity, Their Measurement, and Their Relationship* (Defence Technical Information Center Research Note 85-12). Alexandria, VA.

NTSB (1982) *Air Florida, Inc., Boeing 737-222, N62AF, Collision with 14th Street Bridge near Washington National Airport, Washington, DC, January 13, 1982.* (NTSB Report Number AAR-82/08). Washington, D.C.: NTSB.

Orasanu, J. M. (1993) Decision-making in the cockpit. In E.L. Wiener, B.G. Kanki and R.L. Helmreich (eds.) *Cockpit Resource Management* (pp137–168). San Diego: Academic Press.

Patterson, R. (2000) *The Assertiveness Workbook: How to Express Your Ideas and Stand up for Yourself at Work and in Relationships*. Oakland, CA: New Harbinger Publications.

Pease, A. and Pease, B. (2005) *The Definitive Book of Body Language.* London: Orion.

Reader, T., Flin, R. and Cuthbertson, B. (in press) Factors affecting team communication in the Intensive Care Unit. In C. Nemeth (ed.) *Improving Healthcare Team Communication – Building on Lessons from Aviation and Aerospace.* Aldershot: Ashgate.

Reader, T., Flin, R., Lauche, K. and Cuthbertson, B. (2006) Non-technical skills in the intensive care unit. *British Journal of Anaesthesia*, 96, 551–559.

Reason, J. (1997). *Managing Organizational Accidents*. Aldershot: Ashgate.

Ross, R.R. and Altmaier, E.M. (1994) *Intervention in Occupational Stress. A Handbook of Counselling for Stress at Work.* London: Sage Publications.

Tannenbaum, S.I., Smith-Jentsch, K.A. and Behson, S.J. (1998) Training team leaders to facilitate team learning and performance. In J.A. Cannon-Bowers and E. Salas (eds.) *Making Decisions Under Stress: Implications for Individual and Team Training* (pp247–270): American Psychological Association.

Townsend, A., DeMarie, S.M. and Hendrickson, A.R. (1998) Virtual teams: Technology and the workplace of the future. *Academy of Management Executive* 12, 17–29.

Tushman, M. (1977) Special boundary roles in the innovation process. *Administrative Science Quarterly*, 22, 587–606.

US Chemical Safety and Hazard Investigation Board (2005) *Fatal Accident Investigation Report: Isomerisation Unit Explosion Final Report, Texas City, Texas, Usa.* Washington, D.C.: US Chemical Safety and Hazard Investigation Board.

Vrij, A., Akehurst, L., Soukara, S. and Bull, R. (2004) Detecting deceit via analyses of verbal and nonverbal behaviour in children and adults. *Human Communication Research*, 30, 8–41.

Volpe, C.E., Cannon-Bowers, J.A. and Salas, E. (1996) The impact of cross-training on team functioning: An empirical investigation. *Human Factors*, 38, 87–100.

Weick, K. (1990). The vulnerable system: An analysis of the Tenerife air disaster. *Journal of Management*, 16, 571–593.

West, M.A. (2004) *Effective Teamwork. Practical Lessons From Organisational Research* (2nd ed.). Leicester: BPS Blackwell.

Wiener, E., Kanki, B. and Helmreich, R. (1993) (eds.) *Cockpit Resource Management.* San Diego, CA: Academic Press.

Zuckerman, M., DePaulo, B.M. and Rosenthal, R. (1981) Verbal and non-verbal communication of deception. In L. Berkowitz (ed.) *Advances in Experimental and Social Psychology* (Volume 14, pp1–59), New York: Academic Press.

Chapter 5

Team Working

Introduction

This chapter focuses on the skills relating to working in a team. Teams are increasingly a feature of organisations, as work often involves people with different expertise who have to co-operate on the same tasks. What makes a team perform well has been the focus of psychological research in business, sport and military settings for many years (Hackman, 2002; 2003) and there are hundreds of books on how to manage and develop teams (e.g. Clutterbuck, 2007; West, 2004), including virtual teams (Brown et al., 2006). More recently teams in high-risk work settings have become a key topic for researchers, given the importance of effective team working for the maintenance of workplace safety (Awad et al., 2005; Turner and Parker, 2004; Glendon et al., 2006). This chapter presents some of the theoretical background on team working skills, especially relating to safety outcomes, including group behaviour, team development and effectiveness, as well as decision making in teams, team working problems, training techniques and the effects of stress on team working.

Despite the widespread use of teams in organisations, they do not always live up to expectations, due to factors such as inadequate communication, lack of role clarity or even team members not being prepared to operate as a team (Lencioni, 2005). Teams, however, can have a major impact on the success and functioning of organisations. There is increasing demand in the workplace for teams composed of distributed experts and cross-functional specialists to enhance flexibility and responsiveness (Kozlowski, 1998). Teams are seldom constant, and in modern organisations team members frequently change out over time. A core concept of crew resource management (CRM) training is not necessarily to strengthen any particular team but rather to make individuals more effective in whichever team they are working in.

This chapter presents an overview of team working skills, including skills for decision-making in teams. The type of team being discussed is an operational level group (typically 2–10 members), working on a shared task/s, and who are located in the same place. An individual team member needs the abilities to perform their tasks, as well as the abilities to work effectively in a team (West, 2004). Different types of training for team working skills are described (see also Chapter 10).

What is a team?

The definition of a team, according to Salas et al. (1992: p4), is:

> a distinguishable set of two or more people who interact, dynamically, interdependently, and adaptively toward a common and valued goal/objective/mission, who have each been assigned specific roles or functions to perform, and who have a limited life-span of membership.

The team itself has performance goals, as do the individuals in it. West (2004) describes advantages for implementing team-based working in organisations as including:

- enacting organisational strategy by creating consistency between changing organisational environments, strategy and structure
- developing and delivering products and services quickly and cost-effectively
- enabling organisations to learn and to retain learning more effectively
- promoting innovation due to cross-fertilisation of ideas
- integrating and linking together in ways that individuals cannot to ensure that information is processed effectively in the complex structures of modern organisations
- staff who work in teams report higher levels of involvement and commitment, as well as lower stress levels than staff who do not work in teams.

The main generic elements of individual team working skills are shown below.

Table 5.1 Category and elements of team working

Category	Element
Team working	Support others. Solve conflicts. Exchange information. Co-ordinate activities.

Team working is central to most work settings but is especially important in higher-risk industries, such as offshore oil and gas production, petrochemical, as well as aviation, medicine and maritime. Teams typically must function effectively from the moment they are formed to achieve their team task, therefore team members must have a common understanding of how they will be expected to work together from the outset (CAA, 2004). For instance, in the nuclear power industry, effective control-room operations rely on high-level team performance involving skills such as communication, co-ordination, co-operation and control (Stanton, 1996).

For many years, psychologists have been studying how people behave when they are in the company of other people, and there is an extensive literature on group behaviour (Baron and Kerr, 2003; Brown, 1999). More recently, significant progress has been made towards understanding work group performance and increasing current knowledge about team composition and team skills. Drawing on the psychology of group behaviour, a basic process model of team performance is shown in Figure 5.1. This illustrates how individual factors (e.g. experience, skills), leadership and team structures (e.g. roles, norms and status differences) affect group dynamics (e.g. processes like communication, co-ordination, decision-making), that then influence outcomes such as productivity, safety or satisfaction. Leadership is discussed in the following chapter. The structural factors relate to the size of the group (an optimum size for most tasks usually being about five members), as well as the norms or rules the group establish, the status differences or hierarchy (see Reader et al., in press), the roles team members take or are given (Belbin, 2003, discussed below) and the overall cohesion or attractiveness of the group to the team members.

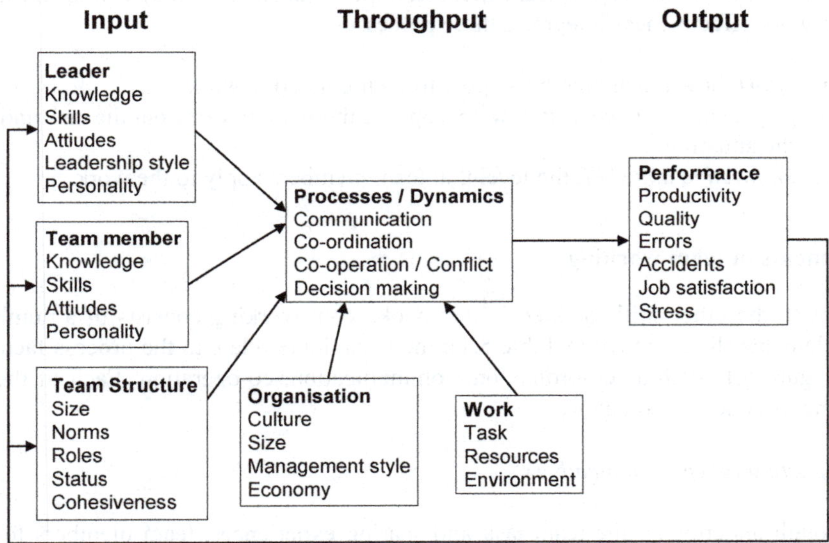

Figure 5.1 Process model of team performance (adapted from Steers, 1988; Unsworth and West, 2000)

As shown in Figure 5.1, the roles adopted by team members also influence the effectiveness of the team. This not only relates to the actual roles and accountabilities of team members, but also to the balance of personality types within a team. Belbin's (2003) team role self-perception inventory describes team roles at work in terms of the role types that he suggests are required for team effectiveness. He identified nine roles (including co-ordinator, shaper, implementer, completer-finisher) that team members need to fulfil if the team is to be successful. Teams do not have to consist of nine people – each team member can fulfil more than one role – but a balance of roles is necessary. Team members have preferred roles that they can take on, and the purpose is to respect the characteristics and strengths of each team member. Belbin's approach emphasises that effective teams require people with different outlooks and strengths; however, diverse teams may have values and expectations of behaviours that are difficult to reconcile, thus the integration of the team may suffer. Although there is limited empirical evidence that it enhances team working (Furnham et al., 1993), Belbin's approach receives much support from managers as a tool to facilitate discussion within a team about how the team members can work together.

The effectiveness of work teams is enabled by features such as leadership ability, well-designed tasks, appropriate team composition and a work context that ensures the availability of information resources and rewards (Hackman, 1987). The outputs from a group working together may be measured in terms of productivity or quality; in relation to safety, the outcome measures are usually errors or accidents. There are also individual effects such as the level of job satisfaction or work stress experienced due to being a member of this group.

In relation to individual team member input, Hackman (2003) comments that teamwork effectiveness integrates three processes:

a. *effort*: how much members apply to their collective work
b. *performance strategies*: how appropriate these are to carry out the task and to the situation
c. *knowledge and skill*: the level that team members apply to the work.

Elements of team working

As with the other skills defined in this book, team working consists of a number of elements (listed above in Table 5.1); the behaviours relate to the process factors in Figure 5.1, such as co-ordination, communication, co-operation. Each of these elements is described below.

Supporting other team members

Through undertaking the team task and sharing experiences, team members form social bonds and strong attachments, creating an emotional background to team working. Benefits of team working can include improved emotional well-being of team members (West, 2004). Effective team working includes providing support to other team members by, for example, sharing workload when appropriate, accepting individual responsibility, maintaining good working relationships and establishing openness. Social support also relates to informational support, such as providing advice and information to assist team members to carry out their tasks, which improves team co-ordination.

Conflict resolution

Although conflict typically has negative connotations and can lead to poor team performance or even the break-up of the team, constructive conflict in teams can be valuable, becoming a source of excellence, quality and creativity. West (2004) lists the skills for conflict resolution as:

- fostering useful debate, while eliminating dysfunctional conflict
- matching the conflict management strategy to the source and nature of the conflict
- using integrative (win–win) strategies rather than distributive (win–lose) strategies.

Conflict can arise due to the task (*what to do?*), team processes (*whose job is it?*) and interpersonal differences (*I don't like...*). Clarity of roles and responsibilities reduces team process conflict, and retaining an objective, non-emotional focus minimises the potential for interpersonal and task conflicts (Lingard et al., 2002).

Managing interpersonal conflicts in a team setting also involves assertiveness. Assertiveness means protecting your own rights without violating the rights of others, and communicating with respect. Examples of assertive speech include:

- 'So what you're saying is…'
- 'I can see that this is important to you, and it is also important to me. Perhaps we can talk more respectfully and try to solve the problem.'
- 'I think… I feel… I believe that…'
- 'I would appreciate it if you…'

Factors underlying assertiveness include persistence (staying focused on the issue), being objective (focusing on the problem rather than the emotions surrounding it), accepting criticism as being feedback rather than an attack, validating the truth and remaining factual, and using humour as appropriate to reduce tension. However, group structure factors influence this kind of behaviour, as indicated in Figure 5.1. It has been frequently demonstrated that team members who perceive themselves to be of lower status in the team are less likely to be fully involved in discussion or to speak up to challenge others or ask for help (Edmondson, 2003; Reader et al., in press). Therefore assertiveness training is important to ensure that team members have the skills and confidence to speak out when appropriate. Chapter 4 also discusses assertiveness.

Exchanging information

Effective exchange of information across all levels is essential for team working (CAA, 2006). Communication is discussed in greater detail in Chapter 4. Both spoken and written communication appropriate to the task and the context are required to ensure that team members are able to perform their required tasks and meet their goals. Analyses of communications from videotaped surgical operations of gall bladder removal showed that while surgeons might have good situation awareness, this was not always shared with other team members (Dominguez et al., 2004). West (2004) lists the skills for teamwork communication as:

- employing communication that maximises an open flow
- using an open and supportive style of communication
- using active listening techniques
- paying attention to non-verbal messages
- taking advantage of the interpersonal value found in greeting other team members, engaging in appropriate small talk, etc.

Co-ordination

Co-ordination results in better team working as opposed to work simply being undertaken by a group of skilled individuals (CAA, 2006). Poor co-ordination can result in breakdowns in communication, increasing errors and conflicts. Co-ordination is improved by the workload being equally distributed among the team to avoid any one team member being overloaded, monitoring of each other's performance, and, as described above, effective exchange of information and team members supporting one another.

What makes a good team?

What makes a good team, or even a great team? An example of effective team working in aviation is described in Box 5.1, where the pilot and co-pilot work together to resolve a novel situation.

Box 5.1 Effective team working in aviation

'There was one instance of a so-called Gimley Glider, where an aircraft, a [Boeing] 767, was flying across Canada and ran out of fuel for all the usual reasons; they had imperial and metric and all kinds of systemic stuff…
…they ran out 60 miles from Winnipeg and there wasn't a way of getting to Winnipeg without fuel. But two things happened: one, the 1st Officer recognised that down there on the edge of the lake was a deserted airstrip, which he flew from when he was a National Service pilot, and the Captain happened to be a very good glider pilot, so between them, they vectored themselves onto this tiny little airstrip, and the Captain said to the 1st Officer, "Shall I sideslip?" (kick it sideways, so you lose height without gaining speed, and then kick it straight at the last minute…)
… It's absolutely amazing, because they couldn't do it on simulator afterwards.'
Radio National (Australia) interview with Professor James Reason, May 2005.

The minimum characteristics of high-performance teams have been identified in a variety of domains (Salas and Cannon-Bowers, 1993):

- individual task proficiency (team members should be individually competent not only at their own tasks but must also have the necessary team working skills)

- clear, concise communication
- task motivation (the group need to feel like a team and be motivated to perform well)
- collective orientation (how team members view themselves in relation to other team members)
- shared goal and mission.

Additional requirements for enhanced performance also include:

- shared understanding of the task and of other team members' roles and responsibilities (shared mental models) to anticipate each other's needs
- team leadership
- collective efficacy (sense of 'teamness', i.e. the team members' belief in their team's ability or competence to achieve desired outcomes)
- anticipation (getting ahead of the 'power curve')
- flexibility (adjusting the allocation of resources to fit task and altering strategies to suit task)
- efficient, implicit communication (driven by an awareness of each other's needs)
- monitor own performance (self-correcting).

Advanced team performance can be developed through training, however this demands focused and directed interventions based on a clear understanding of the complexities and dynamics of the team and team functioning. A team is more than just a collection of individuals, and high-performance teams are made, not born, which raises the issue of the importance of team training (see Chapter 10). Training of individuals working in a team is based upon developing the competence, knowledge, skills and attitude of individuals, and training objectives, methods and tools. To minimise the possibility of team working problems (see below), teams should develop an 'identity' where team members learn their own roles and tasks, and understand those of other team members, either through training, shared experiences or by reflecting on how the team functions. Teams need to develop more automatic ways of co-ordinating effectively (Klein, 1998). Team members also need to know how things are done in their team, i.e. following accepted processes.

Team working problems contributing to accidents

Good teamwork is particularly important to reduce error and maintain safety in the workplace. There are many documented cases where teamwork failures have contributed to accidents, especially in aviation (see Wiener et al., 1993), and, as discussed earlier, this was the trigger for the development of crew resource management and other non-technical skills training. The example of good team working shown in Box 5.1 can be contrasted with the failure of team working in a naval incident described in Box 5.2 below.

Box 5.2 Failure in team working on board a nuclear submarine

On the afternoon of 9 September 2001, the *USS Greeneville* (a fast-attack nuclear submarine) collided with the *Ehime Maru* (a Japanese fishing and training vessel) off the coast of Oahu, Hawaii. At the time of the collision the *Greeneville* was performing an emergency surfacing manoeuvre for a group of civilian guests on board the submarine. As the submarine rose to the surface, it struck the *Ehime Maru*, causing the ship to sink in less than 10 minutes. Of the 35 Japanese crew, instructors and students, 26 were rescued and 9 perished.

The National Transportation Safety Board (2005) determined that the probable cause of the collision was the Commanding Officer's overly directive style, and a failure in team working and communication of key watchstanders. An outline of the main failures in non-technical skills that led to the accident are described below:

- As the submarine was behind schedule, the Commanding Officer (CO) rushed his crew, resulting in the truncation of recommended steps for safe operation.
- Both the sonar supervisor and the fire control technician of the watch made assumptions about sonar contacts in the area, which later resulted in their providing incomplete and erroneous information to the CO.
- As opposed to overseeing the inexperienced Officer of the Deck-2 (OOD-2) by having him direct vessel movements that the CO could verify as being correct, the CO essentially took over the conn (directing the steering of the submarine) without acknowledging that he was doing so. He ordered specific depths and turns, which the OOD-2 repeated to the diving officer and the helmsman.
- Before the Greeneville rose to periscope depth, the CO announced to the personnel in the control room that he had a 'good feel' for the contacts. 'In effect, the CO's flawed situational awareness regarding the proximity of vessels at the surface reinforced and influenced the watchstanders' own limited situational awareness which had the effect of discouraging backup from his crew' (p47).

The failure of the two-way communication necessary for effective bridge resource management in the *Greeneville* control room was further compromised by the CO's management of, and interaction with, the visitors. The 16 civilian visitors created both physical and communication barriers for the personnel in the control room.

NTSB (2005); Roberts and Tadmore (2002).

Reviews of other high-profile accidents – Pan Am Flight 401 crash (1972), the explosion of the chemical process plant at Flixborough in England (1974), the Pan Am and KLM collision at Tenerife (1977), the Three Mile Island accident in the USA (1979), and the *USS Vincennes* incident (1988) – identified three main teamwork problems (Rouse et al., 1992), namely:

- *Roles not clearly defined* – a lack of clearly, and appropriately, delineated roles was found in all instances. For example, during the Pan Am Flight 401 accident in 1972 (see Box 5.3), a troubleshooting task dominated the crew's attention and flight control was ignored.

Box 5.3 Flight 401 incident (1972)

A Lockheed L-1011 with 176 people on board (163 passengers and 13 crew members) crashed 19 miles north-west of the Miami International Airport on 29 December 1972. In total, 99 passengers and 5 crew members were fatally injured in the incident. The flight from New York to Miami was turning in to land when the crew noticed that only two of the three main landing gear lights had illuminated but the nose wheel light had not.

The NTSB (1973) attributed the crash to 'the failure of the crew to monitor the flight instruments during the final four minutes of flight, and to detect an unexpected descent soon enough to prevent impact with the ground.' Furthermore, the report commented on the 'preoccupation with a malfunction of the nose landing gear position indicating system [which] distracted the crew's attention from the instruments and allowed the descent to go unnoticed.'

The crew, consisting of the captain, first officer and flight engineer, all became involved in dealing with the nose wheel light, such that no one was monitoring the plane's flight.

- *Lack of explicit co-ordination* – conflicting goals were not balanced properly in the operations team in two of the incidents listed above. A contributory cause of the release at the Three Mile Island nuclear power plant was due to the focus on the goal of protecting pumps rather than shutting down, which resulted in losing the plant. At Flixborough, the focus was on maintaining operations rather than safety, and again the result was losing the plant.
- *Miscommunication/communication* – the plane crash at Tenerife is a classic example that has been used earlier to illustrate failures in non-technical skills (Box 1.1). In Box 5.4, we present another aspect of this accident to illustrate communication problems within and between teams. Although English is the international language of aviation, there can be difficulties for non-native speakers relating to accents and language differences. The lack of communication, as well as miscommunication, contributed to the collision between the two airliners.

Additional analysis of the formal records of investigations into other aircraft accidents indicated incidents where crew co-ordination failed at critical moments (Helmreich and Foushee, 1993). For example:

- A co-pilot, concerned that take-off thrust was not properly set during a departure in a snow storm, failed to get the attention of the captain, resulting in the aircraft stalling and crashing into the Potomac River in Washington, D.C. (see Box 4.5).
- A crew, distracted by non-operational communications, failed to complete checklists and crashed on take-off because the flaps were not extended.
- A breakdown in communications between a captain, co-pilot and air traffic control regarding fuel state led to a crash following complete fuel exhaustion.

Box 5.4 Tenerife disaster (1977)

The following is the final section of the transcripts from the cockpit voice recorder of the KLM and Pan Am Boeing 747 collision in Tenerife (27 March 1977) (taken from ICAO Circular 153-AN/56 (pp22–68)):

Note:	**KLM – KLM captain (KLM 2 – co-pilot; KLM 3 – engineer)** **APP – Tenerife tower** **RDO – Pan Am radio communications (co-pilot)**
KLM:	...four eight zero five is now ready for take-off ... uh and we're waiting for our ATC clearance
APP:	KLM eight seven zero five uh you are cleared to the Papa Beacon climb to and maintain flight level nine zero right turn after take-off proceed with heading zero four zero until intercepting the three two five radial from Las Palmas VOR.
KLM:	Ah roger, sir, we're cleared to the Papa Beacon flight level nine zero, right turn out zero four zero until intercepting the three two five and we're now (at take-off).
KLM 2:	We gaan. [We're going]
APP:	OK.
RDO:	No... eh.
APP:	Stand by for take-off, I will call you.
RDO:	And we're still taxiing down the runway, the clipper one seven three six

RDO and APP communications caused a shrill noise in KLM cockpit – messages not heard by KLM crew.

APP:	Roger alpha one seven three six report when runway clear
RDO:	OK, we'll report when we're clear
APP:	Thank you
KLM 3:	Is hij er niet af dan? [Is he not clear then?]
KLM:	Wat zeg je? [What do you say?]
KLM 2:	Yup.
KLM 3:	Is hij er niet af, die Pan American? [Is he not clear that Pan American?]
KLM:	Jawel. [Oh yes. – emphatic]

Pan Am captain sees landing lights of KLM Boeing at approximately 700m. The two planes collide.

While the aviation accidents are particularly well documented, studies of adverse events in health care (Leonard et al., 2004; Reader et al., 2006; Vincent, 2006) and other industries (Turner and Parker, 2004; Glendon et al., 2006) have frequently demonstrated failures relating to communication and co-ordination in teams with status differences and role confusion as key contributing factors.

Team working problems can also arise due to the level of experience within a particular team, i.e. how long and how effectively the team has been working together, irrespective of any individual team member's experience. Members of inexperienced teams show more of a tendency to focus on their own individual tasks and the overall goals. Confusion can also arise in inexperienced teams about roles and responsibilities. Some team members may not fully assume their tasks, work less hard when in the team setting than if they were working alone, and may be insensitive to the needs of other team members, resulting in loss of effort. On the other hand, members of experienced teams who are accustomed to working together focus on the overall status of the team, identify themselves with the whole team, appreciate that team members may need to compensate for others, ask for help and support each other to achieve team goals (Klein, 1998). Related to the issue of experience, are the changes that occur at different stages in a team's lifespan.

Stages of team development

Tuckman (1965) has suggested that teams go through a series of stages from their initial formation through to performing, as shown in Table 5.2.

Table 5.2 Stages of team development (Tuckman, 1965)

Stage	Definition
Forming	Typically characterised by ambiguity and confusion when the team first forms. Team members may not have chosen to work together and may be guarded, superficial and impersonal in communication, as well as unclear about the task.
Storming	A difficult stage when there may be conflict between team members and some rebellion against the task as assigned. Team members may jockey for positions of power and frustration at a lack of progress in the task.
Norming	Open communication between team members is established and the team starts to confront the task at hand. Generally accepted procedures and communication patterns are established.
Performing	The team focuses full attention on achieving the goals. The team is now close and supportive, open and trusting, resourceful and effective.

A final stage – adjourning – was added by Tuckman and Jensen (1977), which relates to the dissolution of the team. This involves termination of roles, completion of tasks and reduction of dependency. Some teams, e.g. project teams who are brought together for a specific task, have well-defined adjournments, whereas for other teams this stage is less well defined. It should be noted that not all teams either go or need to go through these stages of development. For example, air crew, control room teams and medical teams (e.g. accident and emergency) may not undergo the forming and norming stages, as the organisational norms are in place to cover this aspect in terms of the roles and procedures for a team achieving a specific task.

These stages are not necessarily an accurate representation of every team's development but the model is generally accepted, although there is limited empirical support. Teams may go through periods of flux or variation that can affect the team development. Some team members may already know each other and quickly establish interactions within the team. Also, stages may be missed out, demarcation between stages can be unclear and teams may need to repeat stages.

Team effectiveness

Team effectiveness, according to Guzzo and Dickson (1996), is indicated by group-produced outputs (as shown in Figure 5.1). These include the quantity or quality of production, speed, the consequences a team has for its members, or the enhancement of a team's capability to perform effectively in the future. In risky work settings, safety is an important outcome, which may be assessed by errors or accidents. For a team to be effective, the task proficiency of individual team members can be enhanced through selection or training (Tannenbaum et al., 1996). A successful team is a function not only of team members' abilities and available resources, but also of the processes that teams use to interact with each other to accomplish the team task (Marks et al., 2001). That is, due to team member interdependency, team effectiveness can be enhanced through improving the team members' interactions with one another (Salas et al., 1992). Teams such as operating theatre teams, i.e. composed of people with expertise in different disciplines and who may only work together for a limited period of time, need to develop relationships quickly. In project teams, members are also from different disciplines, and this type of team may be formed for a limited duration – nevertheless, this is typically weeks or months, so the team members have time to train together and develop team processes.

Teamwork and taskwork behaviours

We must also emphasise here the distinction between taskwork and teamwork, which is crucial to understanding team performance (Salas et al., 2002).

- 'Taskwork' refers to behaviours that are related to the operations activities that team members must perform, i.e. the technical aspects of the team task. Taskwork behaviours include understanding task requirements, operating procedures and task information.
- 'Teamwork' skills reflect the behavioural interactions and attitudes that team members must develop before they can function effectively as a team. Teamwork behaviours are distinct from taskwork behaviours as they relate to how team members, irrespective of role and task within the team, ensure efficient team functioning.

Effective team performance is achieved through the integration of taskwork and teamwork skills. Teamwork problems, especially in non-routine situations, often occur as a result of poor teamwork skills. McIntyre and Salas (1995) proposed four critical teamwork behaviours and two team performance norms that are interdependent and contribute to effective performance, as shown in Table 5.3. These teamwork behaviours strengthen the quality of functional interactions, relationships, co-operation, communication and co-ordination of team members.

Table 5.3 Teamwork behaviours and performance norms (Salas et al., 2000, p24)

	Dimensions	Definition
Teamwork behaviour	Performance monitoring	Team members observe the behaviour of team members and accept that their behaviour is being observed or monitored as well.
	Feedback	A climate in which feedback is freely offered and accepted among team members.
	Closed-loop communication	A three-step sequence whereby a message is sent by a team member, another team member provides feedback regarding the received message and the originating team member then verifies that the *intended* message was received.
	Backing-up behaviours	Behaviour among team members that indicates both the competence and willingness to help, and be helped by, other team members.
Team performance norms	Team self-awareness	The attitude or value that team members see themselves first as members of the team and as individuals second.
	Fostering of team interdependence	A team member attitude that reflects the extent that individual team members' success depends on the success of other team members as well.

Additional factors include adaptability, shared situational awareness, leadership/team management, interpersonal relations, co-ordination, communication and decision-making.

Various approaches to studying team working have been proposed. These include basic group behaviour models (e.g. McGrath, 1984, or Figure 5.1) and more specific analyses of work teams. Salas and his colleagues (2002; 2004; 2006) have conducted extensive research into military teams. Their integrated team effectiveness model is presented below, as it provides a comprehensive account of relevant factors.

Integrated model of team effectiveness (Salas et al., 2002)

One of the main concepts of the integrated model of team effectiveness is that of the competencies, i.e. knowledge, skills, attitudes (KSAs), required by individuals. These KSAs, once identified, can be improved through training. Although the model describes the characteristics of the individual, the team and the task, as well as the work structure, which impact on team output, and the organisational characteristics in which a team functions (as shown in Figure 5.1), we shall focus on KSAs in this chapter, as these competencies are key to effective performance. This model also defines the co-ordination and communication patterns that develop.

Individual team members who make up a team each generally have specific tasks and responsibilities that are related to their particular position in the team. Examination of the taskwork involved can help to diagnose the root of individual performance outcome problems (Johnston et al., 1997). Individual characteristics include the knowledge, skills, motivation, attitudes and personalities that individuals bring to the team task, which all impact on team performance. Taken together, individual proficiency and motivation to perform are necessary, but not sufficient, conditions for effective team performance.

Knowledge-based competencies: By working towards a shared goal or objective, members of a team use existing knowledge, which allows them to predict and plan performance in future situations, along with knowledge of their own role and responsibilities and those of the rest of the team (Cannon-Bowers et al., 1995). Individuals must possess compatible mental models (i.e. knowledge structures) about their team-mates' roles, the tasks and the situations encountered by the team to be effective. Expectations are created by these knowledge structures, allowing individuals to generate predictions concerning how to perform during routine situations, as well as when the team encounters a novel or stressful situation. In the workplace, this knowledge may be encapsulated in the form of a doctrine or 'how we will operate' document that outlines how the team will work.

Skill-based competencies: These refer to the co-ordination, communication, synchronisation and adaptation of behaviours and actions by individuals in the team to attain goals and to perform effectively and accurately. Skill-based team competencies include adaptability, shared situational awareness, mutual performance monitoring, motivation, decision-making, assertiveness, interpersonal relations and conflict resolution. These skills, employed by team members in a team working situation, should also be brought by individuals to the team, to fulfil goals and missions by performing and acting in an accurate and timely fashion.

Attitude-based competencies: These refer to the cohesion, morale and motivation of members of the team, at an individual and team level, to perform collectively and interdependently. Furthermore, collective efficacy, shared vision, mutual trust and a belief in the importance of teamwork are desirable attitudes. How individuals feel about the task and team members is important, including respect for others, a readiness to listen to the opinions of others, an ability to predict and to take the behaviours of others into account.

Effective team performance does not only depend on the KSAs of individuals but also on the traits of individual team members that facilitate team interaction and functioning, such as learning ability, initiative, risk-taking propensities, adaptability, tolerance for stress and so on (Paris et al., 2000). Leadership qualities (as discussed in Chapter 6) also influence team performance (Klimoski et al., 1995). While appreciating that each team member does not necessarily require exactly the same set of KSAs, it is important that these are compatible. They can be categorised as being either task- or team-specific/-generic (Paris et al., 2000), as shown in Table 5.4. Specific competencies mean that these relate only to that particular team or task, whereas generic competencies relate to any team and to any task. The competencies required by various types of teams are related to the tasks and the make-up of the team.

Table 5.4 Team competencies framework (adapted from Paris et al., 2000, Table 4)

	Team-generic	**Team-specific**
Task-generic	*Transportable* – can be applied to both different task situations and teams – needed where team members work on a variety of teams, as well as on a variety of tasks	*Team-contingent* – specific to a particular set of team members, but not to a particular task – needed when team membership is stable, but the tasks vary
	• taskforces • process-action teams • project teams	• self-managing work teams • management teams • department teams
Task-specific	*Task-contingent* – related to a particular task, but not dependent on a specific set of team members – needed when team members perform a specific team task but do not work consistently with the same team-mates	*Context-driven* – dependent on a particular task and team configuration – needed when team membership is stable and the number of tasks is small
	• commercial aviation crews • specialist emergency teams • medical teams	• nuclear operations teams • offshore oil platform teams • military teams • emergency medical teams • sports teams

The relevant competencies, in terms of KSAs, vary for each of these four team types. For example, members of teams that have a stable set of personnel (that require context-driven and team-contingent competencies) require knowledge of team-mates, whereas this is not as necessary for teams where team members are not constant (requiring task-contingent and transportable competencies).

Team identity

Team identity refers to the extent to which team members conceive of the team as an interdependent unit and operate from that perspective while engaged in the task (Militello et al., 1999) and includes:

- *Engaging all members*: The extent to which the team members participate in the team's work and take responsibility for reaching the team goals, including behaviours that signify self-involvement as well as the encouragement of other members to get involved.
- *Compensating and coaching*: The extent to which team resources are shifted to cover areas in which the team cannot satisfy its roles, functions and responsibilities. For example, team members step appropriately outside their assigned team roles or functions or add to their assigned functions to help the team reach its goals.
- *Interpersonal aspects*: The harmonious or conflicting styles of the team members, as determined by a number of different tests or profiles in use.

Teamwork is the seamless integration of a number of KSAs, both of the individual and the team. Teamwork occurs when the task drives team members to pull from their repertoire appropriate competencies to accomplish an objective, and the KSAs that team members use to adapt and enhance performance. It is important to understand the nature of teamwork because the KSAs that compose it become the target of team training.

Team processes and interventions

Team processes and team interventions have an impact on what the team produces. Team processes include co-ordination, communication and problem-solving, as shown in Figure 5.1 and described below. In high-risk work settings, special emphasis is placed on decision-making in teams (see below) due to the additional complexities – such as different motives, agendas, opinions and the impact of organisational policy – that can emerge when teams are making decisions. The team's ability to operate as an intelligent entity that thinks, solves problems, makes decisions and takes actions collectively so that the task demands are met (Militello et al., 1999) include:

- *Envisioning goals*: The extent to which the goals of the team are understood and used to direct the team's behaviour towards those goals – this entails the team's ability to identify its goals, to ensure that all members share a common definition of the goals and the degree to which this shared understanding contains a workable temporal element.
- *Maintaining dynamic focus*: The team's ability to concentrate its planning and decision-making within an appropriate span of time and on a relevant breadth

of concepts and information. It can also concern the team's ability to attend to the most relevant cues, and to focus ahead an appropriate amount of time to anticipate and act on forthcoming events.

- *Assessing the situation*: The ability to form a shared understanding of the situation with which the team is operating, which involves forming plausible scenarios about what is happening and holding on to alternative hypotheses until ambiguity is resolved. The team must allow for divergence in which views from a variety of team members are sought and information from multiple sources is considered. This includes retaining ambiguous information and uncertainty until it can either be confirmed or rejected. Convergence, which is the team's ability to harness various views to form a situation assessment that is as close as possible to the truth, is also implied. Methods such as the use of maps, diagrams, messages and meetings ensure that relevant team members are viewing the situation similarly and are working from a similar perspective.
- *Articulating expectations*: The extent to which the team articulates its expectations about the progress of a course of action to relevant team members. Making such expectations explicit helps the team to attend to relevant information and serves as an alerting function, which in turn increases the team's ability to grasp the significance of a message and to react quickly.
- *Envisioning and evaluating courses of action*: This concerns the team's ability to use its collective experience to visualise how a plan will be co-ordinated and conducted and to envision where it might run into trouble.
- *Monitoring*: The team's ability to examine itself for signs of both effective and ineffective teamwork behaviours. The team can recognise how it is operating as a team and make adjustments for meeting the team task. Recognising effective behaviours increases the team's ability to use them in the future. The team can also look to the future to anticipate events that might require specific team performance strategies.
- *Adjusting*: The team's ability to change strategies when current ones become less effective or are not working. Also involves the team's ability to anticipate changes that may require different strategies for meeting the demands of a future task or situation.
- *Detecting gaps and inconsistencies*: The extent to which the team actively attempts to discover and fill gaps in the team's information base and assumptions, to recognise and handle inconsistencies or contradictions that might be present, either through eliminating the inconsistencies or noting them until they can be resolved.
- *Time management*: The team's ability to meet goals before deadlines overtake them. Managing time includes sequencing sub-tasks so that output from one task connects where and when it should as input to the next task.

Although a number of additional factors, over and above those described above, that contribute to the overall effectiveness of a team's performance are included, this merely emphasises the complexities involved in effective team performance. Individual team member and leader skills, through communication, co-ordination, decision-making and problem-solving, enable a team to perform and achieve its goals.

This model of team effectiveness provides an integrated approach based on a number of different team performance models, team research and team-driven theory. It involves individual and team competencies that have an impact on team performance, and emphasises aspects such as decision-making, situation awareness, as well as motivation and personality. Team effectiveness research has helped to make significant advances in understanding what teams do, feel and think and have allowed for the development of training interventions to examine team processes (Salas and Cannon-Bowers, 1997).

Team decision-making

One of the key aspects of team working is that of team decision-making. Decision-making by teams refers to the process of reaching a decision undertaken by interdependent individuals to achieve a common goal (Orasanu and Salas, 1993). Team decision-making is distinguished from individual decision-making as more than one information source or task perspective must be combined to make a decision. Team members, although working towards the same goal, may have different agendas, motives, perceptions and opinions that must be combined to reach the decision. In some cases, team members feed information to a team leader who then makes decisions for the team; however, in team decision-making the components of the decision-making process may be distributed across many individuals (Zsambok, 1997). Box 5.5 describes the team problem-solving and decision-making that occurred between the crew and mission control in the Apollo 13 incident (1970).

Box 5.5 Team decision-making in the Apollo 13 incident (1970)

The resolution of the incident involving the crew and controllers of the Apollo 13 mission has been referred to as a 'successful failure'. The famous quote, *'Houston, we have a problem,'* indicated that an explosion had occurred on the spacecraft, resulting in a reduction in the supply of oxygen to the fuel cells that formed the craft's primary power source and crippling the electrical system. The astronauts worked closely with the ground-based team at mission control over a three-day period to resolve the situation.

On the ground, mission controllers, along with specialists from contractors and other experts, set up a network involving computers and simulators to solve the problem of determining course corrections, manoeuvres and equipment resources.

(http://liftoff.msfc.nasa.gov/Academy/History/APOLLO-13/mission-report.html)

Skills for collaborative problem-solving are (West, 2004):

- using an appropriate level of participation for any given problem, and
- avoiding obstacles to team problem-solving, such as domination by some team members, by structuring how team members interact.

Shared situation awareness

Team decision-making involves concepts such as shared situation awareness (Salas et al., 1995) and shared mental models (Orasanu and Salas, 1993):

- *Shared, or team, situation awareness* refers to the extent to which the team members share the same interpretation of ongoing events (Cannon-Bowers et al., 1993; Zsambok, 1993). Shared situation awareness is defined as the ability of team members to gather and use information to develop a common understanding of the task and team environment (Salas et al., 2006). In dynamic environments, it is easy for team members to form differing impressions of the current situation without realising it, and for discrepant views to create difficulties when making decisions. This in turn can create conflict between team members. However, it is not essential that all team members receive all information, only that they receive information relevant to their role. Ensuring that all team members are given the same information to form shared situation awareness may result in information overload. Teams need to consider which team member needs to know what information to carry out their individual tasks, while at the same time maintaining an overview of the team task and how their actions contribute to achieving that task. Analyses of communications from videotapes of gall bladder surgery showed that while surgeons might have good situation awareness, this was not always shared with other team members (Dominguez et al., 2004).

  A tool to assist shared situation awareness developed by Endsley and Jones (2001) is shown in Table 5.5 below. These features generally appear to promote good team functioning, by indicating the importance of sharing basic information, the comprehension of what other team members need to share and anticipation of how the situation may be altered.

Table 5.5 Shared situation requirements (adapted from Endsley and Jones, 2001)

What? (Data)	• job and kit • environment • other team members
Now what? (Comprehension)	• status of own goals • status of others' goals • impact of own actions on others • impact of others' actions on self/goal
So what? (Projection)	• actions of team members and leaders

The basic data stage refers to sharing information about the operational system and its environment. Next, team members need to understand what team members are doing and how this affects their own activities as well as how this affects the overall goal. Finally, team members think ahead about what they will do next but also what other team members will do.

- *Shared mental models* is defined as the 'knowledge structures held by members of a team that enable them to form accurate explanations and expectations for the task, and, in turn, to co-ordinate their actions and adapt their behaviour to demands of the task and other team members.' (Cannon-Bowers et al., 1993: p228). In a team, a shared mental model relates to the team members having the same understanding of the dynamics of key processes (Klein, 2000). Key processes include, for example, the roles and functions of each team member in accomplishing the task, the nature of the task, equipment. Encouraging the development of shared mental models can improve the team's ability to co-ordinate efforts, to adapt to changing demands and to anticipate the needs of the task and other members. Regular review of team functioning helps to improve team adaptability and reflexivity (West, 2000). A four-stage exercise in role negotiation (West, 2004) can be used to help develop a better understanding of roles and functions of team members:

  Step 1: Team members list objectives and principal activities.
  Step 2: Team members examine the information for each role.
  Step 3: Team members note what behaviours each other role should adopt.
  Step 4: Pairs of team members meet to examine the results and reach agreement about the various requests.

In the Apollo 13 incident described in Box 5.5, shared situation awareness was crucial, as the ground-based mission controllers had to make sense of what the problem was and how it could be solved. Klein (1998) comments that mission controllers had to form a 'big picture' of what had happened to the spacecraft, which would then allow them to work out how to save the astronauts. Box 5.6 shows how experts in the oil and gas industry also form a big picture to solve problems and make decisions.

Box 5.6 Team decision-making in the oil and gas industry

'Technical problems are dealt with pretty straightforwardly. We lay out all the options and engage the "bigger brain", look at it and find the right technical solution, and that means including service suppliers or whoever, and our research people if necessary to derive the best solution we can.

With respect to people, we will gather the team leaders and discuss how we can best deal with a particular situation and we will sit down and talk with the individual(s) and try to work through it. We do not let problems linger. That doesn't mean that we don't purposefully park things if we think that it is a much lower priority than other things that need to be dealt with right now, but it is noted, documented, and we will come back to it.' Comments by member of oil and gas drilling team.

Difficulties in team decision-making

Sources of failure in team decision-making, according to Orasanu and Salas (1993), include poor communication, logical errors, inadequate situation assessment and pressure to conform. Compared with individuals, teams have increased cognitive resources and might be expected to perform better than individuals. For example, team members can monitor each other's performance, pool their knowledge or observations, suggest strategies or options, provide alternative viewpoints, reduce workload by sharing tasks, provide feedback to one another and critique each other. However, some difficulties with team decision-making can arise:

- 'Groupthink': Originally identified by Janis (1972), groupthink involves a group suspending its rational judgement in order to maintain group cohesion, often by accepting, without challenge, a proposal by a respected leader. An example of hesitance to contradict figures of authority is that of the Bay of Pigs incident in 1961 (Kramer, 1998) described in Box 5.7.

Box 5.7 Bay of Pigs incident (1961)

In 1961, an invasion of Cuba by a group of Cuban exiles was supported by the CIA with the authorisation of President John Kennedy and his advisers. The invasion was a failure being readily repulsed by the Cuban army. President Kennedy and his advisers were young, optimistic and committed to civil rights and democracy, yet they collectively had considered that this event could be successful. They, led by the President, had formed a cohesive group which appeared to be insulated from outside opinion.

- *Inhibition:* Individual team members may feel inhibited in contributing information relevant to a decision, but instead team members offer information that is already shared by the group.
- *Failure to challenge:* Team members may assume that they share similar goals and operate on that assumption, leading to false consensus or pluralistic ignorance.
- *Poor communication/shared experiences:* Team members may think along similar lines but all can be incorrect. Another problem can be if team members assume that others share understanding of words (e.g. risk, threat, likelihood) when they actually may interpret them differently. In an incident involving delayed treatment to a patient in an intensive care unit (ICU), the staff involved were all using the term 'bumpable' to refer to patients suitable for transfer out of ICU, but their use of the term may have had slightly different meanings depending on their professional group (Cook, 2006).
- *Decision-making:* Teams may make poor decisions due to factors such as hostility within the team, lack of co-operation, lack of motivation.
- *Status:* A higher status team member may reject relevant information offered by a lower status team member.

- *Organisational policy:* The organisational context in which the team operates may affect team decision-making.

Improving team decision-making

One approach to ameliorate the difficulties in team decision-making described above is to adopt the advanced team decision model (ATDM). This was developed by Zsambok et al. (1992) as a training programme for senior military officers to achieve more effective strategic team decision-making. The project focused on providing officers with the skills needed to observe, diagnose and improve decision-making of teams on which they may serve for only a brief period. The ATDM model identifies 10 key behaviours that are possessed by high-performing teams. The model was initially developed for planning teams (Klein et al., 1993; Klein and Miller, 1999) but has since been adapted for action teams (Thordsen et al., 1996). The distinction between planning and action teams is that the job for a planning team is to produce a plan, whereas the job for an action team is to accomplish a task, i.e. by carrying out a plan.

The three core components of the ATDM model are shown in Figure 5.2.

- *Team identity* describes the extent to which team members view the team as an independent unit, and operate from that perspective while engaging in tasks.
- *Team conceptual level* (or team thinking) is concerned with the group mind of the team that thinks, solves problems and takes actions collectively.
- *Team self-monitoring* is a regulatory process for all other processes. It is the ability of the team to observe itself while acting to accomplish its tasks.

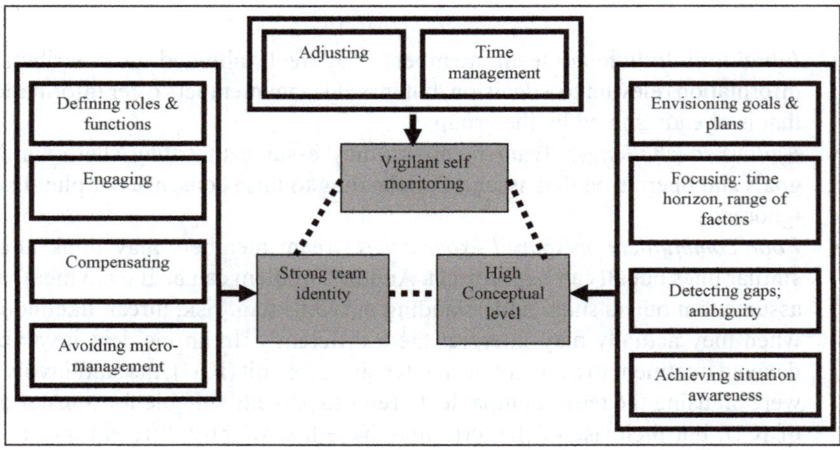

Figure 5.2 Advanced team decision-making (Zsambok et al., 1992)
Reproduced by permission of Klein Associates Inc.

The ATDM is intended to form the basis of both a training and an assessment tool. Full descriptions of the model and examples of the concepts and behaviours should be provided to team members and observers before use, and a questionnaire can be used to assess the team's performance on the 10 key behaviours. The ATDM was developed for strategic as opposed to operational decision-making, therefore it would need to be further examined and modified to suit any alternative circumstances.

Training team decision-making

Team decision-making can be improved by organisational design, system design or formal team training (Orasanu and Salas, 1993). Effective training in team decision-making includes training teams in situation assessment, i.e. rapidly combining information and interpretations from team members, to evaluate options by using mental simulations, and using communication to develop shared mental models. CRM training courses typically train crews to manage information resources and workload, to co-ordinate their actions and communicate effectively. Chapter 10 provides more details on tools and techniques that can be used to train team decision-making.

Team working and stress

Where team members have been trained and are highly motivated, then their team co-ordination should be more resilient to stress (physical, emotional or managerial). High stress has been shown to lead to decreases in communication and increases in errors, such as incorrect decisions. Environmental stressors that affect team performance in an operational environment include (Cannon-Bowers and Salas, 1998):

- multiple information sources
- incomplete, conflicting information
- rapidly changing, evolving scenarios
- requirement for team co-ordination
- adverse physical conditions
- performance pressure
- time pressure
- high work/information load
- auditory overload/interference
- threat.

Interventions aim to reduce the effect of these stressors on team performance, especially team decision-making. Team adaptation and co-ordination is one of a range of training interventions (e.g. event-based training, team leader training, cross-training) and is designed to teach team members about the importance of teamwork and introduce important team working skills (Serfaty et al., 1998). The topic of training team working is discussed in more detail below.

Teams who maintain superior performance under high levels of workload and stress have been shown to employ different co-ordination strategies from low-performing teams (Serfaty et al., 1998). High-performing teams have the ability, when faced with demanding tasks, to adapt their decision-making strategies, co-ordination strategies and even their structure. It has been proposed that effective teams are those who have developed shared situational mental models of the task environment and the task, as well as a mental model of other team members' tasks and abilities. The ability to anticipate changes in the situation and changes in team members' needs contribute to the team's superior performance, especially under stress.

Some of the key lessons learned from team co-ordination training include (Serfaty et al., 1998):

- *Adaptive teams are high performers*: Flexibility in decision-making and teamwork processes is achieved through successful strategies of adaptation to stress. Stressors such as task uncertainty and time pressure stimulate changes in behaviour, strategy and organisational structure.
- *Team adaptive co-ordination is trainable*: Co-ordination and communication strategies can be trained to allow operational teams to cope with stressful conditions. Strategies include adaptive co-ordination strategies and regular situational updates.
- *Significant improvements in team performance can be achieved using a relatively simple training intervention:* Principles of training (i.e. instruction, demonstration, practice and feedback) can produce teamwork skill improvements.
- *Shared knowledge is necessary but not sufficient*: Team co-ordination strategies that operate on, update and maintain the team's shared mental models are necessary. Effective team training depends on the ability to help teams develop shared mental models and the ability to train teams in using these mental models.
- *Knowledge training should be combined with team skills training to maximise performance*: Strengthening shared mental models should have a positive effect on the team and consequently on team performance. Shared mental models can be enhanced through additional interventions such as cross-training (familiarising team members with the roles and tasks of other team members). As team members share a common awareness and understanding of the situation, then team performance is enhanced.

Much of the research described above has been undertaken with naval combat information centre teams under the tactical decision-making under stress (TADMUS) research programme (see Cannon-Bowers and Salas, 1998, for further details). The objective of the research programme was to understand complex team processes and to develop team training strategies. Many of the concepts researched and subsequent conclusions may also apply to control room personnel or operational teams in a range of settings.

Training team working

Team-building interventions for established or fixed teams, such as away-days (e.g. paintballing, abseiling), have been shown to have little effect on team task performance, although they often affect team members' relationships, attitudes to each other and team cohesiveness (Tannenbaum et al., 1996). In terms of developing the skills of team working, however, team-building interventions *per se* may have little impact, unless clear skills development objectives are included. Training team working to enhance team performance therefore relies on specific training objectives, and that training should be continuously focused on team working. Training of established teams also differs from training of individuals to perform better in teams, which CRM training addresses (see Chapter 10).

Salas and his group at the US Naval Air Warfare Centre (USNAWC) during the late 1990s developed and tested a number of different team training techniques designed to improve team performance. This research was a result of the shooting down of an Iranian passenger airliner by the *USS Vincennes* in 1988 after misinterpretation of data from electronic sensors (Orasanu and Salas, 1993). Four examples of team training instructional strategies that were developed from this military research – i.e. cross-training, team self-correction, event-based and team facilitation training – are described.

Cross-training

Volpe et al. (1996) define cross-training as an instructional strategy in which each team member is trained in the duties of his or her team-mates. Salas and Cannon-Bowers (1997) recommend this type of training as being particularly useful where there is a high turnover of personnel. Three levels of cross-training can be identified, which differ in the depth of information provided (Blickensderfer et al., 1998).

- *Positional clarification:* This aims to provide team members with a general understanding of the role and the responsibilities of each team member. Discussions, lectures and demonstrations can be used to provide this knowledge.
- *Positional modelling:* This type of cross-training provides the team members with a greater level of understanding. The duties of each team member are discussed and observed. This technique results in the team members knowing both about the duties of each team member and how these relate to the other members of the team. The training consists of observing team-mates in a simulated situation.
- *Positional rotation:* This is the highest level of cross-training. Team members are given direct, hands-on practice of their colleagues' specific tasks and how the tasks interact. Although the aim is not for team members to become experts in the other roles, they should reach a basic competence level for specific tasks carried out by their team-mates, which demand high degrees of co-operation and interdependency. This type of training is inappropriate for teams where team members have highly specialised jobs, but it is more

suitable where team members may be called upon to undertake other team members' roles if required.

Cross-training has been evaluated and shown to be an effective training strategy (e.g. Volpe et al., 1996). However, organisations must balance between the benefits gained from cross-training and the cost associated with not allowing sufficient time to develop individual-level expertise within the team (McCann et al., 2000).

Team self-correction training

Team self-correction training works on the premise that effective teams review events, correct errors, discuss strategies and plan future events (Salas and Cannon-Bowers, 2000). Therefore, the training simply provides direction on processes that typically occur. This type of training is 'a natural mechanism by which team members correct their attitudes, behaviours, and cognitions' (Blickensderfer et al., 1997: 268). The team members provide the feedback. Peer feedback has been shown to be a highly accurate and effective strategy to help team members to develop skills and understanding (Norton, 1992).

There are a number of stages in team self-correction training. Firstly, the learning objectives need to be clearly stated to highlight the event that will be discussed and to set the tone of the analysis. It is important that team members do not feel that they are being singled out for unnecessary criticism and that the discussion does not develop into an argument. The next step is error identification and problem-solving. This allows the team to discuss any errors that occur and discuss the implications of these for performing the task in the future. Blickensderfer et al. (1997) suggest that the training will most importantly influence the team members' knowledge of non-technical skills (e.g. by building shared mental models). This will then have a positive effect on the skill-based competencies as a result of a better understanding of the roles of the team members, which in turn should foster improved attitudes in cohesiveness and collective orientation.

For feedback to be effective it should be given in a non-threatening manner, showing outcomes of problem behaviour, and give positive suggestions for preventing the problem occurring in the future (Ilgen et al., 1979). For a more detailed outline of how to actually perform the training, see Smith-Jentsch et al. (1998), who outline the application of self-correction training for US naval teams. They provide some evidence of its effectiveness in the form of a case study of a pre-commissioned combat information centre team participating in a simulated test.

Event-based training

'Event-based training is an instructional approach that systematically structures training in an efficient manner by tightly linking learning objectives, exercise design, performance measurement and feedback' (Dwyer et al., 1999: 191). The first part of event-based training (EBT) is to specify the training objectives through performing task analysis and establishing acceptable performance standards.

For each training objective, specific learning objectives are identified for inclusion in the training exercises. They represent components of behaviour that may have always been lacking, subject to skill decay, or so difficult to perform that they are not practised with sufficient frequency for team members to maintain levels of proficiency (Dwyer et al., 1999). Once the learning objectives have been defined, the next stage is to identify 'trigger events' for each learning objective. These events are the stimulus conditions and cues that are embedded in the exercises and require a response by the participants. They provide an opportunity for the participants to demonstrate their ability to perform the tasks associated with each learning objective and they also provide a controlled situation in which performance can be assessed. Ideally, multiple events should be specified for each learning objective. They should vary in difficulty and should be introduced at several points throughout the exercise. Finally, the data from the performance measurement should be used to provide feedback to the participants, which links back to the learning objectives. EBT provides structure and control to training and ensures proper task exposure within an exercise. Dwyer et al. (1999) describe two case studies in which EBT was used with 13 to 19 military personnel. It was found that performance improved over the days of training in all three mission phases (planning, contact point and attack).

Team facilitation training

Facilitation training is designed to help team leaders stimulate learning by creating an effective learning environment, supporting more formal training experiences, and facilitating and encouraging team discussions. Team leaders can contribute to team effectiveness in many different ways, but one key role is to facilitate ongoing team development (Kozlowski et al., 1996). This type of training attempts to provide participants with the skills to be an effective team facilitator. Tannenbaum et al. (1998) identify eight team leader behaviours that should be exhibited during pre-briefs and post-event reviews. These are described in Chapter 4 on communication.

Tannenbaum et al. (1998) carried out training with 14, five-person teams of naval officers that were required to perform as a team in a simulated combat information centre of a navy warship. The leader of each team received a day of training on conducting effective pre-briefs and post-action reviews. When compared with teams with leaders who had not received the training, it was found that trained team leaders exhibited more effective briefing behaviours. In addition, teams with trained leaders displayed more effective team working, and also achieved better performance outcomes.

Assessing team working

Assessment of team working skills can be carried out using a range of different measures (see Stanton et al., 2005, for examples). Methods to assess team working skills include observing a team in the actual workplace, in a simulator, or other group behaviour task (e.g. solving a problem in a tactical decision game). In each of these situations, the evaluation of team working can include an expert's assessment or peer ratings of the team's performance.

Assessment of team working skills may involve a behaviour rating system: this might rate teamwork for a team as a whole (e.g. Undre et al., 2006) or rate the teamwork skills of individual team members (discussed in Chapter 11). In the NOTECHS behavioural marker system developed in aviation (see Appendix 1), individual team working skill is covered by the category 'co-operation', with elements consisting of team-building and maintaining, considering others, supporting others and conflict-solving. In anaesthetics, a behavioural rating system (anaesthetists' non-technical skills (ANTS)) has a team working category, defined as 'working with others in a team context, in any role, to ensure effective joint task completion and team satisfaction; focus is particularly on the team rather than the task' (Fletcher et al., 2004: p168). This has five elements:

- co-ordinating activities with team members
- exchanging information
- using authority and assertiveness
- assessing capabilities
- supporting others.

Peer assessment often consists of the use of a questionnaire where the team members rate aspects such as their own and other members' contribution to the team's task, willingness to take on responsibility, amount of support needed, amount of support offered, ability to promote ideas, ability to solve problems in the team and co-operation. These peer assessments can be used by the team to identify strengths and weaknesses in team interactions and skills. Questionnaires can also be used for team members to comment on aspects of teamwork in their unit or department (Awad et al., 2005; Crichton, 2005; Flin et al., 2004; Reader et al., 2007; West, 2004).

Conclusion

In high-reliability industries, whether a team is brought together and needs to perform quickly or it has had time to form and work together for a while, team co-ordination is critical to overall team performance. A team is more than the sum of its parts, as team members bring a wide range of experience and abilities to the task. Team working is the advantage of a team over a collection of highly skilled individuals. In aviation, benefits of team co-ordination have been shown to include:

- increased safety by redundancy to detect and remedy individual errors
- increased efficiency by the organised use of all existing resources, which improves in-flight management.

Effective teams are organised and supported within the organisation, support and respect each other, and are able to adapt to situational changes. Team members need to share mental models of the situation, of their task and of their goal. Team co-ordination can be improved through targeted training interventions.

Problems with team co-ordination have manifested in incidents in a range of settings, including aviation, health care, military and petrochemicals. Various theories of team co-ordination have been described and guidance for team co-ordination training presented. In sum, effective team co-ordination is not a case of trial and error, or of good luck, but is a result of recognising the importance of team working knowledge skills, and abilities, and targeted training.

Key points

- Teams are increasingly a feature of organisations, especially in today's uncertain, high-work-demand and rapidly changing organisational structures. They are embedded in organisations and are part of the larger social system.
- Team working problems have been reported in high-profile accidents and include: roles not clearly defined; lack of explicit co-ordination; failures to resolve conflict; and miscommunication/communication problems.
- The definition of a team proposed by Salas and colleagues is 'a set of two or more individuals who must interact and adapt to achieve specified, shared, and valued objectives, and whose members have meaningful task interdependencies and task-relevant knowledge'. Team co-ordination is required to achieve the advantage of a team over a collection of highly skilled individuals.
- Team decision-making refers to the process of reaching a decision undertaken by interdependent individuals to achieve a common goal, and relies on concepts such as team situation awareness and shared mental models.
- Team co-ordination training can result in adaptive teams. Adaptive co-ordination is trainable, and significant improvements in team performance can be achieved by relatively simple training interventions.

Suggestions for further information

Key texts

Hackman, J.R. (2002) *Leading Teams. Setting the Stage for Great Performances*. Harvard: Harvard Business School Press.

Swezey, R. and Salas, E. (eds.) (1992) *Teams. Their Training and Performance*. New York: Ablex.

West, M. (2004) *Effective Teamwork: Practical Lessons from Organizational Research* (2nd ed). Leicester: BPS Blackwell.

Websites

Hackman, J.R.: http://www.leadingteams.org
CAA: http://www.caa.co.uk/docs/33/CAP737.PDF, Appendix 7
Team Performance Laboratory, University of Central Florida: www.tpl.ucf.edu

References

Awad, S. et al. (2005) Bridging the communication gap in the operating room with medical team training. *The American Journal of Surgery*, 190, 770–774.

Baron, R. and Kerr, V. (2003) *Group Process, Group Decision, Group Action* (2nd ed). Milton Keynes: Open University Press.

Belbin, R.M. (2003) *Team Roles at Work*. Oxford: Butterworth-Heinemann.

Blickensderfer, E., Cannon-Bowers, J. and Salas, E. (1997) Theoretical bases for team self-correction: fostering shared mental models. *Advances in Interdisciplinary Studies of Work Teams*, 4, 249–279.

Blickensderfer, E., Cannon-Bowers, J. and Salas, E. (1998) Cross training and team performance. In J. Cannon-Bowers and E. Salas (eds.) *Making Decisions Under Stress: Implications for Individual and Team Training* (pp. 299–311). Washington, D.C.: American Psychological Association.

Brown, R. (1999) *Group Processes* (2nd ed). Oxford: Blackwell.

Brown, K., Huettner, B. and James-Tanny, C. (2006) *Managing Virtual Teams: Getting the Most from Wikis, Blogs, and other Collaborative Tools*. London: Wordware.

CAA (2006) *Crew Resource Management (CRM) Training. CAP737. Guidance for Flight Crew, CRM Instructors (CRMIs) and CRM Instructor-Examiners (CRMIEs)*. Hounslow, Middlesex: Civil Aviation Authority.

CAA (2004) *CAP716 Aviation Maintenance Human Factors (EASA/JAR145 Approved Organisations)* Sussex: CAA Safety Regulation Group.

Cannon-Bowers, J. and Salas, E. (1998) Individual and team decision making under stress: Theoretical underpinnings. In J. Cannon-Bowers and E. Salas (eds.) *Making Decisions Under Stress. Implications for Individual and Team Training*. Washington, D.C.: APA.

Cannon-Bowers, J.A., Salas, E. and Converse, S. (1993) Shared mental models in expert team decision making. In J. Castellan (ed.) *Individual and Group Decision Making: Current Issues*. Hillsdale, NJ: LEA, 221–246.

Cannon-Bowers, J.A., Tannenbaum, S.I., Salas, E. and Volpe, C.E. (1995) Defining team competencies and establishing team training requirements. In R. Guzzo and E. Salas (eds.) *Team Effectiveness and Decision Making in Organisations* (pp. 333–380). San Francisco, CA.

Cook, R. (2006) Being bumpable: Consequences of resource saturation and near-saturation for cognitive demands on ICU practitioners. In D. Woods and E. Hollnagel (eds.) *Joint Cognitive Systems*. London: Taylor & Francis.

Crichton, M. (2005) Attitudes to teamwork, leadership and stress in oil industry drilling teams. *Safety Science*, 43, 679–696.

Clutterbuck, D. (2007) *Coaching the Team at Work*. London: Brealey.

Dominguez, C., Flach, J., McDermott, P., McKellar, D. and Dunn, M. (2004) The conversion decision in laparoscopic surgery: Knowing your limits and limiting your risks. In K. Smith and P. Johnston (eds.) *Psychological Investigations of Competence in Decision Making*. Cambridge: Cambridge University Press.

Dwyer, D., Oser, R., Salas, E. and Fowlkes, J. (1999) Performance measurement in distributed environments: Initial results and implications for training. *Military Psychology*, 11, 189–215.

Edmondson, A. (2003) Speaking up in the operating room: How team leaders promote learning in interdisciplinary action teams. *Journal of Management Studies*, 40, 1419–1452.

Endsley, M. and Jones, D.G. (2001) A model of inter- and intrateam situational awareness: Implications for design, training and measurement. In M. McNeese, E. Salas and M. Endsley (eds.) *New Trends in Cooperative Activities: Understanding System Dynamics in Complex Environments* (pp. 46–67). Santa Monica, CA: Human Factors and Ergonomics Society.

Fletcher, G., Flin, R., McGeorge, P., Glavin, R., Maran, N. and Patey, R. (2004) Rating non-technical skills: Developing a behavioural marker system for use in anaesthesia. *Cognition, Technology and Work*, 6, 165–171.

Flin, R., Fletcher, G., McGeorge, Sutherland, A. and Patey, R. (2004) Anaesthetists' attitudes to teamwork and safety. *Anaesthesia*, 58, 233–242.

Furnham, A., Steele, H. and Pendleton, D. (1993) A psychometric assessment of the Belbin team-role self-perception inventory. *Journal of Occupational and Organizational Psychology*, 66, 245–257.

Glendon, I., Clarke, S. and McKenna, E. (2006) *Human Safety and Risk Management*. Sydney: CRC Press.

Guzzo, R. and Dickson, M.W. (1996) Teams in organisations: Recent research on performance and effectiveness. *Annual Review of Psychology*, 47, 307–338.

Hackman, J.R. (1987) The design of work teams. In J. Lorsch (ed.) *Handbook of Organizational Behaviour* (pp. 315–342). Englewood Cliffs, NJ: Prentice-Hall.

Hackman, J.R. (2002) *Leading Teams. Setting the Stage for Great Performances*. Harvard: Harvard Business School Press.

Hackman, J.R. (2003) Learning more by crossing levels: Evidence from airplanes, hospitals, and orchestras. *Journal of Organisational Behavior*, 24, 905–922.

Helmreich, R. and Foushee, H. (1993) Why crew resource management? Empirical and theoretical bases of human factors training in aviation. In E. Wiener, B. Kanki and R. Helmreich (eds.) *Cockpit Resource Management*. New York: Academic Press.

Ilgen, D., Fisher, C. and Taylor, M. (1979) Consequences of individual feedback on behaviour in organizations. *Journal of Applied Psychology*, 64, 531–545.

Janis, I.L. (1972) *Victims of Groupthink*. Boston: Houghton-Mifflin.

Johnston, J.H., Smith-Jentsch, K.A. and Cannon-Bowers, J.A. (1997) Performance measurement tools for enhancing team decision-making training. In M.T. Brannick, E. Salas and C. Prince (eds.) *Team Performance Assessment and Measurement*. Mahwah, NJ: LEA.

Klein, G. (1998) *Sources of Power. How People Make Decisions*. Cambridge, Massachusetts: The MIT Press.

Klein, G. (2000) Cognitive task analysis of teams. In J.M. Schraagen, S.F. Chipman and V.L. Shalin (eds.) *Cognitive Task Analysis*. Mahwah, NJ: LEA.

Klein, G. and Miller, T. (1999) Distributed planning teams. *International Journal of Cognitive Ergonomics*, 3, 203–222.

Klein, G., Zsambok, C.E. and Thordsen, M.L. (1993, April) Team decision training: Five myths and a model. *Military Review*, 36–42.

Klimoski, R. and Jones, R.G. (1995) Staffing for effective group decision making: Key issues in matching people and teams. In R. Guzzo and E. Salas (eds.) *Team Effectiveness and Decision Making in Organisations* (pp. 291–332). San Francisco, CA: JosseyBass.

Kozlowski, S.W.J., Gully, S.M., McHugh, P.P., Salas, E. and Cannon-Bowers, J. (1996) A dynamic theory of leadership and team effectiveness: Developmental and task contingent leader roles. *Research in Personnel and Human Resources Management*, 14, 253–305.

Kozlowski, S.W.J. (1998) Training and developing adaptive teams: Theory, principles and research. In J. Cannon-Bowers and E. Salas (eds.) *Making Decisions Under Stress. Implications for Individual and Team Training* (pp. 115–153). Washington, DC: American Psychological Association.

Kramer, R.M. (1998) Revisiting the Bay of Pigs and Vietnam decisions 25 years later: How well has the groupthink hypothesis stood the test of time? *Organizational Behavior and Human Decision Processes*, 73, 236–271.

Lencioni, P. (2005) *Overcoming the Five Dysfunctions of a Team*. Chichester: Wiley.

Leonard, M., Graham, S. and Bonacum, D. (2004) The human factor: The critical importance of effective teamwork and communication in providing safe care. *Quality and Safety in Health Care*, 13, 85–90.

Lingard, L., Reznick, R, Espin, S, Regehr, G. and De Vito, I. (2002) Team communication in the operating room. Talk patterns, sites of tension and implications for novices. *Academic Medicine*, 77, 232–237.

Marks, M.A., Mathieu, J.E. and Zaccaro, S.J. (2001) A temporally based framework and taxonomy of team processes. *Academy of Management Review*, 26, 356–376.

McCann, C., Baranski, J., Thompson, M. and Pigeau, R. (2000) On the utility of experiential cross-training for team decision making under time stress, *Ergonomics*, 43, 1095–1110.

McGrath, J.E. (1984) *Groups: Interaction and Performance*. Englewood Cliffs, NJ: Prentice Hall.

McIntyre, R.M. and Salas, E. (1995) Measuring and managing for team performance: Emerging principles from complex environments. In R. Guzzo and E. Salas (eds.) *Team Effectiveness and Decision Making in Organisations*. San Francisco, CA: Jossey Bass.

Militello, L., Kyne, M.M., Klein, G., Getchell, K. and Thordsen, J. (1999) A synthesized model of team performance. *International Journal of Cognitive Ergonomics*, 3, 131–158.

National Transportation Safety Board (1973) *Eastern Air Lines, Inc., L-1011, N310EA, Miami, Florida, December 29, 1972*. NTSB report number: MAB-05-01. Washington, D.C.: NTSB.

National Transportation Safety Board (2005). *Collision between the U.S. Navy Submarine USS Greeneville and Japanese Motor Vessel Ehime Maru Near Oahu,*

Hawaii, February 9, 2001, NTSB report number: MAB-05-01. Washington, D.C.: NTSB.

Norton, S. (1992) Peer assessment of performance and ability: An exploratory meta-analysis of statistical artefacts and contextual moderators. *Journal of Business and Psychology*, 6, 387–399.

Orasanu, J. and Salas, E. (1993) Team decision making in complex environments. In G. Klein, J. Orasanu, R. Calderwood and C. Zsambok (eds.) *Decision Making in Action*. Norwood, NJ: Ablex.

Paris, C.R., Salas, E. and Cannon-Bowers, J.A. (2000) Teamwork in multi-person systems: A review and analysis. *Ergonomics*, 43, 1052–1075.

Reader, T., Flin, R. and Cuthbertson, B. (in press) Factors affecting team communication in the Intensive Care Unit. In C. Nemeth (ed.) *Improving Healthcare Team Communication: Building on Lessons from Aviation and Aerospace*. Aldershot: Ashgate.

Reader, T., Flin, R., Lauche, K. and Cuthbertson, B. (2006) Non-technical skills in the Intensive Care Unit, *British Journal of Anaesthesia*, 96, 551–559.

Reader, T., Flin, R., Mearns, K. and Cuthbertson, B. (2007) Interdisciplinary communication in the intensive care unit. *British Journal of Anaesthesia*, 98, 347–352.

Roberts, K. and Tadmore, C. (2002) Lessons learned from non-medical industries: the tragedy of the *USS Greeneville. Quality and Safety in Healthcare*, 11, 355–357.

Rouse, W.B., Cannon-Bowers, J. and Salas, E. (1992) The role of mental models in team performance in complex systems. *IEEE Transactions on Systems, Man, and Cybernetics*, 22, 1295–1308.

Salas, E. and Cannon-Bowers, J. (1993, August) Making of a dream team. Paper presented at the American Psychological Association Conference, Toronto.

Salas, E. and Cannon-Bowers, J.A. (1997) The anatomy of team training. In L. Tobias and D. Fletcher (eds.) *Handbook on Research in Training*. New York: Macmillan.

Salas, E., Rhodenizer, L. and Cannon-Bowers, J.A. (2000) The design and delivery of crew resource management training: Exploiting available resources. *Human Factors*, 42(3), 490–511.

Salas, E., Burke, C.S. and Stagl, K.C. (2004) Developing teams and team leaders: Strategies and principles. In D.V. Day, S.J. Zaccaro and S.M. Halpin (eds.) *Leader Development for Transforming Organizations. Growing Leaders for Tomorrow*. Mahwah, NJ: Lawrence Erlbaum.

Salas, E., Cannon-Bowers, J. and Weaver, J. (2002) Command and control teams: Principles for training and assessment. In R. Flin and K. Arbuthnot (eds.) *Incident Command: Tales From the Hot Seat*. Aldershot: Ashgate.

Salas, E., Dickinson, T., Converse, S. and Tannenbaum, S. (1992) Toward an understanding of team performance and training. In R. Swezey and E. Salas (eds.) *Teams. Their Training and Performance*. New York: Ablex.

Salas, E., Prince, C., Baker, D.P. and Shrestha, L. (1995) Situation awareness in team performance: Implications for measurement and training. *Human Factors*, 37, 123–136.

Salas, E., Wilson, K.A., Burke, C.S., Wightman, D.C. and Howse, W.R. (2006) A checklist for crew resource management training. *Ergonomics in Design*, 14, 6–15.

Serfaty, D., Entin, E.E. and Johnston, J.H. (1998) Team coordination training. In J. Cannon-Bowers and E. Salas (eds.) *Making Decisions Under Stress. Implications for Individual and Team Training*. Washington, D.C.: American Psychological Association.

Smith-Jentsch, K., Zeisig, R., Acton, B. and McPherson, J. (1998) Team dimensional training: A strategy for guided team self-correction. In J. Cannon-Bowers and E. Salas (eds.) *Making Decisions Under Stress: Implications for Individual and Team Training* (pp. 271–298). Washington, D.C.: American Psychological Association.

Stanton, N. (1996) Team performance: communication, co-ordination, co-operation and control. In N. Stanton (ed.) *Human Factors in Nuclear Safety*. London: Taylor & Francis.

Stanton, N., Salmon, P., Walker, G., Baber, C. and Jenkins, D. (2005) *Human Factors Methods. A Practical Guide for Engineering and Design*. Aldershot: Ashgate.

Steers, R.M. (1988) *Introduction to Organizational Behaviour* (3rd ed). Glenview, IL: Scott Foresman.

Tannenbaum, S.I., Salas, E. and Cannon-Bowers, C. (1996) Promoting team effectiveness. In M. West (ed.) *Handbook of Work Group Psychology*. Sussex, England: Wiley and Sons.

Tannenbaum, S.I., Smith-Jentsch, K.A. and Behson, S.J. (1998) Training team leaders to facilitate team learning and performance. In J.A. Cannon-Bowers and E. Salas (eds.) *Making Decisions Under Stress: Implications for Individual and Team Training* (pp. 247–270). American Psychological Association.

Thordsen, M.L., McCloskey, M.J., Heaton, J.K. and Serfaty, D. (1996) *Decision-centred Development of a Mission Rehearsal System* (Contract N61339-95-C-0101 for Naval Air Warfare Center Training Systems Division, Orlando, FL). Fairborn, OH: Klein Associates.

Tuckman, B.W. (1965) Development sequence in small groups. *Psychological Review*, 63, 384–399.

Tuckman, B.W. and Jensen, M.A.C. (1977) Stages of small group development revisited. *Group and Organisational Studies*, 2, 419–427.

Turner, N. and Parker, S. (2004) The effect of teamwork on safety processes and outcomes. In J. Barling and M. Frone (eds.) *The Psychology of Workplace Safety*. Washington: APA Books.

Undre, W., Sevdalis, N., Healey, A., Darzi, A. and Vincent, C. (2006) Teamwork in the operating theatre: cohesion or confusion? *Journal of Evaluation in Clinical Practice*, 12, 182–189.

Unsworth, K. and West, M. (2000) Teams: The challenges of cooperative work. In N. Chmiel (ed.) *Work and Organizational Psychology*. Oxford: Blackwell.

Vincent, C. (2006) *Patient Safety*. Oxford: Churchill Livingstone.

Volpe, C., Cannon-Bowers, J. and Salas, E. (1996) The impact of cross-training on team functioning: An empirical investigation. *Human Factors*, 38, 87–100.

West, M. (2000) Reflexivity, revolution, and innovation in work teams. In M. Beyerlein, D. Johnson and S. Beyerlein (eds.) *Product Development Teams* (pp. 1–29). Stamford, CT: JAI Press.

West, M. (2004) *Effective Teamwork: Practical Lessons from Organizational Research* (2nd ed). Leicester: BPS Blackwell.

Wiener, E., Kanki, B. and Helmreich, R. (1993) *Cockpit Resource Management.* San Diego: Academic Press.

Zsambok, C.W. (1993) Advanced team decision making in C2 settings. In *Proceedings of the 1993 Symposium on Command and Control Research.* McLean, VA: SAIC, 45–52.

Zsambok, C. (1997) Naturalistic decision making research and improving team decision making. In C. Zsambok and G. Klein (eds.) *Naturalistic Decision Making*. Mahwah, NJ: LEA.

Zsambok, C., Klein, G., Kyne, M. and Klinger, D. (1992) *Advanced Team Decision Making: A Developmental Model* (Contract MDA903-90-C-0117) (pp. 111–120). US Army Research Institute for the Behavioral and Social Sciences. Fairborn, OH: Klein Associates.

Chapter 6

Leadership

Introduction

Exploits of leaders such as Hannibal – who in 219 BC led his army with war elephants across both the Pyrenees and the Alps – and Alexander the Great – who conquered lands from Iraq to India around 330 BC – are passed from generation to generation. There are further examples of great leadership from the 20th century: in 1916, Ernest Shackleton ensured the survival of all of his crew – involving an extraordinary journey across the ice that lasted for almost two years – after their ship, the *Endurance*, became trapped in the Antarctic (Shackleton, 2001); Mahatma Ghandi led India to independence from Great Britain in 1947 through mass, non-violent, civil disobedience. In all cases, these leaders have acted as an inspiration for their followers. In the workplace, leadership of crews or teams is equally crucial for effective operations (Yukl, 2005). Team leaders occur in all work settings, including business, industry, health care, public services and the military. One of their main responsibilities is to build an effective team by ensuring safe and efficient team functioning in order to maximise task performance. This chapter explores the concept of leadership and the behaviour of team leaders in the workplace. Leader characteristics, behaviours, competencies and styles, as well as leadership theories, are discussed. Although leadership is often considered to be an innate quality, leadership skills can be trained. Guidance on techniques to improve them are also described.

A team leader is defined as the 'person who is appointed, elected, or informally chosen to direct and co-ordinate the work of others in a group' (Fiedler, 1995: p7). Team leadership is about directing and co-ordinating the activities of team members; encouraging them to work together; assessing performance; assigning tasks; developing team knowledge, skills and abilities; motivating; planning and organising; and establishing a positive team atmosphere (Salas et al., 2004). The thoughts and behaviour of others in the team are influenced by the team leader's ideas and actions. Leaders come in various forms. Some are task specialists who owe their position to being good at the job. Others are good with people. A few leaders are good at both. Some are good at neither. A team leader acts as a linchpin for a team. The role of team leader can usually be described in a job description, but leadership, on the other hand, is not as easy to define. Leadership refers to the personal qualities, behaviours, styles and strategies adopted by the team leader, and other team members, and influences how and whether a team achieves its objective. Table 6.1 identifies the elements associated with the skill of leadership.

Table 6.1 Leadership elements

Category	Elements
Leadership	Use authority Maintain standards Plan and prioritise Manage workload and resources

These leadership skills, however, not only apply to the team leader but sometimes may apply to other team members. Within a team setting, team members may also use leadership skills in terms of the elements described above; for instance use authority when necessary, maintain standards, plan, prioritise, and manage workload and resources. Leadership can therefore be expanded beyond the role of the team leader to consider leadership as it emerges within a team and is drawn from teams as a result of people working together to accomplish shared work (Day et al., 2004).

Leadership and safety

Box 6.1 Examples of poor leadership: South Canyon fire (1996)

Sub-optimal decisions have been identified as being made by the leader of a wildland firefighter crew at South Canyon, Colorado. Typically, when fighting large-scale wildland fires, one incident commander takes overall responsibility for the effective and safe execution of activities by the multiple crews on duty. Due to under-preparation in leadership decisions, combined with acute stress (e.g. time pressure, multiple tasks) and ambiguous authority, decisions were taken by a smoke-jumper team leader during the management of this incident that resulted in the deaths of 12 firefighters.

The smoke-jumper team leader on duty that day recognised the danger that the firefighters were in after they had been clearing a fireline for several hours. This fireline was the crew's only exit point, but the decision was taken too late and only six of the crew managed to reach the safe point; 12 other crew members were overtaken by the fire. Conditions had gradually become more and more challenging during the two days that the fire had been raging. Lack of leadership training meant that the smoke-jumper team leader violated several standard operating procedures: he did not post a lookout to observe the fire's activities below their post; and also he was overconfident about their ability to control the fire. In addition, a problem with ambiguous authority emerged as the management team comprised three leaders and three crews with firefighters from five states, causing confusion in orders and delegation.

Consequently, the smoke-jumper team leader suffered from a lack of situation awareness (e.g. not assigning fire lookouts, loss of radio contact with other firefighters and personal location), as a result of poor leadership, as well as lack of information (especially critical weather forecasting that did not reach him).
(Useem et al., 2005)

While we know that good leadership can influence team performance, it is also clear that poor team leadership skills can contribute to accidents, as the example in Box 6.1 demonstrates.

Effective leadership has been shown to be crucial for maintaining safe performance in the workplace (Hofmann and Morgeson, 2004; Glendon et al., 2006). For example, leaders influence key worksite safety behaviours such as compliance with rules and procedures (Thompson et al., 1998) and may have to manage critical incidents or emergencies (Flin, 1996). The term 'safety leadership' is now being used in industry and this refers to managers' and supervisors' leadership behaviours in relation to safety outcomes. Flin and Yule (2004) listed examples as follows:

- monitoring and reinforcing workers' safe behaviours;
- participating in workforce safety activities;
- being supportive of safety initiatives;
- emphasising safety over productivity.

In the UK oil and gas industry, the Step Change in Safety programme is a safety forum that draws together representatives from across the industry and from all management levels. One of the activities of the programme has been the introduction of safety leadership training. Safety leaders are expected to create an environment in which unsafe acts are challenged and safe behaviours promoted to achieve a workplace without harm (Step Change in Safety, 2006). North Sea divers have reported that their confidence in the supervisor's ability to manage accident risk was the most important factor in preventing accidents (Osman et al., 2003). Thus, the supervisor's ability to manage risk and the attitudes and leadership skills of those in the supervisory position are crucial to the safety and effectiveness of a diving operation. Leadership skills are also key to managing novel, risky incidents, as described in Box 6.2 below.

Box 6.2 Example of effective leadership in oil and gas drilling industry

An incident management team (IMT) was established to manage a novel incident, namely that the 6,000ft drilling riser (a pipe that connects the drilling rig to the sea bed) parted at approximately 3,200ft. The IMT consisted of the incident manager and a number of team leaders, each of whom took responsibility for one part of the incident. In terms of decision-making, each team leader had to make often difficult and challenging decisions, and accept accountability for those decisions. Team leaders talked through issues, discussed expectations and set deadlines. Team leaders also had to delegate tasks to appropriate members and check that these tasks were appropriately undertaken. Finally, team leaders needed to focus on the major issues involved in managing the incident and were not distracted by other tasks or demands. This involved prioritisation of activities and actions, with shedding of tasks that were not relevant at that particular time. This did not mean that these tasks were overlooked, but were left until a more appropriate period, i.e. when time was available. (Crichton et al., 2005)

Safety research has shown that the most effective leaders of multi-disciplinary teams, whose team members must co-ordinate action in risky, uncertain, dynamic situations, are those who communicate a motivating rationale for change and minimise concerns about status differences. This helps their team members to speak up and engage in more proactive co-ordination. Edmondson (2003), studying teams in cardiac surgery, found that where the lead surgeon behaved in this way, the teams learned more quickly to use a new technique.

What are the skills of leadership?

Leadership has been defined as the art of getting others to do (and want to do) something that the leader believes should be (must be) done, involving interpersonal influence, goal-setting and communication (Furnham, 2005). The power wielded by leaders can vary across different types, e.g. coercive (power to punish), reward (power to reward), legitimate (positional power), expert (special skill or knowledge power) and referent (power of a follower's identification). Leaders can also vary in terms of different styles, ranging from democratic through to autocratic. A good team leader needs to both focus on accomplishing the task and create an enabling performance environment for the development of team member knowledge, skills and attitudes.

Effective leadership processes represent a critical factor in the success of teams in organisations (Zaccaro et al., 2001). In terms of non-technical skills, effective leadership in the workplace helps to achieve safe task completion within a motivated, full-functioning team, through co-ordination and persuasiveness. Even self-managed teams have someone who influences the team, although this may change depending on the situation. In aviation, a leader on an aircraft is defined as 'a person whose ideas and actions influence the thought and the behaviour of others. Through the use of example and persuasion, and an understanding of the goals and desires of the group, the leader becomes a means of change and influence' (CAA, 2006: Appendix 7, p. 3). This involves the elements listed in Table 6.1, and described in more detail below.

Use of authority and assertiveness

This refers to the ability to create a proper challenge and response atmosphere, by balancing assertiveness and team member participation and being prepared to take decisive action if required by the situation. The leader must also know when to apply his or her authority to achieve safe task completion.

Providing and maintaining standards

This relates to compliance with essential standards, e.g. standard operating procedures (SOPs and others) for task completion, as well as supervision and intervention that may be required due to deviations from standards by other team members.

Planning and prioritising

This element describes how leaders apply appropriate methods of planning and prioritising for organised task management and delegation to achieve the best performance. This also involves co-ordination by communicating plans and intentions.

Managing workload and resources

Leaders must manage not only their own workload and resources but also that of the team. This involves understanding the basic contributors to workload and developing the skills of organising task-sharing to avoid workload peaks and dips. Causes of high workload include unrealistic deadlines and under-resources.

There are a number of different accounts of the skills required by leaders and these can differ depending on whether they are derived from research with military, industry or elsewhere. Several versions are presented below. For instance, Zaccaro et al. (2001) suggest that leadership aspects that affect team performance include:

- active participation by the team leader and other team members
- defining team directions and organising the team to maximise progress along such directions
- gaining respect from the team members
- knowledgeable about their own speciality areas and willing to respect other team members who are experts in their own speciality areas
- encouraging open communications, including discussion of the team goals and performance expectations, which promotes commitment to the team and consensus within the team.

Flin's (1996) study of incident commanders showed that the most effective leaders:

1. diagnose the situation (the task/problem, the mood, the competence and motivation of the team),
2. have a range of styles available (e.g. delegative, consultative, coaching, facilitating, directive), and
3. match his or her style to the situation.

From studies of military teams, Salas and Cannon-Bowers (1997) found that the following skills were essential:

1. Define the social structure, encourage open communications and exhibit self-disclosure to develop team cohesion.
2. Use effective communications and inform the other team members about matters affecting team performance.
3. Plan, structure and co-ordinate the team.
4. Maintain the team focus on their task.

5. Ask for input from other team members and openly discuss potential problems.
6. Maintain coherence within the team by managing situation awareness.
7. Provide feedback to the other team members, the degree of successful feedback depending on the leader's style.
8. Adjust their role to match team progress.
9. Define and encourage team goals and performance to promote commitment and consensus.

Leaders carry a great deal of responsibility for ensuring that their team achieves its goals and one aspect of this is to solve complex problems. Salas et al. (2004) proposed four dimensions of a team leader's behaviours for problem-solving (see Table 6.2).

Table 6.2 Team leader problem-solving behaviours (Salas et al., 2004)

Dimension	Sub-dimension
Information search and structuring	Acquiring information Organising and evaluating information Feedback and control
Information use in problem-solving	Identifying needs and requirements Planning and co-ordinating Communicating information
Managing personnel resources	Obtaining and allocating personnel resources Developing personnel resources Motivating personnel resources Utilising and monitoring personnel resources
Managing material resources	Obtaining and allocating material resources Maintaining material resources Utilising and monitoring material resources

These behaviours show what needs to be done to encourage adaptive team performance, and it was suggested that this can be applied in any team context.

In summary, skilled leadership may be required to handle a range of situations and depends on a number of different skills and competencies discussed elsewhere in this book, such as decision-making, communication, teamwork and situation awareness. Situational demands affect how a leader can perform or is viewed. Different tasks may require different leader characteristics and/or behaviours. Thus the context or organisational factors can support or frustrate a leader's behaviours and actions. The next section describes various leadership theories that have attempted to explain leadership effectiveness.

Theories of leadership

Many different theories have been proposed to explain leadership performance, including trait theories, style theories, contingency (situation) theories and, more recently, charisma and transformational theories (see Northouse, 2006, or Yukl, 2005, for an overview). A brief summary of some of these theories is presented below.

1. Trait theory

Trait theory suggested that distinctive physical and psychological characteristics account for effective leadership, for example:

- physical characteristics (age, height)
- social background (education, social status)
- intellectual ability (intelligence quotient (IQ), verbal fluency)
- personality (self-confidence, stress tolerance)
- task orientation (achievement need)
- social skills (personal competence, tact).

These traits were considered to distinguish leaders from non-leaders, and also distinguish successful leaders from unsuccessful leaders (Stodgill, 1948). Similarly, according to Arnold et al. (2004), characteristics of leaders, compared with non-leaders, tend to be higher on:

- intelligence
- dominance/need for power
- self-confidence
- energy/persistence
- knowledge of the task.

Box 6.3 Characteristics of leadership in drilling teams

> Members of an oil industry drilling team, taking part in a project to identify non-technical skills in the drilling industry, described their team leader as 'passionate', 'listens open-mindedly', 'engenders a good environment in which to work', and 'very approachable'. He [team leader] was 'honest, very supportive, provides direction, delegates responsibilities and tasks, but doesn't micro-manage – empowers, so team members feel trusted'. (Crichton and Flin, 2004)

Although certain traits may influence the willingness to take a leadership role, there is no clear evidence that certain traits, or characteristics, predict who will be an effective leader in every situation. One problem is that the role of team members is overlooked in this approach. Moreover, the impact of the situation was minimised (Sadler, 1997).

The lists of traits are often gathered from surveys of leaders after they have been seen to be successful. However, it is uncertain whether the traits make the leaders or whether the leadership role makes the traits. Trait theory is descriptive in that it describes the characteristics that leaders show, but it does not explain how, when and why these traits are necessary or sufficient for effective leadership (Furnham, 2005). For example, 'For every fire-breathing orator like General George Patton, you can find a leader like Mahatma Gandhi, who led with compassion, or like Martin Luther King Jr., whose moral convictions inspired his followers' (Fortune, 2003).

2. Style theory

This approach focuses on the management style of leaders rather than their traits. In the early theories, descriptors of the most common styles were developed based on observing leaders, categorising their behaviours and determining which styles were most and least effective. In one analysis, three basic leadership styles were identified (Lewin et al., 1939):

- *Authoritarian* (Directive): centralised authority, dictate work methods, make unilateral decisions, limit employee participation.
- *Democratic* (Participative): involve employees in decision-making, delegate authority (empowerment), encourage participation in deciding work methods and goals.
- *Laissez-faire*: give employees complete freedom in decision-making with minimum leader participation, provide materials (when requested) and answer questions (when asked).

In another line of work, researchers focused on whether the team leader was more concerned with getting the task completed or keeping members of the team happy. Blake and Mouton (1964) combined these two styles – concern for production and concern for people – in their 'managerial grid'. Concern for production related to output, cost-effectiveness and concern for profits. Concern for people related to promoting friendships, helping subordinates with work and paying attention to issues of importance to employees. This resulted in five distinct styles of leadership, shown in Table 6.3.

Table 6.3 Styles of leadership (Blake and Mouton, 1964)

Style	Concern	
	Production	People
Country club	Low	High
Impoverished	Low	Low
Organisation person management	Moderate	Moderate
Authority–obedience	High	Low
Team	High	High

No clear evidence was found to support one style of leadership as enabling effective team output more than another. However, both participative and people-centred leadership appeared to increase satisfaction among subordinates or team members.

This style category approach was subsequently considered to be too simplistic and did not describe all observed leader behaviours. Similar to trait theories, this approach paid little attention to situational context and did not explain which style is most effective in which situation. On the other hand, this approach focused more on what leaders did rather than what they were. Training to make leaders more effective was emphasised and some trainers still use versions of the Blake and Mouton model.

3. Contingency/situational theory

The contingency or situational approach to leadership emerged in the 1970s and proposed that there is no single leadership style that is effective in all situations – the style would need to change to match the situation. The optimal leadership style depends on constraints such as the size of the team, adaptation to the environment, technology and strategies. A number of theories exist, including Fiedler's (1967) contingency theory, Vroom and Yetton's (1973) leadership styles theory, and Hersey and Blanchard's situational theory, which is briefly described below.

Hersey and Blanchard (1977; Hersey et al., 2000) argued that leader effectiveness depends on matching leader style to the level of follower maturity (in terms of competence and commitment). Four different leadership styles were identified that vary on two dimensions, namely consideration and initiating structure. Consideration relates to the welfare of people in the team, concerns for interpersonal relationships, and encouraging friendliness and approachability in the team. Initiating structure relates to concern for the task, especially roles, actions and activities. In combination this produces four styles:

- *Telling* (high task/low relationship behaviour): Characterised by giving a great deal of direction to subordinates and by giving considerable attention to defining roles and goals. Most applicable for dealing with new staff who have low competence but high commitment, or where the work was menial or repetitive, or where things had to be completed within a short time span.
- *Selling* (high task/high relationship behaviour): Most of the direction is given by the leader, but some attempts are made at encouraging people to take ownership of the task, as they have some competence but variable commitment.
- *Participating* (high relationship/low task behaviour): Decision-making is shared between leaders and followers/team members – the main role of the leader is to facilitate and communicate. This style should be used when team members have competence but variable commitment.
- *Delegating* (low relationship/low task behaviour): The leader identifies the problem or issue, but the responsibility for carrying out the response is given to followers, when they have both high competence and high commitment.

To effectively apply this approach, leaders need to have good diagnostic skills and a repertoire of styles that they can match to team needs. Yukl (2005) summarised the key factors influencing situational leadership, some of which are appropriate for first-line managers or supervisors:

- The lower managers are in the hierarchy, the less likely they are to use participative leadership and the more likely they are to focus on technical matters and to monitor subordinate performance.
- Production managers tend to be more autocratic and less participative than sales or staff managers.
- Leaders are more directive and less participative as the task becomes more structured, as this type of behaviour is easier to exercise on less complex tasks. As tasks become more complex/less structured, leaders depend more on subordinates.
- Leaders are less participative and more autocratic as the number of their subordinates increases, and are less likely to show consideration or support for subordinates.
- As the stress in a situation increases (e.g. time pressure, risk), the leader becomes more directive and task-orientated and less considerate.
- As subordinate performance and competence decline, leaders are more likely to react with increasingly close, directive, punishing and structuring behaviour and show less consideration and participation.

Underpinning the contingent leadership theory, Klein et al. (2006) refer to adaptive leadership, identified as a hierarchical, de-individualised system of shared leadership. The authors describe this form of leadership in extreme action teams, i.e. teams where 'highly skilled members co-operate to perform urgent, unpredictable, interdependent and highly consequential tasks while simultaneously coping with frequent changes in team composition and training their teams' novice members' (p. 590). Extreme action teams operate in situations of urgency, uncertainty and high consequence; for example, they studied trauma teams in emergency medicine. Adaptive leadership is characterised by dynamic delegation, in that leadership is rapidly and repeatedly delegated to and withdrawn from more junior leaders in the team by senior leaders, resulting in improvements in the action team's reliability of performance while also building up the skills of novice team members, i.e. they learn by doing.

This form of leadership in dynamic situations illustrates how the most appropriate leadership style varies in response to the characteristics of the task, the subordinates or the leader. In situations of low urgency and low novelty, and where the subordinates' skills and behaviour are deemed appropriate, the senior leaders delegate to the subordinates. On the other hand, in more urgent and highly novel situations, a senior leader is more likely to retain the leadership role.

These approaches emphasise the effect of the situation on leadership and different leadership styles. For example, a more directive style may be required where a quick response or decision is necessary. Nevertheless, these theories have been criticised, in part, for their North American bias – that the effects of culture are overlooked, and also that they focus mainly on the relationship between leaders and immediate subordinates and pay limited attention to organisational structure or politics (Bolman and Deal, 1997).

4. Transformational leadership theory

Leaders are often perceived as being inspirational or charismatic figures who can unite and motivate followers by offering shared visions and goals (Burns, 1978). A more recent version of this leadership theory is that of transactional and transformational leadership (Bass, 1990; Bass and Riggio, 2006). Transactional leaders offer some sort of exchange to subordinates, i.e. the leader rewards good performance and maintains existing work methods unless performance goals are not being met. Bass (1990) also refers to the concept of 'laissez-faire' leadership, where the leader abdicates responsibilities and avoids making decisions. This leadership style is neither transformational nor transactional. Transformational leaders articulate a clear vision and mission, while treating individuals on their own merits and encouraging free thinking. These leaders use charisma to motivate subordinates by raising their awareness about key outcomes, encourage subordinates to place team or organisational goals above their own, motivate subordinates to have stronger drive for responsibility, challenge and personal growth (Bass, 1985). Bass and Avolio (1990) propose that these two styles of transformational and transactional leadership, described in Table 6.4, lead to very different interactions within the team, but when combined result in enhanced performance.

Table 6.4 Transactional and transformational leadership styles (Bass and Avolio, 1990)

Transactional leadership	Transformational leadership
Contingent reward: Contracts exchange of rewards for effort, promises rewards for good performance, recognises accomplishments	*Charisma*: Provides vision and sense of mission, instils pride, gains respect and trust
Management by exception (active): Watches and searches for deviations from rules and standards, takes corrective action	*Inspiration:* Communicates high expectations, uses symbols to focus efforts, expresses important purposes in simple ways
Management by exception (passive): Intervenes only if standards are not met	*Intellectual stimulation:* Promotes intelligence, rationality and careful problem-solving
	Individualised consideration: Treats employees individually, coaches, advises

The two styles are not mutually exclusive, and Bass argues that leaders should use a combination of both styles. Transactional leadership is the basis of all management and results in expected levels of performance. Transformational leadership builds on this foundation to achieve increased motivation and performance beyond expectations, as shown in Figure 6.1.

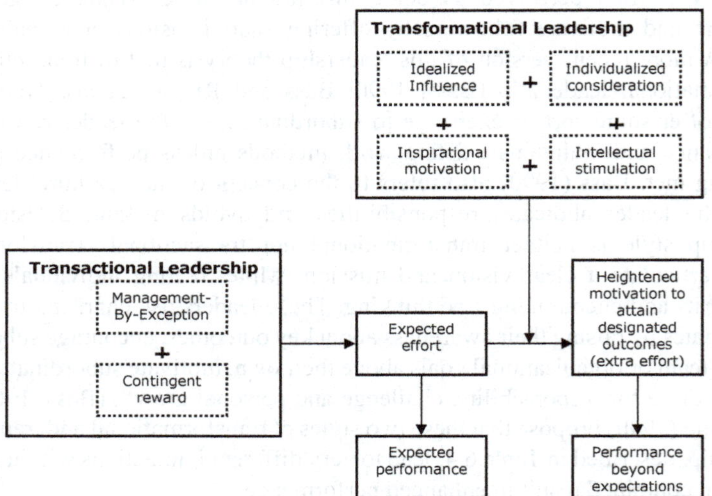

Figure 6.1 The augmentation effect of transformational on transactional leadership (adapted from Bass and Avolio, 1990, p237)
Reprinted with permission of Elsevier (JAI Press).

Transformational leaders are also viewed more positively by the team than those who show less of these behaviours. Transformational leadership has been shown to be effective in relation to subordinate satisfaction, motivation and performance (Yukl, 2005). This theory has also been extensively tested in relation to safety in the workplace, with studies consistently indicating that team leaders who show a more transformational style result in fewer unsafe behaviours and accidents in the workplace (Zohar, 2003).

A criticism of this approach is that transformational leadership may depend on how subordinates support the idea of finding new ways of doing and seeing things (Howell and Avolio, 1993). A further limitation is that the essential behaviours for charismatic and transformational leadership may be ambiguous, with both of these types of leadership – which are often treated as synonymous but do include plausible differences – rarely occurring at the same time (Yukl, 2005).

Summary of leadership theories

Leadership is acknowledged as being critical to effective team performance and leadership skills should be developed by all team members through training. As the CAA (2006) states, leadership involves teamwork, and the quality of a leader depends on the success of the leader's relationship with the team. As the various theories described above indicate, effective leadership involves a combination of traits, behaviours, styles and charisma. These theories also illustrate that, for the individual, existing leadership skills can be enhanced through specific training courses and on-the-job training. Salas et al. (2004) distinguish between leadership, in terms of traditional leadership theories, as described above, and a functional approach to team leadership, as outlined in Table 6.5.

Table 6.5 Traditional and functional leadership (Salas et al., 2004)

Traditional leadership	Team leadership
Fixed to a situation	Dynamically varies within the situation
Loosely connected subordinate roles and linkages	Tight interdependencies and co-ordination requirements
Fit the leader to the situation, task, or subordinates	Structuring and regulating team processes to meet shifting internal and external contingencies

A functional approach to leadership is proposed in that a team leader's primary directive is to get things done (McGrath, 1962). Leadership is a form of social problem-solving that promotes co-ordinated, adaptive team performance by helping to define and attain goals, by identifying and diagnosing problem situations, generating solutions, and implementing the chosen solution.

Leadership under stress

Team leaders in high-risk organisations are not unique, in that most of the time they are simply managing a process, just like leaders in a low-risk organisation. However, occasions can arise where managers and supervisors in high-risk organisations may be required to lead in an emergency, i.e. take on an incident command role. The knowledge, skills and abilities of the leader in this role can play a major part in managing an incident (Flin, 1996). An example from the *Piper Alpha* disaster in 1988 is described in Box 6.4.

Whether every workplace leader is capable of being a capable incident commander is debatable. Ullman (1994), an oil industry offshore installation manager, describes the shift expected from being a platform manager to becoming an incident commander for an offshore emergency. He suggested that what the oil industry required was

> …somebody like Clark Kent, who will sit in his office chair carrying out mind-numbing admin for 364 days with a wry grin on his granite hewn features, only to disappear into the closet on hearing the general alarm and to appear in a trice with flowing cape, a pair of tights, complete with modesty knickers over them and a huge OIM emblazoned on the

chest. Ready to take on the world. Where they are going to find him, God only knows and when they do I want to be there to read the job description and see the salary grade.' (p5).

Box 6.4 Piper Alpha disaster (1988)

Occidental Petroleum (Caledonia) Ltd's *Piper Alpha* oil and gas production platform suffered an explosion on the production deck of the platform on 6 July 1988. This explosion was, it has subsequently been suggested, caused by the ignition of a cloud of gas condensate leaking from a temporary flange. The fire spread rapidly and was followed by a number of smaller explosions. A major explosion occurred 20 minutes later, caused by the rupturing of a pipeline carrying gas to the *Piper Alpha* from a nearby platform (the *Tartan*, owned by Texaco).

Referring to the management of the incident, Sefton (1992) observed that the crisis on the *Piper Alpha* could have been managed more effectively, commenting 'It seems the whole system of command had broken down.' (p6). In his report for the official inquiry, Lord Cullen observed that 'The failure of the OIMs [offshore installation managers – the most senior manager on an oil platform] to cope with the problems they faced on the night of the disaster clearly demonstrates that conventional selection and training of OIMs is no guarantee of ability to cope if the man himself is not able in the end to take critical decisions and lead those under his command in a time of extreme stress' (Cullen, 1990: 20.59, p353).

Under stressful conditions, characterised by time pressure, risk, dynamic conditions, high information load and uncertainty, team performance has been linked to the team leader's effectiveness (Burgess et al., 1992). Command is often considered to be synonymous with leadership, yet others argue that command is distinct from leadership (Arbuthnot, 2002). The exercise of command requires good leadership, including aspects such as team- and morale-building, as well as developing mutual respect and understanding. Leadership skills based on routine, as opposed to non-routine, situations therefore act as the basis for command.

Brunacini (2002) describes different incident commander styles and their effects on team members when he was a member of the Phoenix Fire Department in the 1950s. He explains how his station was under the area command of Battalion Chief 1, but often responded to multiple unit events under the command of Battalion Chief 2. The two battalion chiefs had different approaches to managing incidents, as described in Box 6.5.

Box 6.5 Different approaches used by fire incident commanders (US)

'Battalion Chief 1 was a traditional, 1960s-type fire officer. He was an assertive, autocratic, intelligent and very experienced (30+ years) micromanager. Listening only occurred on those rare occasions when he wasn't talking (actually lecturing). Battalion Chief 2 was a younger guy, sort of quiet, laid back with a good sense of humour. He treated people with a "light hand" and did not over-manage.'

When responding to an incident, Brunacini explained the effect of the two different approaches shown by the battalion chiefs on the firefighters.

'Upon his arrival [Battalion Chief 1] he would abandon his vehicle, circle the incident at a dead run, screaming conflicting orders to anyone and everyone he encountered. His typical response to an expanding fire was to order the first wave of responders to inside positions with hand lines, call for more resources and then assign the second-wavers to outside positions where they would blast the insiders with deck guns. The result of this lack of effective command… was that we burned up a lot of property that was saveable and unnecessarily beat up a lot of firefighters…

'…Battalion Chief 2 would arrive, get out of his car and lean (his body language always looked sort of casual) against the door. He would direct companies to come to his vehicle (no portable radios in those days) where he would assign them face to face. Many times he would calmly engage the companies by asking "whatdaya think?" – then he would actively listen and together he and the assignee would develop and agree on a work place/function. He would never raise his voice and generally would smoke his pipe (another casual signal) as he evaluated conditions and observed the effect of operational action. He always had a clipboard with an old-time yellow pad and as he made assignments he would note with a stubby little pencil the who/where/what of their task. I never saw him leave his vehicle while a fire fight was under way and I never saw him get excited.' (Brunacini, 2002: p. 60)

Mentoring, as described later, is often proposed as a useful training mechanism for leaders, and, as Brunacini explains, can provide both negative and positive examples. The leadership styles of the two incident commanders in Box 6.5 resulted in totally different outcomes for the firefighters and the incident. Characteristics and competencies regarded as important for leading under stress are shown in Table 6.6.

Leaders of emergency response teams must be able to change leadership style in response to a fast-moving situation. That is, they must have the ability to switch to a more directive style rather than consultative in response to situational demands. Such changes also depend on the leader's personality, organisational culture, team's expertise and expectations, and operational management structure (Flin, 1996). Core incident command skills have been identified as leadership, team management, stress management, situation (risk) assessment and decision-making (Flin and Arbuthnot 2002).

Table 6.6 Summary of characteristics and competencies for leading under stress (Flin, 1996, pp42–44)

Topic	Description
Leader characteristics	• willingness to take a leadership role • emotional stability • stress resistance • decisiveness • controlled risk-taking • self-confidence • self-awareness
Leader competencies	• leadership ability • communication skills, especially briefing and listening • delegating • team management • decision-making, under time pressure and especially under stress • evaluating the situation (situation awareness) • planning and implementing a course of action • remaining calm and managing stress in self and others • pre-planning to prepare for possible emergencies

Team leader behaviours under stress often differ from those of team leaders when in normal work situations. Under stressful conditions, team leaders typically initiate almost all team communications, consisting of commands, suggestions, observations and statements of intent (Oser et al., 1991), and, although team leaders may be more receptive to task inputs from team members, they can be less likely to defer to those inputs (Driskell and Salas, 1991). Effective team leader principles, skills and knowledge, for stressful situations, are listed in Table 6.7 (Burgess et al., 1992).

These team leader principles can then be used as a basis for guidelines for training team leaders to deal with stressful situations. Box 6.6 describes the impact of the behaviour of a London Metropolitan Police Inspector on management of the King's Cross fire.

Table 6.7 Team leader principles, skills and knowledge, under stress (Burgess et al., 1992)

Principle		Team leader skill	Team leader knowledge
1	Giving immediate feedback to team members	Relays information pertaining to performance of team member tasks	
2	Collecting performance information	Requests performance information from team members	Double-checks team member performance and provides feedback Ensures team member skill competence
3	Structuring the team	Communicates specific and unambiguous role expectations Discusses role strategies in advance of potential crisis situations Directs team members to communicate information in proper sequence	
4	Co-ordinating the team to work together	Discusses member role inter-relation Discusses team member interdependencies Communicates the effects of team interaction on team performance Advises team members to communicate with each other when necessary	Structures the team according to knowledge of team member abilities and interaction styles Ensures that team members align their individual goals to those of the team
5	Using strategic and concise communication		Ensures that team members focus on the task at hand
6	Preparing and planning for upcoming crises	Educates team members in recognising unexpected events Educates team members in actions necessary in the event of emergency	Implements strategies in handling possible crisis situations
7	Assisting team members in attaining goals	Provides team members with specific goals relative to the mission/task Provides team members with motivational guidance	Removes obstacles that may inhibit team member motivation
8	Encouraging personal involvement in meeting team goals	Communicates information pertaining to issues that affect team performance	
9	Remaining aware and responsive to the needs of the team members		Perceives team member limitations and strengths Efficiently identifies potential problems due to knowledge of team member positions
10	Being perceived as approachable and unintimidating to team members		Establishes trust and opportunities for co-operative interaction Provides support when team members make mistakes Communicates to team members that it is acceptable to correct team leader if they think team leader is wrong Communicates to team members that team leader will remain open to constructive criticism

Box 6.6 King's Cross fire (1987)

On 18 November 1987, a small fire on an escalator was reported by a passenger and was inspected by a member of staff on duty at the underground station. Not normally being based at this station, this staff member did not report the incident to either the station manager or the line controller. The London Fire Brigade was called by a Metropolitan Police Officer who was fortunately in the underground station at the time. The fire, however, escalated while passengers were being evacuated and engulfed the ticket hall. In total, 31 people were killed, including a London Fire Brigade Station Officer (Fennell, 1988). The police inspector who responded was commended, as he '...was able to initiate with speed after a prompt reconnaissance. In the result an efficient and effective back-up was available to deal with the results of the disaster.'
(Fennell, 1988: p. 90)

Followership

Followership is defined as leadership influence of a manager on subordinates (Conger et al., 2000). Leadership is not something a leader possesses, so much as a process involving followership. Without followers, there are no leaders or leadership (Hollander, 1993; Wiener et al., 1993). Followership, like leadership, is a skill that can be learned (CAA, 2006), but for a follower, the skills are exercised in a supporting role that does not attempt to undermine the leader. This can be noted by how the leadership behaviour of the manager or team leader influences subordinates' reactions towards the manager as a leader, and towards themselves as subordinates and their task efforts.

Leadership and followership are intertwined, yet they are not equally examined. Leadership receives much attention, but this also requires an understanding of leader–member relations in teams and groups involving active followership (Hollander and Offerman, 1997). The topic of followership has been examined in relation to charisma theories of leadership. Charismatic leadership has been found to produce higher performance levels among followers, as well as more motivated and satisfied followers (Shamir et al., 1993). According to Conger et al. (2000), followers of charismatic leaders develop reverence for their leaders, in that they perceive their leader to be extraordinary. These followers also have higher collective identity as the leader employs vision to set goals and to shape a collective identity for followers, and have better understanding and have more realistic assessments of what tasks can actually be accomplished collectively by the team.

Perceptions of leaders can vary depending on whether things go well or not, and this is sometimes called 'the romance of leadership'. If things go really well, this is attributed to the leader's skill, but if things go badly, the leader's ineptitude is blamed (Meindl et al., 1985). However, if things are moderate, then it is unlikely that this is attributed to average leadership.

Training leadership skills

Leadership skills training courses have been specifically developed for first-line managers in the UK by the Institute of Leadership and Management (ILM), and courses typically include modules on leadership and team-building, communication, planning change, managing stress and problem-solving. Self-report questionnaires are often used in training to gauge the leader's perception of their style, for example the *Multifactor Leadership Questionnaire* by Bass and Avolio (1991) (see Mindgarden website) measures transformational, transactional and laissez-faire leadership. This can also be given to team members or peers to rate the leader's style.

Team leaders, and team members, should also be encouraged to seek out, engage in and craft their own experiences to develop more effective skills and knowledge for adaptive performance and overall team effectiveness (Salas et al., 2004). Team leaders drive team performance and effectiveness by communicating a clear direction, creating an enabling performance environment and providing process coaching, as well as training and developing team members' individual and team-level competencies (Hackman and Walton, 1986; Kozlowski et al., 1996).

In aviation, some CRM courses include a section on 'management, supervision and leadership'. This relates to developing leadership skills for appropriate personnel and would include aspects such as delegation, prioritisation of tasks and leadership styles – use of authority or assertiveness (CAA, 2004). Pilots also have to undertake specific training in command skills as part of their development to be promoted to captain.

Designing effective leadership training

The content of the leadership training can either focus on a sub-set of skills for particular tasks/situations or generic skills and behaviours. Effective training relies on effective design. Designing a training programme for developing leadership should be based on theories of both leadership and learning. Guidance for designing an effective leadership training programme includes (Yukl, 2005):

- *Clear learning objectives*: These objectives should describe the behaviours, skills and knowledge that trainees are expected to acquire.
- *Clear, meaningful content*: The content should build on a trainee's prior knowledge and focus on important aspects.
- *Appropriate sequencing of content*: Training activities should be organised and sequenced to facilitate learning, moving from simple to more complex ideas.
- *Appropriate mix of training methods*: Training methods should be appropriate for the knowledge, skills, attitudes and behaviours to be learned.
- *Opportunity for active practice*: The skills to be learned should be practised, for example practising behaviours, recalling information from memory.
- *Relevant, timely feedback*: Feedback should be obtained from a variety of available sources and should be constructive.
- *Trainee self-confidence*: Trainee self-efficacy and expectations should be enhanced by the instructional processes, and trainees should experience progress and success.

- *Appropriate follow-up activities*: Learning of complex skills should be enhanced by, for example, holding follow-up sessions at an appropriate interval after the training to review progress and problems.

Salas et al. (2002) suggest that skills training for team leaders should include:

- developing knowledge of other team members' roles to facilitate co-ordination, communication and team performance
- critical thinking skills, and
- promoting continuous learning within teams.

Box 6.7 Training metacognition in naval combat information centre (CIC) officers

The training content of a metacognitive skills training course developed for naval CIC officers (Cohen et al., 1998) consisted of scenario-based exercises that used interactive simulation and feedback. The content of the training exercises included:

- *Creating, testing and evaluating 'stories'*, i.e. decision-makers enhance their understanding of a situation by creating a story based on their assessment of the situation, which is then tested by comparing expectations against what is known or observed. If too many unreliable assumptions are used to form the story, then alternative assessments and stories can be generated.
- *Hostile-intent stories* can be used to teach trainees by practice and example to identify typical story components.
- *Critiquing stories* to evaluate their plausibility by identifying hidden assumptions in a story and generating alternative interpretations of the evidence.
- *When to think more*. It may not always be appropriate to use critical thinking, but it is necessary to evaluate the time available before making a decision. In a command environment, decisions can be delayed if a) the risk of delay is acceptable; b) the cost of an error if one acts immediately is high; c) the situation is non-routine or problematic in some way.

Feedback was provided by instructors in the form of hints throughout the scenario, as required, and by class discussion at the end of the exercise.

Studies with groups of five or six officers using this metacognitive training have shown that critical thinking skills can be taught and can improve decision-making processes as well as decision outcomes. These skills are important in tactical decision-making, as these skills help experienced decision-makers to handle uncertainty effectively. Critical thinking skills were also shown to be extended to teams by helping to provide team members with a shared mental model of the situation. Moreover, a shared metacognitive model of ongoing uncertainties was formed, which prompts team members to volunteer relevant information.

Although the trainees in this study were naval officers, this training strategy has also been used with Army battlefield command staff (Cohen et al., 1997a), and with Rotorcraft pilots (Cohen et al., 1997b).

Critical thinking skills, also known as metacognition, put simply could be defined as 'thinking about thinking'. This refers to the skills that involve reflecting upon and regulating one's own thinking, including memory, comprehension and performance (Flavell, 1979), and involves both explicit, conscious, factual knowledge as well as implicit, unconscious knowledge. Metacognitive skills describe awareness of the cognitive demands and specific strategies associated with different tasks. Cohen et al. (1998) comment that training of metacognition initially comprises a cognitive task analysis (see Chapter 9) to identify thinking strategies, ways of organising information and decisions. A syllabus can then be developed using information-based (e.g. lectures) and practice-based sessions (e.g. scenario-based exercises; see Box 6.7) to train leaders in metacognitive skills.

Leadership training techniques

Training for team leaders often consists of lectures, one-on-one coaching, on-the-job training, feedback from subordinates, superiors and peers, mentoring, and outdoor challenges (Bass and Riggio, 2006; Yukl, 2005). Effective leadership development involves a combination of these practices as well as the following specific techniques that focus on leadership skills training:

- *Behaviour role modelling*: This technique is based on demonstration and role-playing to enhance interpersonal skills for leaders. After being shown the effective behaviours, either by the trainer modelling the behaviour or on a video, trainees practise the necessary behaviours and receive constructive feedback. This technique is useful for concrete behaviours that are effective in particular circumstances but may not be as suitable for teaching flexible adaptive behaviours or cognitive knowledge.
- *Case studies:* This technique uses descriptions of events in an organisation, ranging from detailed descriptions of events over time or brief descriptions of specific incidents from a leader's career. Detailed cases about how specific situations or events were dealt with (e.g. operations or maintenance activities) may be used to practise analytical and decision-making skills for leadership. After the individual, or group, develop recommendations, these are compared with what the organisation or leader actually did. A benefit of case studies is that they increase understanding about situations that leaders can encounter, and different ways that can be considered for dealing with the situation.
- *Simulations:* Simulations require trainees to analyse complex problems and make decisions while leading their team; however, unlike case studies, trainees have to consider and deal with the consequences of their decisions as the scenario unfolds. Simulations can either be large-scale, with a large number of participants involved in role-playing, or smaller scale, e.g. a single team in a high-fidelity simulator. Low-fidelity simulation games can also be used, such as tactical decision games (see Chapter 10). Trainees receive directed feedback on their interpersonal skills (leadership, communication, teamwork), and cognitive skills such as decision-making and situation awareness.

The aim in all of the above techniques is to allow team leaders to observe, learn and practise leadership skills. Learning improves as trainees are provided with the opportunity to learn to deal with challenging and varied situations. Trainees also receive guided feedback about their performance and are encouraged to use this to reflect on their own experiences and performance to increase their learning.

Day and Halpin (2001) reviewed best practices in industry for developing leaders, including leadership training practices such as formal training, executive coaching, mentoring and outdoor challenges. Having examined leadership development in these organisations and the different types of leadership development training, they presented best practice principles:

- Successful leadership development efforts require an influential champion (preferably a senior manager).
- Leadership capacity is everywhere; leadership development initiatives should be orchestrated throughout the organisation.
- The most effective leadership development practices are tied to specific business imperatives.
- Leadership development is used to socialise managers on key corporate values and build a strong, coherent culture.
- Leadership development is a systemic process and not a single event.
- Successful leadership development depends more on consistent implementation than on using innovative practices.
- An important job of leaders is to make more leaders. High-potential leaders make for effective leadership preceptors in designing and delivering the curriculum.
- Leadership development is about creating entrepreneurial change agents who provide creative solutions in ambiguous situations.
- Leadership development is an investment in the future. Like most investments, it may take years before the dividends are realised.

Leadership development is related to organisational culture and strategy, and best practice organisations recognise that this training is an investment in the future.

Assessing leadership skills

There are various methods by which leadership skills can be assessed, such as appraisal and upward appraisal (see Yukl, 2005). As with other non-technical skills, assessment of leadership skill can be undertaken using observations and behavioural rating, which are described in more detail in Chapter 11. One example, the NOTECHS behavioural markers framework for pilots (Flin et al., 2003), includes a category and elements on leadership skills (see Table 6.8).

Table 6.8 NOTECHS Behavioural markers for leadership skills

Category	Elements	Example behaviours
Leadership and managerial skills	Use of authority and assertiveness	Takes initiative to ensure involvement and task completion
	Maintaining standards	Intervenes if task completion deviates from standards
	Planning and co-ordinating	Clearly states intentions and goals
	Workload management	Allocates enough time to complete tasks

These elements are specific to the setting in which they have been developed, i.e. the above elements and behaviours are drawn from aviation. Domain-specific categories and elements have been developed in other settings, such as the non-technical skills for surgeons system (NOTSS), where the elements of leadership are setting and maintaining standards, supporting others and coping with pressure (Yule et al., 2006).

Conclusion

This chapter has described leaders and leadership and discussed the competencies, skills and knowledge required by leaders. Over the past century, different theories of leadership have been proposed, focusing on the personality traits of leaders, observable behaviours of leaders and the influence of the situation. The theories appear to have some overlap, but none provides a single explanation of the concept of leadership. The functional approach to team leadership described the role of team leader in terms of enhancing team performance through developing the knowledge, skills and attitudes of team members.

Guidance for leadership has been discussed, such as the competencies required by team leaders, the knowledge, skills and attitudes required by team leaders, leadership under stress, and the interaction between team leaders and team members. Team leaders and team members do not operate in a vacuum, but are part of an overall organisational context. Leadership skills and knowledge has been emphasised rather than the tasks expected of leaders.

It is acknowledged that much of the current knowledge about leadership is based on managers in industrial settings or on military teams. There appears to be less empirical research reported into the behaviours of leaders at the 'sharp end' in other types of organisation. Nevertheless, much of this understanding does appear to apply to any operational team leader who has a direct responsibility for safe and effective performance in the workplace. An organisation's effectiveness depends on the behaviours of its team leaders and team members. This requires the consistent implementation of leadership development practices and the encouragement of leaders to reflect on and develop their own leadership skills.

Within high-reliability organisations, leaders are critical to safe and effective team performance, and it is the interaction between leader and team members that makes the difference. It is proposed here that leaders can be made, that leadership competencies can be trained through guided training to not only lead but to lead well.

Key points

- Leadership is primarily relevant to team leaders, but other team members should also show leadership skills as well as followership skills.
- Key tasks for team leaders are: to articulate a vision, to provide direction to the team, to create conditions for the team to work in, to build and maintain the team, and to coach and support the team to achieve success.
- Different theories of leadership have been proposed, such as trait theory (distinctive physical and psychological characteristics account for effective leadership), behavioural theory (observable leader behaviours), contingency/situational theory (leadership style changes to match the situation), and transformational leadership theory (charisma and motivation).
- Team leaders need to be able to vary their leadership style (e.g. delegative, consultative, democratic, directive) in response to different situations – there is no single leadership style that is effective in all situations.
- In emergency situations, the team leader may take on the role of commander, which requires specific skills for leading in stressful conditions.
- Team leaders need to develop their leadership competence through specific training and guided feedback, as well as by reflecting on their own experiences in terms of behaviours, skills, knowledge and attitudes.

Key texts

Flin, R. and Arbuthnot, K. (eds.) (2002) *Incident Command: Tales from the Hot Seat*. Aldershot: Ashgate.

Hackman, J.R. (2002) *Leading Teams. Setting the Stage for Great Performance*. Harvard: Harvard Business School Press.

Bass, B. (1990) *Bass and Stodgill's Handbook of Leadership: Theory, Research and Managerial Applications* (3rd ed). New York: Free Press.

West, M. (2004) *Effective Teamwork. Practical Lessons from Organizational Research*. (2nd ed). Oxford: BPS Blackwell

Yukl, G.A. (2005) *Leadership in Organisations*. (6th ed.). Upper Saddle River, NJ: Prentice Hall.

Websites

Hackman, J.R: http://www.leadingteams.org

CAA: http://www.caa.co.uk/docs/33/CAP737.PDF, Appendix 7
Step Change in Safety (2006): http://stepchangeinsafety.net/stepchange
Mindgarden: www.mindgarden.com for MLQ transformational leadership test

References

Arbuthnot, A. (2002) Key issues in incident command. In R. Flin and A. Arbuthnot (eds.) *Incident Command: Tales from the Hot Seat* (pp. 10–31). Aldershot: Ashgate.

Arnold, J., Silvester, J., Patterson, F., Robertson, I., Cooper, C. and Burness, B. (2004) *Work Psychology. Understanding Human Behaviour in the Workplace* (4th ed.). Harlow: Pearson Education.

Bass, B.M. (1985) *Leadership and Performance Beyond Expectation.* New York: Free Press.

Bass, B. (1990) *Bass and Stodgill's Handbook of Leadership: Theory, Research and Managerial Applications* (3rd ed). New York: Free Press.

Bass, B. and Avolio, B. (1990) The implications of transactional and transformational leadership for individual, team, and organisational development. In R. Woodman and W.A. Passmore (eds.) *Research in Organizational Change and Development* (pp. 231–272). Volume 4. Greenwich, Conn: JAI Press,

Bass, B.M. and Avolio, B.J. (1991) *The Multifactor Leadership Questionnaire.* Palo Alto, CA: Consulting Psychologists Press.

Bass, B. and Riggio, R. (2006) *Transformational Leadership.* (2nd ed.) Mahwah, NJ: Prentice Hall.

Blake, R.R. and Mouton, J.S. (1964) *The Managerial Grid.* Houston, TX: Gulf.

Bolman, L.G. and Deal, T.E. (1997) *Reframing Organizations. Artistry, Choice and Leadership* (2nd ed.) San Francisco: Jossey-Bass.

Brunacini, A. (2002) Incident command functions. In R. Flin and K. Arbuthnot (eds.) *Incident Command: Tales from the Hot Seat.* Aldershot: Ashgate.

Burgess, K.A., Salas, E., Cannon-Bowers, J.A. and Hall, J.K. (1992) Training guidelines for team leaders under stress. Paper presented at the 36th Annual Meeting of the Human Factors Society, Atlanta, Georgia.

Burns, J.M. (1978) *Leadership.* New York: Harper and Row.

CAA (2006) *Crew Resource Management (CRM) Training. CAP737. Guidance for Flight Crew, CRM Instructors (CRMIs) and CRM Instructor-Examiners (CRMIEs).* Hounslow, Middlesex: Civil Aviation Authority.

CAA (2004) *CAP716 Aviation Maintenance Human Factors (EASA/JAR145 Approved Organisations)*. Sussex: CAA Safety Regulation Group.

Cohen, M.S., Freeman, J.T. and Thompson, B. (1998) Critical thinking skills in tactical decision making: A model and a training strategy. In J. Cannon-Bowers and E. Salas (eds.) *Making Decisions Under Stress: Implications for Individual and Team Training.* Washington, D.C.: American Psychological Association.

Cohen, M.S., Parasuraman, R., Serfaty, D. and Andes, R. (1997a) *Trust in Decision Aids: A model and Training Strategy.* Arlington, VA: Cognitive Technologies.

Cohen, M.S., Thompson, B. and Freeman, J.T. (1997b) Cognitive aspects of automated target recognition interface design: An experimental analysis. Arlington, VA: Cognitive Technologies.

Conger, J.A., Kanungo, R.N. and Menon, S.T. (2000) Charismatic leadership and follower effects. *Journal of Organisational Behavior*, 21, 747–767.

Crichton, M. and Flin, R. (2004) Identifying and training non-technical skills of nuclear emergency response teams. *Annals of Nuclear Energy*, 31, 1317–1330.

Crichton, M., Lauche, K. and Flin, R. (2005) Incident command skills in the management of an oil industry drilling incident: A case study. *Journal of Contingencies and Crisis Management*, 13, 116–128.

Cullen, The Hon. Lord (1990) *The Public Inquiry into the Piper Alpha Disaster*, vols I and II (Cm 1310). London: HMSO.

Day, D.V. and Halpin, S.M. (2001) *Leadership Development: A Review of Industry Best Practices* (Technical report 1111). Alexandria, VA: US Army Research Institute for the Behavioral and Social Sciences.

Day, D.V., Gronn, P. and Salas, E. (2004) Leadership capacity in teams. *The Leadership Quarterly*, 15, 857–880.

Driskell, J.E. and Salas, E. (1991) Overcoming the effects of stress on military performance: Human factors, training and selection strategies. In R. Gal and A.D. Mangelsdorff (eds.) *Handbook of Military Psychology*. Chichester: Wiley.

Edmondson, A.C. (2003) Speaking up in the operating room: How team leaders promote learning in interdisciplinary action teams. *Journal of Management Studies*, 40, 1419–1452.

Fennell, D. (1988) Investigation into the King's Cross Underground Fire. Department of Transport, London: HMSO.

Fiedler, F. (1967) *A Theory of Leadership Effectiveness*. New York: McGraw-Hill.

Fiedler, F.E. (1995) Cognitive resources and leadership performance. *Applied Psychology: An International Review*, 44, 5–28.

Flavell, J.H. (1979) Metacognition and cognitive monitoring: A new area of cognitive-developmental inquiry. *American Psychologist*, 34, 906–911.

Flin, R. (1996) *Sitting in the Hot Seat: Leaders and Teams for Critical Incident Management.* Chichester: Wiley.

Flin, R. and Arbuthnot, K. (eds.) (2002) *Incident Command: Tales from the Hot Seat.* Aldershot: Ashgate.

Flin, R., Martin, L., Goeters, K., Hoerman, H., Amalberti, R., Valot, C. and Nijhuis, H. (2003) Development of the NOTECHS (non-technical skills) system for assessing pilots' CRM skills. *Human Factors and Aerospace Safety*, 3, 95–117.

Flin, R. and Yule, S. (2004) Leadership for safety: Industrial experience. *Quality and Safety in Health Care*, 13, Suppl 1, i45–i51.

Fortune Small Business magazine (2003) The Best Bosses. http://www.fortune.com/fortune/smallbusiness/managing/articles/0,15114,487500,00.html

Furnham, A. (2005) *The Psychology of Behaviour at Work. The Individual in the Organization.* (2nd ed). Hove: Taylor & Francis.

Glendon, I., Clarke, S. and McKenna, E. (2006) *Human Safety and Risk Management.* (2nd ed.) London: Taylor & Francis.

Hackman, J.R. and Walton, R.E. (1986) Leading groups in organisations. In Goodman and Associates (eds.) *Designing Effective Work Groups* (pp. 72–119). San Francisco: Jossey-Bass, .

Hersey, P. and Blanchard, K.H. (1977) *The Management of Organisational Behaviour* (3rd ed.). Upper Saddle River, NJ: Prentice Hall.

Hersey, P., Blanchard, K.H. and Johnson, D.E. (2000) *The Management of Organisational Behaviour* (8th ed.). Upper Saddle River, NJ: Prentice Hall.

Hofmann, D. and Morgeson, F. (2004) The role of leadership in safety. In J. Barling and M. Frone (eds.) *The Psychology of Workplace Safety.* Washington: APA Books.

Hollander, E. (1993) Legitimacy, power, and influence: A perspective on relational features of leadership. In M. Chemers and R. Ayman (eds.) *Leadership Theory and Research. Perspectives and Directions* (pp. 29–47). San Diego: Academic Press.

Hollander, E.P. and Offerman, L.R. (1997) *KLSP: The Balance of Leadership and Followership*. Maryland: Academy of Leadership Press.

Howell, J. and Avolio, B. (1993) Transformational leadership, transactional leadership, locus of control, and support for innovation: Key predictors of consolidated business unit performance. *Journal of Applied Psychology*, 78, 891–902.

Klein, K.J., Ziegert, J.C., Knight, A.P. and Xiao, Y. (2006) Dynamic delegation: Shared, hierarchical and deindividualised leadership in extreme action teams. *Administrative Science Quarterly*, 51, 590–621.

Kozlowski, S.W.J., Gully, S.M., Salas, E. and Cannon-Bowers, J. (1996) Team leadership and development: Theory, principles, and guidelines for training leaders and teams. In M. Beyerlein, S. Beyerlein and D. Johnson (eds.) *Advances in Interdisciplinary Studies of Work Teams: Team leadership.* Volume 3. Greenwich, CT: JAI, 253–292.

Lewin, K., Lippett, R. and White, R. (1939) Patterns of aggressive behavior in experimentally created 'social climates'. *Journal of Social Psychology*, 10, 271–299.

McGrath, J.E. (1962) *Leadership Behavior: Requirements for Leadership Training.* Washington, DC: US Civil Service Commission Office of Career Development.

Meindl, J.R., Ehrlich, S.B. and Dukerich, J.M. (1985) The romance of leadership. *Administrative Science Quarterly*, 30, 78–102.

Northouse, P. (2006) *Leadership: Theory and Practice.* (4th ed.) London: Sage.

Oser, R.L., Prince, C., Morgan, B.B. and Simpson, S.S. (1991) *An analysis of Aircrew Communication Patterns and Content*. Technical report 90-009. Orlando, FL: Human Factors Division, Naval Training Systems Center.

Osman, L. Adie, W. and Cairns, J. (2003) *Attitudes to Safety Culture among Professional Divers And Offshore Workers.* Suffolk, UK: HSE Books.

Sadler, P. (1997) *Leadership*. London: Kogan Page.

Salas, E. and Cannon-Bowers, J.A. (1997) The anatomy of team training. In L. Tobias and D. Fletcher (eds.) *Handbook on Research in Training*. New York: Macmillan.

Salas, E., Burke, C.S. and Stagl, K.C. (2004) Developing teams and team leaders: Strategies and principles. In D. Day, S. Zaccaro and S.M. Halpin (eds.) *Leader*

Development for Transforming Organizations. Growing Leaders for Tomorrow. Mahwah, NJ: Lawrence Erlbaum.

Salas, E., Cannon-Bowers, J. and Weaver, J. (2002) Command and control teams: Principles for training and assessment. In R. Flin and K. Arbuthnot (eds.) *Incident Command: Tales from the Hot Seat*. Aldershot: Ashgate.

Sefton, A. (1992, April) Introduction to the first offshore installation management conference: Emergency command, Aberdeen.

Shackleton, E. (2001) *South!: The story of Shackleton's Last Expedition 1914–1917.* Santa Barbara, CA: Narrative Press.

Shamir, B., House, R. and Arthur, M.B. (1993) The motivational effects of charismatic leadership: A self-concept based theory. *Organisational Science*, 4, 387–409.

Step Change in Safety (2006) http://stepchangeinsafety.net/stepchange.

Stodgill, R. (1948) Personal factors associated with leadership: A review of the literature. *Journal of Psychology*, 24, 35–71.

Thompson, R.C., Hilton, T.F. and Witt, L.A. (1998) Where the safety rubber meets the shopfloor: A confirmatory model of management influence on workplace safety. *Journal of Safety Research*, 29, 15–24.

Ullman, M. (1994) A funny thing happened to me on the way to the forum. In Collected Papers of the Third Offshore Installation management conference, Robert Gordon University, Aberdeen, April.

Useem, M., Cook, J. and Sutton, L. (2005) Developing leaders for decision making under stress: Wildland firefighters in the South Canyon fire and its aftermath. *Academy of Management Learning and Education*, 4, 461–485.

Vroom, V.H. and Yetton, P.W. (1973) *Leadership and Decision Making*. Pittsburgh: University of Pittsburgh Press.

West, M. (2004) *Effective Teamwork. Practical Lessons from Organizational Research*. (2nd ed). Oxford: BPS Blackwell.

Wiener, E., Kanki, B. and Helmreich, R. (1993) *Cockpit Resource Management*. San Diego: Academic Press.

Yukl, G. (2005) *Leadership in Organisations* (6th ed.). Upper Saddle River, NJ: Prentice-Hall International.

Yule, S., Flin, R., Paterson-Brown, S., Maran, N. and Rowley, D. (2006) Development of a rating system for surgeons' non-technical skills. *Medical Education*, 40, 1098–1104.

Zaccaro, S.J., Rittman, A.L. and Marks, M.A. (2001) Team leadership. *The Leadership Quarterly*, 12, 451–483.

Zohar, D. (2003) The influence of leadership and climate on occupational health and safety. In D. Hofman and L. Tetrick (eds.) *Health and Safety in Organizations: A Multilevel Perspective* (pp. 201–230). San Francisco, CA: Jossey-Bass.

Chapter 7

Managing Stress

Introduction

Most people know what it feels like to be under stress. Many different definitions of stress exist. Stress can be mechanical, engineering, linguistic or psychological. One of the most widely accepted definitions of psychological stress by stress researchers is that of Lazarus and Folkman (1984). They define stress as 'a particular relationship between the person and the environment that is appraised by the person as taxing or exceeding his or her resources and endangering his or her well-being' (p19). Similarly, the UK Health and Safety Executive (HSE) define stress as 'the adverse reaction people have to excessive pressure or other types of demand placed on them' (HSE, 2005a).

Within the context of high-risk work environments, stress can have both acute and chronic effects. Acute stress – often known as emergency stress or critical incident stress – is sudden, novel, intense and of relatively short duration. It disrupts goal-oriented behaviour and requires a proximate response (Salas et al., 1996). Acute stress occurs, at its most extreme, when the individual is suddenly exposed to a threatening situation, such as a life-endangering event or traumatic scene, and experiences a pronounced physiological and psychological reaction (Flin, 1996a). Acute stress reactions in humans can be related to the classic fight, flight or freeze response in animals (Cannon, 1929).

Chronic stress is related to conditions in the workplace and the individual's reaction to these, usually over a protracted period of time. Baum et al. (1993) define chronic stress as 'the persistent negative experience or exposure of threat or excessive demand' (p274). In an increasingly high-tech and complex world, the levels of stress are escalating, and chronic stress is rife across many types of workplace. A secretary working in a bank, a control room operator in a nuclear power plant, or a soldier patrolling the streets of a foreign country must be able to cope with periods of chronic stress. However, individuals working in high-risk industries must also be able to function effectively under shorter periods of acute stress, and may need to make critical decisions under extreme pressures and demands. Therefore, an understanding of both chronic and acute stress is crucial to maintain, or improve, job performance.

Stress has been linked to safety outcomes, such as accident involvement (Cooper and Clarke, 2003). Therefore, the ability to recognise and manage stress in self and others is an important non-technical skill. In this chapter, a theory of stress is presented followed by a discussion of both chronic stress and acute stress. The skills related to identifying causes, symptoms, effects and methods of addressing the two types of stress are described. The basic elements of coping with stress are shown in Table 7.1

Table 7.1 Elements of stress management

Category	Elements
Managing stress	Identify causes Recognise symptoms and effects Implement coping strategies

Theory of stress

Theories of stress tend to emphasise the role of individual appraisal in the human stress response (Cooper et al., 2001). These theoretical models of stress can be portrayed as a balance mechanism (see Figure 7.1). A *demand*, or *stressor*, is something that causes stress. It may be a single event (e.g. meeting a specific deadline at work), or an ongoing event (e.g. excessive work pressures). *Resources* are an individual's means to address these demands, such as prior training, experience and skills. Another aspect that affects the balance is an individual's *mediating* factors (e.g. personality, fitness, coping strategies and social support). Mediators can reduce or exacerbate the stress effects experienced by an individual.

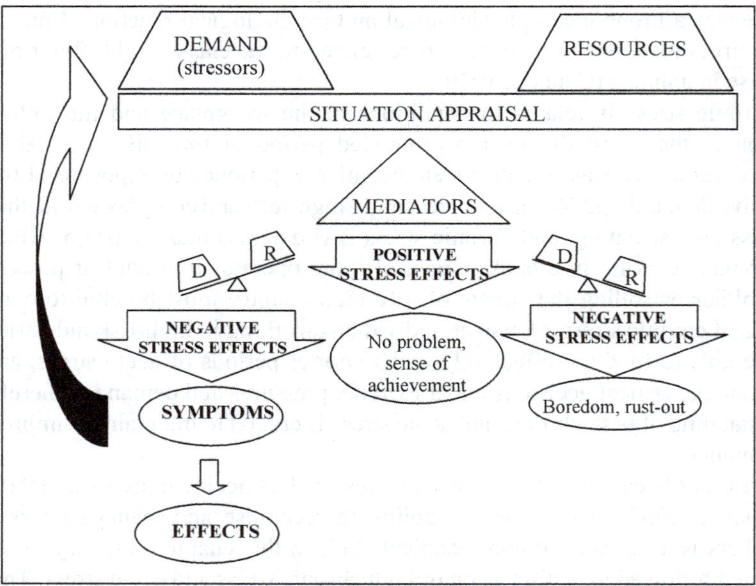

Figure 7.1 The balance model of stress (based on Cox, 1993)

What is critical in this model is the role of situation appraisal. From Figure 7.1 it can be seen that when the available resources are judged to be equal to the demands, the individual feels in control and comfortable. In addition, while in this state, moderate increases in demand may actually increase motivation and performance, as low levels of pressure can have a beneficial effect on performance. Further, if an individual's resources far exceed the demands, then they may experience 'rust-out' or boredom. We need some level of demand to ensure we are not bored and to be motivated (see Figure 7.1). However, for the purpose of this chapter, the term stress will be taken to represent negative, unpleasant effects. Where the stressors outweigh the perceived resources to cope with the demands, then the stress reactions begin to occur. Initially the reaction leads to *symptoms* or stress. These reactions are a complex and interacting package of responses with behavioural, emotional, somatic (physical) and thinking (cognitive) effects. These reactions then feed back into the individual's *appraisal of the situation*. Finally, the symptoms lead to *effects* or disease, which have detrimental effects on the performance and health of the individual, team and organisation.

It is important to emphasise that it is the individual's *perceived* awareness of the demands and of his or her resources to meet these demands that are crucial. The *absolute* level of the demand or resource does not appear to be so important (Cox, 1985). Rather, it is the discrepancy that exists between the individual's perception of the demands and perceived ability to cope. Consequently, one emergency response team member faced with an incident may feel calm, confident and totally in control, while another in the same situation may be uneasy, irritable and losing a grasp of the situation (Flin, 1996a).

Chronic stress

As described above, chronic stress is related to how an employee reacts to stressors (or demands) in the workplace over an extended period of time. In the developed world, work-related stress is one of the greatest challenges to the health of working people and to the healthiness of their work organisations (Cox et al., 2002). A survey of the European Union member states found that 28% of employees reported stress-related illness or health problems (this equates to 41 million workers, European Foundation for the Improvement of Working Conditions, 2000). Further, those in the health care sector are one of the groups identified as most at risk to chronic stress (Houtman, 2005).

Based on a survey of 8,000 people in the Bristol area of the UK, one in five people reported being very or extremely stressed at work (Smith et al., 2000). In the UK, stress is the second most commonly reported work-related health problem (HSE, 2005b) and during 2004/2005, stress, depression and anxiety accounted for an estimated 12.8 million lost working days (HSE, 2005c). In the USA, workers who suffer from a stress-related illness are off work for a median of 25 days, compared with a median of six days for other non-fatal injuries at work (National Institute for Occupational Safety and Health (NIOSH), 2004). NIOSH also quote a survey carried out by Northwestern National Life in which 40% of respondents report that their job is very or extremely stressful. In a study by Yale University, 29% of respondents perceived themselves to be 'quite a bit' or 'extremely' stressed at work (NIOSH, 1999).

Figure 7.2 Model of *chronic* stress (adapted from Cooper et al.'s 1988 model of stress at work)

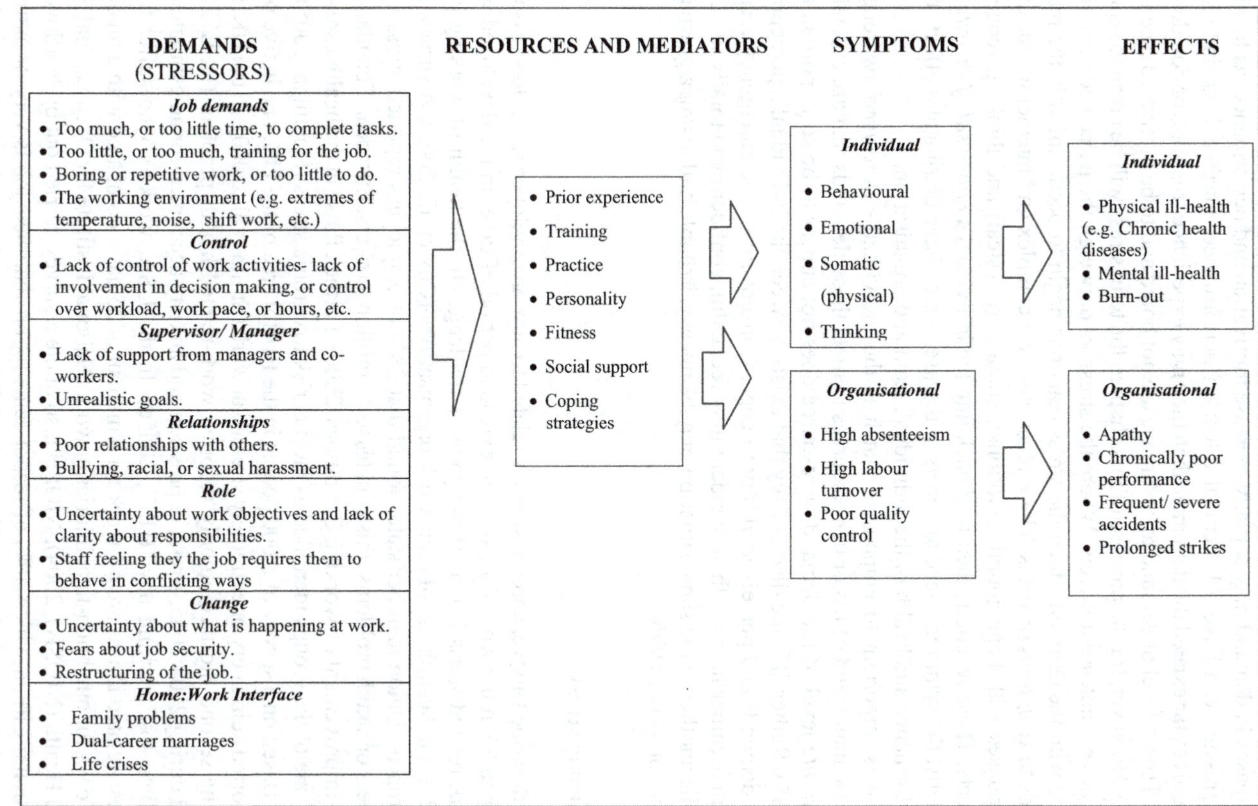
DEMANDS
(STRESSORS)
Job demands
• Too much, or too little time, to complete tasks.
• Too little, or too much, training for the job.
• Boring or repetitive work, or too little to do.
• The working environment (e.g. extremes of temperature, noise, shift work, etc.)
Control
• Lack of control of work activities- lack of involvement in decision making, or control over workload, work pace, or hours, etc.
Supervisor/ Manager
• Lack of support from managers and co-workers.
• Unrealistic goals.
Relationships
• Poor relationships with others.
• Bullying, racial, or sexual harassment.
Role
• Uncertainty about work objectives and lack of clarity about responsibilities.
• Staff feeling they the job requires them to behave in conflicting ways
Change
• Uncertainty about what is happening at work.
• Fears about job security.
• Restructuring of the job.
Home:Work Interface
• Family problems
• Dual-career marriages
• Life crises
RESOURCES AND MEDIATORS
• Prior experience
• Training
• Practice
• Personality
• Fitness
• Social support
• Coping strategies
SYMPTOMS
Individual
• Behavioural
• Emotional
• Somatic (physical)
• Thinking
Organisational
• High absenteeism
• High labour turnover
• Poor quality control
EFFECTS
Individual
• Physical ill-health (e.g. Chronic health diseases)
• Mental ill-health
• Burn-out
Organisational
• Apathy
• Chronically poor performance
• Frequent/ severe accidents
• Prolonged strikes

Chronic Stressors

Researchers have categorised sources of workplace stress, or stressors, into a number of taxonomies. Six categories of workplace stressors have been identified by the UK HSE (2005c), as shown in Figure 7.2.

Job demands include issues such as workload, shift patterns and the work environment. The level of noise, temperature, hazards and other working conditions can cause both physical and mental discomfort. Many research studies have shown increased workload leads to a degradation in performance (Morgan and Bowers, 1995). Environmental conditions such as noise have also been shown to be a stressor (Smith, 1991).

Lack of control relates to the amount of control a person has in the way they do their work. One of the dominant models of stress (Karasek and Theorell, 1990) suggests that it is the relationship between work demands and the amount of control that can be exercised over them that is most critical. In a review by Schnal et al. (1994), 17 out of 24 studies found significant associations between low job control and cardiovascular disease.

Relationships relates to bullying, harassment and other poor relationships with people at work. Workplace bullying was found to be associated with an increase in the sickness absenteeism of hospital staff (Kivimaki et al., 2000).

Change is concerned with how organisational change is managed and communicated in the organisation. Little et al. (1990) found that pilots who worked for airlines with histories of corporate instability reported significantly more stress and depression symptoms, and a greater accumulation of symptoms, than pilots employed by more stable companies.

Role relates to how well the employees understand the role they perform in the organisation and whether they have conflicting roles. Kahn et al. (1964) found that workers who suffered from role ambiguity were more likely to experience a greater incidence of job-related stress.

Supervision or managers are recognised as a source of stress. Hogan et al. (1994) state that 60–70% of American employees regard their immediate supervisor as the most stressful aspect of their job. Lack of support of the individual includes colleagues, managers and people outside work. This is often a mediator. De Longis et al. (1988) found that individuals with low social support are vulnerable to illness and mood disturbance when their stress levels increase, even if they generally have little stress in their lives.

The *home/work interface*, particularly in terms of time commitment, can also be an added stressor. With nearly 70% of married women working, and 40% of working women with pre-school children in some European countries, the dual-career couple is becoming increasingly prevalent (Cooper and Cartwright, 2001). People cannot simply forget about stressors they may have in their personal life (e.g. poor relationship with spouse, new child, sick parent, etc.) when they get to work. Holmes and Rahe (1967) developed the Social Readjustment Scale, which consists of a list of 43 life-changing events with numerical values attached to them. Using this scale allows an estimation of the level of stress experienced by an individual as a result of events such as changing job, moving house, a death in the family, etc.

However, the above list of stressors is not exhaustive. Every workplace is likely to have its own particular set of stressors. For example working on an offshore oil platform has a unique combination of stressors (e.g. remote location, absence from family and friends, shift and rotation patterns, social density and helicopter travel; Flin, 1996b) that are probably not shared by other workplaces. There are questionnaires available to measure sources of stress in a particular workplace (see below).

Resources and mediators

Mediating factors are the 'lens' through which the effects of stressors are amplified or reduced. These factors moderate the relationship between the causes and effects of stress by exerting influence upon the ability of the individual to cope with the perceived stressors.

Prior experience, training and practice If a person has received adequate training to perform a job, and performed the job effectively in the past, then they are less likely to suffer from symptoms of stress than a less experienced or less well-trained individual. In addition, if someone failed to adequately perform a task effectively in the past, then they are more likely to suffer from symptoms of stress the next time they have to do the same task than if they successfully completed the task. Therefore, employees should ensure they are adequately trained and have the necessary experience to perform their job.

Personality Studies of occupational stress have repeatedly shown that there are marked individual differences in how people experience and react to stress. Research findings as to the effect of personality as a mediating factor in the experience of stress are not clear cut. However, three aspects of personality that have been shown in some studies to have an effect are outlined below:

- Psychological hardiness includes a belief in one's ability to influence the situation, being committed to or fully engaged in one's activities, and having a positive view of change. Kobasa et al. (1982) found business executives who are hardy are less likely to suffer the ill effects of stress than less hardy individuals. Further, individuals high in both hardiness and exercise remain healthier than those high in one or the other only.

- Studies of patient personality and heart disease 30 years ago suggested that there were two basic types of person:
    - Individuals with a Type A disposition tend to be competitive, hard-driving, time-driven, achievement oriented, and impatient.
    - In contrast, Type B individuals are more easy-going, patient and relaxed.

  A Type A personality was found to be significantly correlated with daily stress, tension, anger symptoms and ambitiousness (Haynes et al., 1978). A study of North Sea offshore workers found that those who were Type A were more likely to experience stress than those who were Type B (Flin, 1996b). Although the idea that Type A personality might be linked to an increased risk of heart disease was never proven, there is evidence that the hostility and anger aspects of Type A may be linked to heart disease (Geipert, 2007).
- Neuroticism can be defined as an enduring tendency of an individual to experience negative emotional states. Low neuroticism is associated with good coping skills, favourable mental health, emotional stability and resistance to stress. On the other hand, individuals with high neuroticism tend to be more emotionally unstable, prone to distress and have a greater likelihood of 'snapping' under stressful conditions. Gunthert et al. (1999) found that when compared with low-neurotic college students, high-neurotic college students reported more interpersonal stressors and reacted with more distress in response to those stressors.

Fitness Fitness and general well-being is also a good mediator against stress. It is well known that feelings of tiredness or aliments such as cold or flu can increase sensitivity to stress. So good diet, exercise and sufficient sleep are effective mediators in coping with stressors (see Chapter 8 on fatigue). A study of executives by Kobasa et al. (1982) found that exercising was associated with an increased ability to cope with stress.

Social support In almost all models of occupational stress, social support is a mediating variable. This can come from friends, family, work colleagues or from professional counsellors. Boyle (1997) found that the majority of ambulance staff admitted that they would find it difficult to do their job without the support of their spouse. La Rocco and Jones (1978) found that support from colleagues was the primary social support used to reduce the effects of stress for ambulance workers.

Coping strategies The extent to which people experience stress can be dependent on the coping strategies they use. There are two forms of coping strategies: problem-focused and emotion-focused coping.

Problem-focused coping can be defined as 'efforts often directed at defining the problem, generating alternative solutions, weighing alternatives in terms of costs and benefits and choosing among them, and acting' (Lazarus and Folkman, 1984: 156). Carver et al. (1989) identified six examples of problem-focused coping strategies:

- active coping – taking active steps to address a stressor
- planning – thinking about how to cope with a stressor
- suppression of competing activities – putting other tasks or events aside to concentrate on dealing with the stressor
- restraint coping – waiting for an appropriate opportunity to act
- seeking social support for instrumental reasons – seeking advice, assistance or information
- seeking social support for emotional reasons – obtaining moral support, sympathy or understanding.

Emotion-focused strategies aim to reduce, or manage, the negative feelings induced by the stressor. Carver et al. (1989) identified seven types of these strategies:

- focus on, and venting, emotion – the tendency to focus on the distress or upset and to vent those feelings
- behavioural disengagement – reducing one's effort to address the stressor
- mental disengagement – distracting one's self with another activity to take one's mind off the stressor
- positive reinterpretation and growth – the focus is on managing distressing emotions rather than dealing with the stressor *per se*
- denial – a refusal to admit the stressor exists
- acceptance – admitting that the stressor exists
- turning to religion – the tendency to turn to religion in times of stress.

Generally, problem-focused strategies are most effective when people have a realistic prospect to change features of their situation to reduce stress. Emotion-focused strategies may be more useful as a short-term strategy. To illustrate, US naval aviators are encouraged to compartmentalise stressors when they are flying (e.g. if an aviator has had a fight with their spouse before coming to work, then they should not think about this while they are flying). However, the danger of this method in the longer term is that the stressor may not ever actually be addressed. Emotion-focused strategies can also be helpful in reducing one's arousal level prior to performing problem-focused coping strategies and can help people when few problem-focused coping options are available (e.g. the death of a spouse).

Symptoms of chronic stress

In the UK and in many other countries, it is the duty, by law, of the employer to ensure that employees are not made ill by their work (HSE, 2000). There is no single way of identifying whether a colleague is suffering from chronic stress. Often people may not wish to admit to themselves, or others, that they are experiencing distress, or indeed may not even be aware that they are suffering from chronic stress. This can then lead to a much more catastrophic effect than if the problem was identified earlier. Therefore, an important element of managing stress is to be able to identify symptoms of stress in oneself, and others.

Unless a colleague tells one of their team-mates they think they are suffering from chronic stress, the most obvious signs are a change from their normal behaviour. If a colleague is always grumpy, then this is not necessarily a sign that they are suffering from chronic stress. However, if an outgoing and happy colleague becomes insular and depressed, these could be indicators that they are suffering from chronic stress.

Individual symptoms

It is possible to classify the indicators of chronic stress effects into four categories: behavioural, emotional, somatic (physical) and thinking. An acronym by which to readily remember these categories is 'BEST'. These symptoms are discussed in detail below.

***B**: Behavioural indicators.* The effects of stress on behaviour are generally the most readily observable by work colleagues. There is not a finite list of behavioural effects, but the defining feature is a change in the individual's normal behaviour pattern (see Table 7.2; Flin, 1996b). For example, a placid worker becomes irritable or a colleague who generally has a neat appearance starts to take less interest in how he looks. As the behaviour change can be subtle (although there may be other clues that a team member is not performing his or her job properly), then the better the leader knows his or her team members the more likely that any behaviour changes will be noticed.

Table 7.2 Behavioural indicators of chronic stress

• absenteeism	• apathy
• abuse of drugs, e.g. increased alcohol use or smoking	• reduced productivity
	• distracted
• hostile behaviour	• careless errors

***E**: Emotional indicators.* A wide range of emotional responses can occur as part of the stress response (see Table 7.3). Through apprehension and anxiety, emotional indicators can lead to loss of emotional control, where the individual becomes irritable or depressed.

Table 7.3 Emotional indicators of chronic stress

• anxiety, feelings of hopelessness	• depression
• cynicism and resentfulness	• irritability

***S**: Somatic (physical) indicators.* As with behavioural indicators, the physical effects and associated symptoms of chronic stress may not be readily observed by others. The physical indicators of stress are often concealed by the sufferer, who may engage in long-term self-medication to manage them. The sufferer may also not recognise or wish to acknowledge that these are being caused by stressors (see Table 7.4). People who have frequent minor health complaints may be suffering from chronic stress. Stress has been shown to be a strong predictor of both the frequency and severity of physical ill health (Wyler et al., 1968).

Table 7.4 Somatic indicators of chronic stress

• decline in physical appearance	• health complaints such as headaches, chest pains, or stomach complaints
• chronic fatigue	
• frequent infections	

***T**: Thinking (cognitive) indicators.* As with acute stress, the cognitive, or thinking, indicators of chronic stress can lead to poor decision-making and difficulty in concentration (see Table 7.5). This is particularly relevant for those working in safety-critical jobs where performance is dependent on their thinking skills.

Table 7.5 Thinking indicators of chronic stress

• lack of concentration	• impaired decision-making
• reduced attention	• failures in planning
• difficulty in remembering	

Organisational indicators

Cox (1993) states that there have been suggestions that if 40% of workers in any group (department or organisation) are having stress-related difficulties, then the group or organisation can manifest symptoms of stress (see Table 7.6). The organisational indicators of stress in the lead up to the *Columbia* space shuttle disaster are described in Box 7.1.

Table 7.6 Organisational indicators of chronic stress

• high staff turnover	• increase in client complaints
• absenteeism	• increase in employee compensation claims
• poor time-keeping	• more near-misses and accidents
• decreased productivity	

Box 7.1 The effects of chronic stress: The Columbia space shuttle disaster

The Columbia STS-107 mission lifted off on 16 January 2003 for a 17-day science mission. Upon re-entering the atmosphere on 1 February 2003, the Columbia orbiter suffered a catastrophic failure due to a breach that occurred during launch when falling foam struck the underside of the left wing. The orbiter disintegrated, killing its seven crew members approximately 15 minutes before Columbia was scheduled to touch down at Kennedy Space Center.

A number of chronic stressors can be identified from Chapters 5 and 6 of Volume 1 of the Columbia accident investigation report (Columbia Accident Investigation Board, 2003).

Demands of the job

- There was pressure to meet an ambitious launch schedule. 'I wasn't convinced people were being given enough time to work the problems correctly' (p134).
- Time was not available for flight controllers to complete recertification requirements.
- Work was being scheduled on holidays, crew rotations were drifting beyond 180 days. In 2001, an experienced observer described the workforce as 'the few, the tired' (p118).

Control

- Management felt that the launch schedule to support the Space Station was not too ambitious, but the workforce disagreed.

Change

- Uncertainty about when the shuttle might be replaced.
- A hiring freeze during the 1990s resulted in a 25% reduction in the NASA workforce. 'Five years of buyouts and downsizing have led to serious skill imbalances and an overtaxed workforce. As more employees have departed, the workload and stress [on those] remaining have increased, with a corresponding increase in the potential for impacts on operational capacity and safety' (p110).
- The launch schedules were increasingly demanding. There was no longer any padding in the system.

Role

- Uncertainty about work objectives and lack of clarity about responsibilities.
- There was concern about how all the work was to be completed. 'I would like to think that the technical issues can take priority over any budget issues or scheduling issue' (p134).

Support and the individual

- Pressure from management to meet an increasingly ambitious launch schedule.
- Fatigue due to lack of holidays and intense work pace.
- NASA had a 'can-do' culture. No one wanted to stand up and say they would be unable to meet a target.

Chronic stress effects

Once the symptoms of stress are present, these can then lead to longer-term effects, which are sometimes labelled disease. Cooper et al. (1988) separates the diseases into those that have an effect upon the individual, and those that effect the entire organisation.

Individual

Stress-related problems are the second most commonly reported case of occupational ill health, after musculoskeletal disorders (Rick et al., 2002). Cox (1993) identifies a number of diseases that have been associated with stress, although they may often be due to other factors. Those diseases identified include: bronchitis, coronary heart disease, mental illness (e.g. depression), thyroid disorders, skin diseases, types of rheumatoid arthritis, obesity, headaches and migraines, ulcerative colitis, and diabetes.

Another effect caused by stress is burnout. Maslach (1978) defines burnout as the result of repeated emotional pressure related to involvement with people and is characterised by emotional exhaustion, depersonalisation and reduced personal accomplishment.

Organisation

Work-related stress contributes to high levels of work absence, a high turnover of staff and reduced performance (HSE, 2005a).

There is also evidence to suggest that individuals who are experiencing stress are more likely to be involved in an accident (Cooper and Clarke, 2003). To illustrate, a number of retrospective studies of US military pilots have linked stressors such as career strain, financial difficulties and interpersonal problems to aircraft mishaps (Alkov et al. 1980, 1982, 1985). Li et al. (2001) compared the responses from petrochemical workers in Taiwan who had been involved in industrial accidents and those who had not. It was found that the accident group were more likely to report experiencing stress, and with a more severe reaction, than the non-accident group. Being able to identify the effects of stress is important when attempting to prevent or mitigate these effects through stress management techniques.

Prevention

The first step prior to implementing any intervention to reduce chronic stress is to identify the scope of the problem. Most workplace regulators now recommend taking a risk management approach to chronic stress (e.g. HSE, 1998). Risk management in health and safety tends to adopt systematic, evidence-based problem-solving (Cox et al., 2002). The approach is founded on the premise that before a problem can be addressed it must be analysed and understood, and an assessment made of the risks (see Figure 7.3).

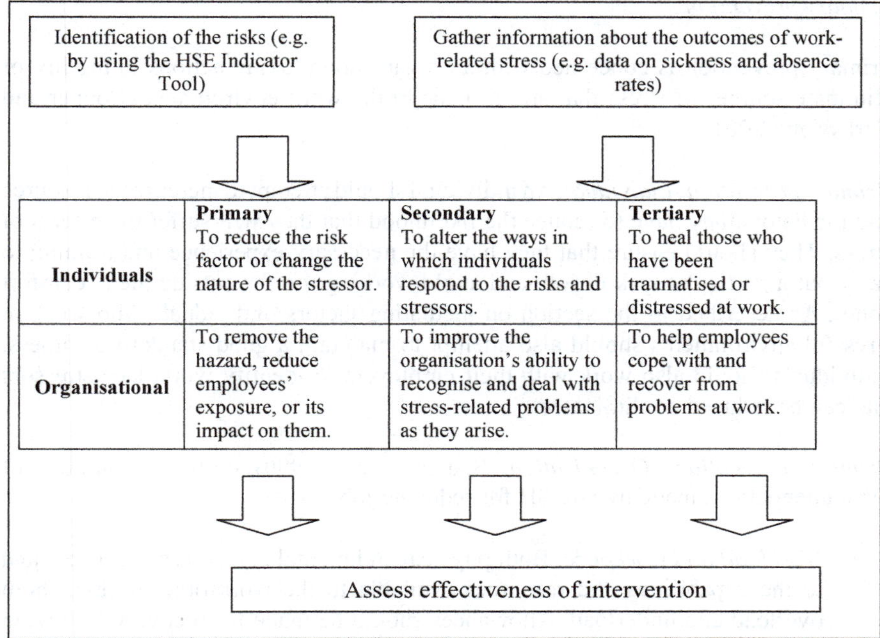

Figure 7.3 Process for management of chronic stress

To aid in the identification of the stressors affecting the employees in an organisation, the HSE has developed a questionnaire to identify workplace stressors. The Indicator Tool is a 35-item questionnaire relating to the six primary stressors identified in Figure 7.1. The tool, and guidance on its use and analysis of the results, are free of charge from the HSE work-related stress website (www.hse.gov.uk/stress/index.htm). The questionnaire aids in the identification of the stressors that are relevant to the organisation and distinguishes any particular groups that are experiencing particularly high levels of stress. It is recommended that the Indicator Tool is not used as the only measure of stress within an organisation. The results from the Indicator Tool survey should be established by discussing the findings with employees and also by considering other data that are available within the organisation, such as sickness absence rates and employee turnover.

Once the stressors have been identified, the next stage for an organisation is to take measures to reduce the levels of stress. Traditionally, chronic stress prevention has been categorised into three levels, consisting of primary, secondary and tertiary (see Figure 7.3). Stress management interventions provide a framework for both individual and organisational involvement in minimising stressful events and reactions. The interventions have been separated into those that can be carried out by the individual and those that can be made by the organisation.

Primary prevention

Primary prevention is concerned with an organisation taking actions to modify or eliminate sources of stress that are intrinsic to the work environment (Cooper and Cartwright, 2001).

Primary prevention: Individual. An individual should attempt to increase the resource and mediators they have to reduce the likelihood that they will suffer the effects of stress. They should ensure that they have the necessary experience and training to carry out a particular task or job, and avoid working too far outside their 'comfort zone'. As described in the section on mediating factors, individuals who work in stressful environments should also attempt to maintain a good standard of fitness. Individuals should also work with their employers to identify workplace stressors that can be reduced or eliminated.

Primary prevention: Organisation. Sauter et al. (1990) outline a number of recommendations made by NIOSH for reducing job stress:

- *Workload and workpace*: Both physical and mental demands should be equal to the capabilities and resources available to the workforce, limiting both overload and underload. Allowances should be made for recovery following periods of high workload.
- *Work schedule*: The work schedule should be compatible with the demands and responsibilities outside the job. This could include flexitime, a compressed work week and job-sharing. If there is a necessity for shift work, the rate of rotation should be stable and predictable and in a forward direction (day to night) (see Chapter 8 on fatigue for more information).
- *Work roles*: The roles and responsibilities of workers should be well defined. The employees should have a clear understanding of their duties, and conflicts in terms of job expectations should be avoided.
- *Job future*: Attempts should be made by the organisation to reduce ambiguity with regard to job security and opportunities for career development. Employees should be aware of promotional opportunities and means for improving skills or professional growth.
- *Social environment*: Organisations should provide mechanisms for personal interaction both for purposes of emotional support and for actual help as needed in accomplishing assigned tasks (e.g. mentoring programmes, team-building activities). The military make extensive use of 'buddy-buddy' systems to enhance emotional support under battle conditions, and the emergency services rely heavily on teamwork in their operational approach, which mediates the exposure to stressors for the personnel involved (Flin, 1996a, and see Chapter 5).
- *Content*: Jobs should be designed to provide meaning, stimulation and an opportunity to use skills. For those jobs in which this may be difficult to achieve, job rotation or increasing the scope of work (enlargement or enrichment) could be attempted.

- *Participation and control*: Individuals should be given the chance to have input on decisions or actions that affect their jobs and the performance of their tasks – for example, as discussed in the chapter on fatigue, involving workers in the design of shift schedules.

It is difficult to assess the relative effectiveness of different primary intervention techniques. This is due to a much smaller body of research evaluating primary interventions than at the secondary or tertiary levels. However, in a review of 18 studies of primary-level stress intervention strategies, evidence suggests that targeted interventions that focus on one or a few stressors that are experienced by a high proportion of employees (e.g. excessive workload, poor supervisor–subordinate communications, rigid work hours) are more successful than those that focus on more general work characteristics such as demands or control (Parker and Sparkes, 1998).

Secondary prevention

This level of prevention is concerned with improving the prompt detection and management of stress. It generally takes the form of stress education and stress management training.

Secondary prevention: Individual. Individuals should be aware of the symptoms of stress so that they can recognise them both in themselves, and in other team members. An individual should examine the coping strategies they use when responding to stressors, and attempt to apply problem-focused techniques to address workplace stressors (see the earlier section on resources and mediators). They should also take advantage of any stress management training provided by their employer.

Secondary prevention: Organisation. Murphy (1996) reviewed 64 workplace stress management studies. He found that a variety of stress-management techniques were used. These included:

- Muscle relaxation – this involves tensing (for 5–10 seconds) and releasing one muscle group at a time in a specific order (it generally starts with the lower extremities and finishes with muscles of the face, abdomen, and chest).
- Meditation – the purpose of meditation is to quiet the mind, emotions and body.
- Biofeedback – this is a training technique in which an individual learns to control the physiological reactions to stress (e.g. increased heart rate and muscle tension).
- Cognitive-behavioural stress management – this involves changing the way the individual thinks about stress. The aim is to help the person recognise negative, or inaccurate, thoughts and to alter the behavioural responses to these thoughts.

Murphy (1996) found the most common techniques used were muscle relaxation, cognitive-behavioural skills, and that combinations of two or more techniques produced the most positive results. It was found that in general, a combination of these techniques (e.g. muscle relaxation plus cognitive-behavioural skills) seemed to be more effective across outcome measures than single techniques. However, none of the stress interventions was consistently effective at an organisational level in relation to outcomes such as absenteeism or job satisfaction.

Tertiary prevention

Tertiary measures are concerned with the treatment, rehabilitation and recovery of individuals who have suffered, or are suffering, from ill health as a result of stress (Cooper and Cartwright, 2001). Interventions at this level tend to involve the provision of counselling services for work or personal problems.

Tertiary prevention: Individual. Employees must identify when levels of stress have become so great that they are having a detrimental effect on health and they require professional help. Being knowledgeable about the cause, symptoms and effects of chronic stress will aid in the early identification of stress-related problems. The help may initially be through a visit to a general practitioner or counsellor.

Tertiary prevention: Organisation. Employers may provide an in-house counsellor, or an employee assistance programme (EAP). The employer pays an annual per capita charge to the EAP company, which in return provides confidential counselling services (telephone and face to face) on demand from all staff, and sometimes also for their immediate family members. Based on reports published in the United States, these services have shown a return on investment of between 3:1 to 15:1 (Cooper and Cartwright, 1994). In an evaluation of an EAP introduced by the UK Post Office, there was a significant improvement in the mental health and self-esteem of employees and a reduction in absenteeism in one year of approximately 60% (Cooper and Sadri, 1991).

Cooper and Cartwright (2001) state that secondary and tertiary levels of intervention are likely to be insufficient in maintaining the health of employees without carrying out interventions at the primary level. These are 'band-aid' approaches and do nothing about reducing or removing causes of stress. Therefore, it is recommended that a multi-level approach is employed. However, the effect of the interventions should be evaluated and re-evaluated on a regular basis to ensure that the organisation continues to get a good return on investment.

Acute stress

Stressors

Individuals working in high-risk settings are not only at risk of the chronic stressors described above, but can also be subjected to acute stressors that can be brought on by periods of high workload, emergencies, attempts to diagnose an unusual problem, or high costs of failure. Figure 7.4 is a summary of the stressors, mediating factors, symptoms and effects of acute stress. It can be seen that the structure of the model is the same as that for the model of chronic stress outlined in Figure 7.2. The psychological balance mechanism (Figure 7.1) and the fact that it is the individual's appraisal of the stressors and of their coping resources that are important are also applicable.

As with chronic stress, different jobs will have a particular set of acute stressors. The potential causes of stress in emergency situations in a nuclear power plant control room were identified by Mumaw (1994). However, although developed for nuclear personnel, the acute stressors he identified are equally applicable to other high-reliability industries. They are classified into three groups: novelty and uncertainty, environmental stressors, and task-related stressors (see Figure 7.4). It can be seen that many of these acute stressors are similar to the chronic stressors identified in Figure 7.2. However, the acute stressors are associated with high-pressure abnormal events, rather than those associated with general day-to-day working conditions.

The most significant stressors in acute stress situations are event uncertainty, workload management, time pressure, fatigue and performance anxiety. Novel events, events in which expectations are violated, events where critical information is missing or when more than one goal must be considered, and events in which plans are not implemented successfully, also occur to create stressful situations that affect the responses of the emergency response personnel.

Mumaw (1994) also identified a number of event characteristics that can cause stress by creating uncertainty in coping with emergencies:

- events that are novel because the physical phenomenon is not well understood, the progression of the event is not known, or appropriate actions are not immediately known
- events in which expectations are violated because parameters fail to trend in the expected direction, certain indications do not fit into current understanding or systems fail to work as expected
- events in which critical information is missing and, therefore, prediction is more difficult and outcome less certain (e.g. critical sensors are lost, the status of critical safety systems is unknown)
- events in which more than one goal must be considered – major accident or complex scenarios present situations in which it is critical to address more than one goal or in which the actions required to achieve the goal are in conflict with a second goal. When operators are required to select among or prioritise multiple goals, uncertainty is introduced about which actions are appropriate
- events in which plans of action are not implemented successfully – plans and procedures provide expectations about the sequence of events. Thus, when implementation fails, expectations are violated and uncertainty is introduced.

Figure 7.4 Model of *acute* stress (based on Mumaw, 1994)

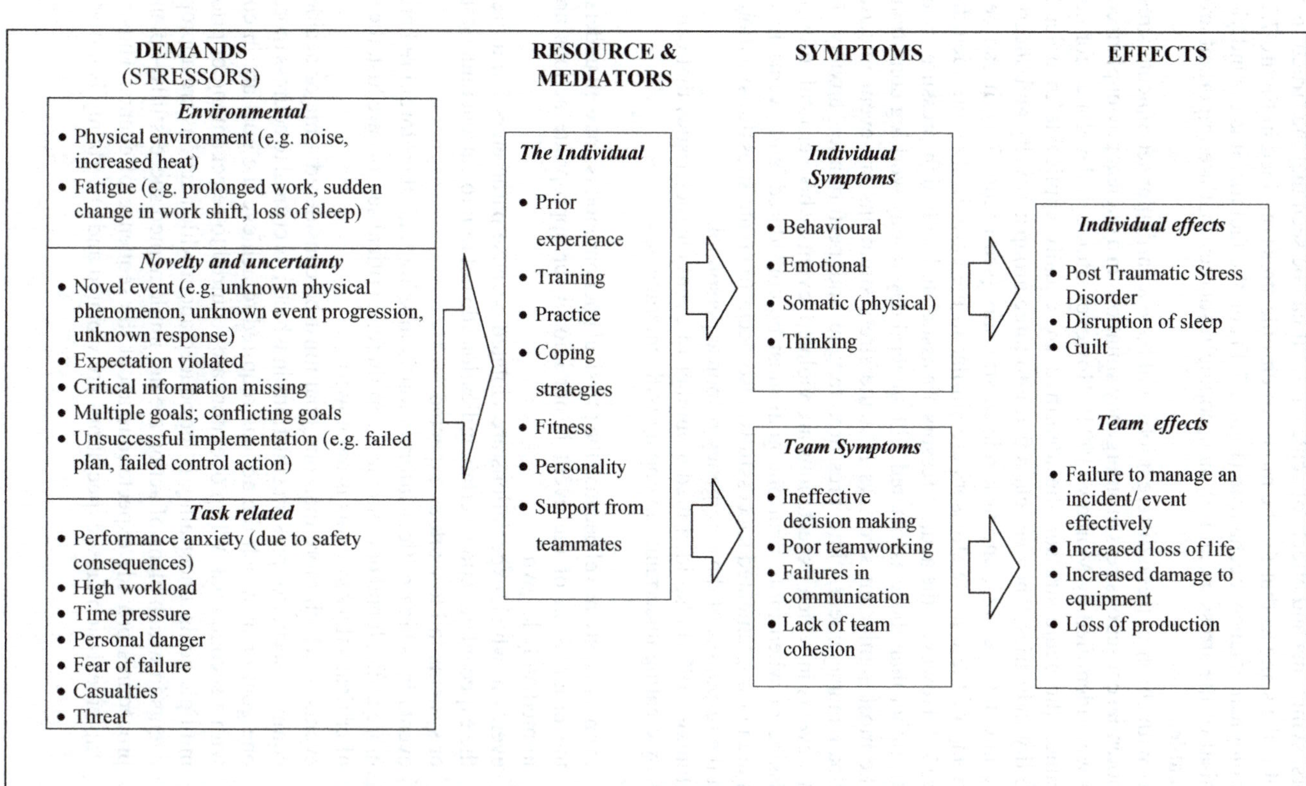

Other additional stressors that may be experienced in an emergency include dealing with casualties, personal danger and fear of failure. Emergency service personnel report experiencing stress due to the fear of personal danger (Hytten and Hasle, 1989). Winkler (1980) reported that approximately 50% of ambulance personnel admitted that they could not stop thinking about incidents until a long time after, and that child trauma was particularly upsetting.

Resources and mediators

As with chronic stress, mediators can reduce or exacerbate the effects of the stressors. It can be seen that these mediating factors for acute stress (see Figure 7.4) are similar to those for chronic stress (see Figure 7.2). As with chronic stress, it seems likely that there are some individuals who are simply better able to cope with acute stress situations. However, apart from people who have a history of mental illness or impairment, there have not been any robust research findings to date that have identified the personality traits that are associated with ability to cope with acute stress (Flin, 1996a).

Individuals are better equipped to cope with stress if they have support from other team members who have shared the experience (Flin, 1996a), and if they have been trained together. Studies of military stress show that support from one's team or unit is a key factor in the mitigating of combat stress (Noy, 1991) and that group cohesion may improve performance under demanding conditions (Orasanu and Baker, 1996).

Acute stress symptoms

As with chronic stress, there are particular symptoms associated with acute stress.

Individual symptoms

It is important to preface this section by emphasising that exposure to stressors does not necessarily produce negative effects, particularly for experienced personnel. There may be immediate positive effects, such as increased motivation and energy, faster reactions, clearer thinking and improved memory retrieval (Charlton, 1992; Orasanu and Baker, 1996). Referring back to Figure 7.1, it is only when the perceived level of challenge begins to exceed the individual's judged ability to cope with the stressors that the symptoms of distress become predominant (Flin, 1996a), so only negative effects are discussed below.

Figure 7.4 depicts a comprehensive categorisation of acute stressors. These may not be applicable or of equal intensity for all high-demand situations. In addition, a wide range of effects of stress exist, causing a complex pattern of behavioural, emotional, somatic (physical) and cognitive, or thinking, reactions (see Tables 7.7–7.10). As with the discussion of indicators of chronic stress, the 'BEST' acronym has been used to categorise the acute stress indicators.

***B**: Behavioural indicators*. The effects of acute stress on behaviour are generally the most readily observable. As Flin (1996a) pointed out for acute stress and discussed earlier for chronic stress, there is not a finite list of behavioural effects, but the defining feature is a change in the individual's normal behaviour pattern. Again, the behaviour change can be subtle (although there may be clues that a key team member is not performing his or her job properly), then the better team members know each other, the more likely that any behaviour changes will be noticed. Table 7.7 outlines a number of behavioural symptoms.

Table 7.7 Behavioural indicators of acute stress

fight/flight:	freeze:
• hyperactivity	• withdrawn ('switch off')
• anger	• detached
• argumentativeness	• apathetic
• irritability	• disengaged from surrounding activities
• jumpiness	
• aggressiveness	
• swearing	
• emotional outbursts	

***E**: Emotional indicators*. As with the behavioural indicators, different emotional responses can occur as part of the acute stress response. Indicators can range from apprehension and anxiety, fear, to loss of emotional control, where the individual becomes aggressive or visibly distressed (e.g. crying; see Table 7.8).

Table 7.8 Emotional indicators of acute stress

• fear	• fear of failure
• anxiety	• vulnerability
• panic	• loss of control

***S**: Somatic (physical) indicators*. This is a pronounced physiological adaptive response that prepares the individual to fight or flee when the brain perceives a threat in the immediate environment (for a list of symptoms see Table 7.9). Any situation perceived by an individual to be very demanding or challenging can produce these effects.

Table 7.9 Somatic indicators of acute stress

• energy surge • increasing heart rate • sweating • dry mouth	• muscle tension (trembling) • heightened sensitivity (e.g. to noise) • effects on digestion (butterflies in stomach)

T: *Thinking (cognitive) indicators*. Stress has a wide range of detrimental effects on thinking. Cognitive activities such as perception, memory, decision-making and task-planning have previously been found to be negatively affected by stress (Mumaw, 1994). Table 7.10 provides a description of some of the negative effects that stress can have on memory, concentration and decision-making. However, it is useful to note that some stress responses may produce some performance-enhancing effects. Task-shedding may be an adaptive function if tasks are shed in an optimal sequence, with the least important tasks abandoned first. Also, when under stress, people prefer to use well-learned techniques to deal with situations, so can revert to familiar rules or recognition-primed decision-making (see Chapter 3). This is known as the availability bias (Tversky and Kahneman, 1974). These techniques can be brought quickly into consciousness and require little cognitive effort. Thus pilots are given extensive training to deal with situations such as an engine failing when the plane is taking off, as they have to take the correct sequence of actions very quickly with no time to consult a manual or deliberate on the appropriate response.

Table 7.10 Thinking (cognitive) indicators of acute stress

Indicators	Descriptions
Impairment of memory	• prone to distraction • confirmation bias (tend to ignore information that does not support following a particular chosen course of action or model) • information overload • task-shedding (the abandonment of certain tasks when stress or workload make it difficult to concentrate on all of the tasks simultaneously)
Reduced concentration	• difficulty prioritising • preoccupation with trivia • perceptual tunnelling (attention becomes narrowly focused on salient cues)
Difficulty in decision-making	• availability bias (resort to familiar routines and not consider plans that are not immediately available in memory) • 'stalling thinking' – mind blank

Box 7.2 provides an example of how the effects of acute stress led to a failure in leadership during the *Piper Alpha* oil platform disaster.

Box 7.2 An example of the effects of acute stress: the *Piper Alpha* disaster

Occidental's *Piper Alpha* platform was situated in the North Sea 110 miles northeast of Aberdeen, Scotland. On 6 July 1988, there was an explosion on the production deck of the platform. The resulting fire spread rapidly and was followed by a series of smaller explosions. At about 22:20 there was a major explosion caused by the rupturing of a pipeline carrying gas from the nearby Texaco *Tartan* platform. Over the next three hours a high-pressure gas fire raged, punctuated by a series of explosions. Of the 226 men on board, only 67 survived.

'The explosion on the Piper Alpha *that led to the disaster was not devastating. We shall never know, but it probably killed only a small number of men. As the resulting fire spread, most of the* Piper Alpha *workforce made their way to the accommodation where they expected someone would be in charge and would lead them to safety. Apparently they were disappointed. It seemed the whole system of command had broken down'* (Sefton, 1992: p6).

Evidence from the Cullen (1990) report into the accident showed the inability of the offshore installation manager (OIM; the most senior member of staff on the installation) – the man responsible for organising the response to the emergency on the *Piper Alpha* – to *'take critical decisions and lead those under his command in a time of extreme stress'* (para 20.59). It was clear that the OIM on *Piper Alpha* was suffering from many of the symptoms of acute stress outlined above.

'The OIM had gone a matter of seconds when he came running back in what appeared... to be a state of panic... The OIM made no specific attempt to call in helicopters from the Tharos [a rescue vessel] *or elsewhere, or to communicate with the vessels around the installation; or with the shore or other installations; or with personnel on* Piper*'* (para 8.9).

'The OIM did not give any other instruction or guidance. One survivor said that at one stage people were shouting at the OIM and asking what was going on and what procedure to follow. He did not know whether the OIM was in shock or not but he did not seem to be able to come up with an answer' (para 8.18).

Furthermore, the radio operator said that *'he himself was also panicking and the message* [that the platform was to be abandoned] *was haphazard'* (para 8.9).

Team symptoms

As with individuals, stressors can also affect overall team performance. When under stress, team members are more likely to become focused on their own tasks, resulting in a decline in team performance. When people are under high stress, there is a tendency to regress to more basic skills where the individual is comfortable operating rather than maintaining an overview (Charlton, 1992). Further, it may be that only one team member needs to be affected by stress for the team's performance to be significantly degraded. Therefore, the symptoms of stress at the team level are failures in teamworking, communication and decision-making (see earlier chapters for a discussion).

Acute stress effects

Individual

The common negative experiences of acute stress that occur after an emergency include: fatigue, sadness, guilt, recurring images, intrusive thoughts. If they last up to four weeks, the individual may be suffering from acute stress disorder. However, if these symptoms last for more than a month, then the individual may be diagnosed as suffering from post-traumatic stress disorder (PTSD) (see Box 7.3).

Box 7.3 PTSD and war

Bower et al. (2005) told the story of Jim Batchelor, a wounded US army infantryman who, despite surviving a gunshot wound between the eyes, was crippled by PTSD. His symptoms included: nightmares and flashbacks, emotional numbness, irritability, difficulty sleeping, migraines, depression and inexplicable fits of rage. Every night he had a recurrent nightmare about the intense firefight in which he was wounded only four days after arriving in Baghdad.

A study by Hodge et al. (2004) of 1,709 US soldiers who had returned from serving in Iraq or Afghanistan found that, just as in the Vietnam and first Gulf war, PTSD was the most prevalent mental illness suffered by US soldiers. The prevalence of PTSD was directly related to the level of fighting to which the soldiers had been exposed. Those soldiers who had not been in an exchange of weapons fire had a PTSD incidence of 4.5%. If the soldiers had been involved in one or two firefights, the incidence doubled to 9.3%, three to five firefights resulted in an incidence of 12.7%, and more than five firefights produced an incidence of 19.3%.

PTSD and acute stress disorder can occur following an individual's involvement in a critical incident. A critical incident is an event that is out with the usual range of experience, and challenges one's ability to cope (e.g. exposure to violent death) (Everly and Mitchell, 2000). PTSD is a syndrome that is medically recognised as a mental disorder (American Psychiatric Association, 2000). The criteria for a diagnosis of PTSD and acute stress disorder are:

1. The person has been exposed to a traumatic event that involved actual or threatened death or serious injury, and the person's response involved intense fear, helplessness or horror.
2. The traumatic event is persistently re-experienced (e.g. dreams, recurrent recollections, flashbacks).
3. Persistent avoidance of things associated with the trauma (e.g. thought, activities) and numbing (e.g. feeling of detachment, unable to show loving feelings, lack of interest in the future).
4. Persistent symptoms of increased arousal (not present before the trauma), as indicated by two (or more) of the following: difficulty falling or staying asleep, irritability or outbursts of anger, difficulty concentrating, hypervigilance, or exaggerated startle response.

Team

As described above, the symptoms of acute stress at a team level are failures in the team members to work effectively together. The effects of poor teamworking will depend upon the task the team is performing. In an emergency response situation this may lead to a failure to manage an incident effectively, resulting in increased loss of life. In a production process (e.g. nuclear power generation, offshore oil production), a failure to manage an incident may lead to increased damage to equipment, loss of production and higher risks to the crew. See Chapter 5 for more information on the effects of failures in teamworking.

Acute stress prevention

As was the case with chronic stress, it is possible to distinguish between primary, secondary and tertiary stress preventative methods for acute stress.

Primary prevention

Many jobs in high-reliability industries involve, at least occasional, periods of high stress, and removing the stressors may be impossible. For example, when one of the authors of this book was interviewing an anaesthetist about his job, he was told that it is 95% complete boredom and 5% sheer panic. Therefore, as was the case with the primary prevention of chronic stress, people who perform jobs where they may be exposed to acute stress must have adequate professional training and experience, so that they are less likely to experience acute stress.

Secondary prevention

Secondary acute stress prevention is concerned with the prompt detection and management of the symptoms and effects of stress. Diagnosis that a debilitating stress reaction is being experienced by oneself or by other team members is a fundamental starting point in stress management. Self-awareness of stress reactions is as essential as recognition of stress reactions in others. All members of any organisation where stressful responses may arise should therefore be trained to take the necessary action of drawing the team leader's attention to any other member who exhibits symptoms of stress that are causing distress or are detrimental to his or her performance. This includes telling the leader when he or she is not coping well with a stressful situation.

Personnel exhibiting a severe negative response reaction should be switched to a non-essential task and kept under observation, but should be treated as if he or she were having a normal reaction and, if possible, kept with the team (Hodgkinson and Stewart, 1991).

Training individuals for stress reactions can take the form of general exercises, or more specific techniques for coping with stress reactions. Building experience, particularly of stressful situations, creates greater self-confidence and therefore reduces the likelihood of stress occurring in the first place.

The importance of training for stress-proofing personnel cannot be overemphasised. Realistic exercises and simulator sessions, i.e. being prepared through practical exercises, are a major stress-reduction mechanism (Hytten and Hasle, 1989). Training of the required emergency response procedures can be employed to ensure effective performance under stress conditions, as well as the use of case studies and presentations by experienced personnel. Knowledge and awareness of the potential causes of emergencies and stressors, and how to deal with them, introduced in training sessions, will assist in minimising negative stress reactions.

Johnston and Cannon-Bowers (1996) describe a three-stage training process designed to improve team performance in stressful situations called stress exposure training. In the first phase the trainees are provided with knowledge about the causes and reactions to acute stress (as outlined earlier in this chapter). In the second phase the trainees are given training designed to help the participants cope with stress through practice and feedback. Driskell et al. (2001) outline a variety of stress management approaches that could be used to reduce the negative effects of stress on performance, both for the individual and team. The most appropriate training technique (or techniques) is dependent on the task and a training needs analysis can be used to identify the areas in which training is required:

- *Cognitive control techniques*. The purpose of this type of technique is to train individuals to regulate emotions (e.g. worry) and regulate distracting thoughts to allow them to maintain concentration on the task. This may include stepping away from the situation for a couple of minutes if they start to feel overwhelmed, or providing a procedure to regulate stress reactions, for example:
  - STOP – *S*tand back, *T*ake stock, *O*verview, *P*rocedures
  - STAR – *S*top, *T*hink, *A*ct, *R*eview
  - or for decision-making, DODAR – *D*iagnose, *O*ptions, *D*ecide, *A*ssign and *R*eview).
- *Physiological control techniques*. This technique attempts to provide methods for regulating the negative physiological reactions to stress. It attempts to

allow an individual to be calm, relaxed and under control when faced with an emergency. Okray and Lubnau (2004) describe a fire department chief who teaches firefighters to be aware of their heart rate. He believes that if the firefighter's heart is beating 90–110 beats per minute, with no exertion, then he or she needs to take a few deep breaths and think about what is happening. If his or her heart is beating at greater than 110 beats per minute, without exertion, he or she needs to take a few steps back, calm down and regain concentration. Singapore Police use several 'stress blasting techniques' for combating operational stress, including 'tactical breathing', which involves slow, deep breaths.

- *Modelling*. In this method the trainees are given the opportunity to observe or model a team responding to a high-stress situation. It is postulated that this gives the trainees the opportunity to observe good, or bad, examples of key behaviours in a realistic setting.
- *Overlearning*. In this training method the trainees are deliberately over-trained beyond a level of proficiency that would normally be required for a particular task. For this method to be effective in a real-life situation, Driskell et al. (2001) recommend that the task trained must be the same as that which would be performed in a high-stress situation, and it should be practised in a simulated stressful environment.
- *Attentional training*. This type of training educates the trainees as to when, how and why attention may be distracted when performing a task during a highly stressful situation. The aim is to overcome the effects of perceptual tunnelling and distractions that occur in a stressful environment.
- *Training time-sharing skills*. Often in emergency situations, it is necessary for the team members to carry out multiple tasks simultaneously. Therefore, trainees are given the opportunity to perform the tasks and provided with skills to prioritise the importance of the tasks.
- *Decision-making training*. This training can be given for skills that are particularly vulnerable to the effects of stress, e.g. decision-making and communication. The inclusion of material on decision strategies, how to build shared mental models, communication and command and control in training, using some type of scenario-based practice strategy, could provide a vehicle to demonstrate and reinforce desired behaviours (see Chapter 10).
- *Enhancing flexibility*. As outlined in Table 7.9, stress leads individuals to restrict attention to a subset of particularly salient cues. However, flexibility leads to more efficient performance under complex conditions in which more than one solution is possible, or in novel task conditions (Driskell et al., 2001). To promote flexible thinking, training material must be presented in a number of different contexts, from different perspectives, and with diverse examples. The practice of a narrow set of skills in training will result in the use of a narrow set of skills in the real world.

In the third phase of training, the participants are given the opportunity to practise the skills through gradual exposure to stress in a training environment. The stressful environment does not necessarily require a high-fidelity simulation. Driskell et al.

(2001) suggest that normal training exercises could be adapted by incorporating stressors such as increased time pressure or noise. Although there are difficulties in creating appropriate levels of stress in such training (Flin, 1996a), the primary outcome associated with the successful completion of the training is improved cognitive and psychomotor performance under stress (Weaver et al., 2001). In a review of 37 articles concerned with stress exposure training, Saunders et al. (1996) found the majority supported the effectiveness of this type of training. Further, there is evidence that skills learned in stress exposure training in a particular type of stressful situation as applied to a particular task, can then be generalised to novel settings (Driskell et al., 2001).

Tertiary prevention

Table 7.11 Core components of CISM (adapted from Everly and Mitchell, 1999)

Intervention	Objectives	Timing	Format
Pre-crisis preparation	Establish expectations. Provide information on stress management. Develop coping skills.	Pre-crisis	Groups/ organisations
Demobilisation and consultation	Inform and discuss. Allow psychological decompression (e.g. talk about experiences). Manage preliminary stress.	Following shift	Groups/ organisations
Defusing	Assess. Reduce acute signs and symptoms. Begin closure. Triage.	Within 12 hours post-crisis	Small groups
Critical incident stress debriefing	Facilitate closure. Mitigate signs and symptoms. Refer as required.	Within 1–10 days; 3–4 weeks after a mass disaster	Small groups
Crisis intervention	Return to level of function of before the crisis. Mitigate signs and symptoms. Refer as required.	Anytime	Individual/ groups
Family critical incident stress management	Foster support. Mitigate signs and symptoms. Refer as required.	Anytime	Families
Follow-up	Determine status. Ensure closure. Refer as required.	Anytime	Individual/ families

Tertiary prevention is concerned with treating team members who have been exposed to an extremely stressful situation. Critical incident stress management (CISM) is an intervention strategy that has been used to prevent PTSD following an unexpected critical event. The goal of CISM is to restore people to their usual state of mental health by mitigating the effects of traumatic stress. CISM can be applied to individuals, groups or organisations. It can be made available after a particular disaster (e.g. the terrorist attack on the World Trade Center; Hammond and Brooks, 2001; the *Piper Alpha* disaster; Alexander, 1993), or it may be available to emergency service personnel who frequently have to deal with traumatic events (e.g. the Los Angeles County Fire Department has conducted more than 500 CISMs since its implementation in 1986; Hokanson and Wirth, 2000). A framework for a particular type of CISM that has been widely used is outlined in Table 7.11.

Flannery and Everly (2000) state that with correct training and an effective assessment procedure, then even a limited array of crisis intervention procedures are effective.

Conclusion

Chronic and acute stress are relevant to individuals working in high-risk work environments. A failure to cope with stressors can result in work errors, reduced productivity, feelings of discomfort or ultimately even illness on the part of the individual, and poor performance for the team or organisation. However, through the identification and reduction of stressors, and effective training, it is possible to minimise the effects of stress on individual and team performance.

Key points

- There are two types of stress experienced by individuals in high-reliability work environments: chronic stress and acute stress.
- Chronic stress is related to conditions in the workplace and the individual's reaction to these, usually over a protracted period of time.
- Acute stress is sudden, novel, intense and of relatively short duration, disrupts goal-oriented behaviour and requires a proximate response.
- It is the individual's *perception* of the demands being placed upon them, and the *perception* of the resources they have available to cope with the demands, that dictates whether the individual feels under stress.
- To mitigate the effects of chronic and/or acute stress in the workplace, it is necessary to understand the stressors, mediators or resource, symptoms and effects of stress on an individual, team or organisation.
- Stress prevention techniques can be primary (prevent stress from occurring), secondary (the prompt detection and management of stress) and tertiary (the treatment of the effects of stress).

Suggestions for further reading

Chronic stress

Cooper, C. and Clarke, S. (2003) *Managing the Risk of Workplace Stress: Health and Safety Hazards*. London: Routledge.

Hancock, P.A. and Desmond, P.A. (eds.) (2001) *Stress, Workload, and Fatigue*. Mahwah, NJ: Lawrence Erlbaum Associates.

Health and Safety Executive stress web page: www.hse.gov.uk/stress/standards/index.htm

National Institute for Occupational Health stress website: www.cdc.gov/niosh/topics/stress/

Acute stress

Cannon-Bowers, J.A. and Salas, E. (eds.) (1998) *Making Decisions Under Stress. Implications for Individual and Team Training*. Washington, D.C.: American Psychological Association.

Driskell, J.E. and Salas, E. (eds.) (1996) *Stress and Human Performance* (pp. 223–256). Mahawah, NJ: Erlbaum.

Flin, R. (1996). *Sitting in the Hot Seat: Leaders and Teams for Critical Incident Management.* Chichester: John Wiley & Sons.

References

Alexander, D.A. (1993) Stress among police body handlers: A long term follow-up. *British Journal of Psychiatry*, 163, 806–808.

Alkov, R.A. and Borowsky, M.S. (1980) A questionnaire study of psychological background factors in U.S. Navy aircraft accidents. *Aviation, Space, and Environmental Medicine*, 51, 860–863.

Alkov, R.A., Borowsky, M.S. and Gaynor, J.A. (1982) Stress coping and the U.S. Navy factor mishap. *Aviation, Space, and Environmental Medicine*, 53, 1112–1115.

Alkov, R.A., Borowsky, M.S. and Gaynor, J.A. (1985) Pilot error as a symptom of inadequate stress coping. *Aviation, Space, and Environmental Medicine*, 56, 244–247.

American Psychiatric Association (2000) *Diagnostic and Statistical Manual of Mental Disorders, Fourth Edition, Text Revision.* Washington, DC: American Psychiatric Association.

Baum, A., Cohen, L. and Hall, M. (1993) Control and intrusive memories as possible determinants of chronic stress. *Psychosomatic Medicine*, 55, 274–286.

Bower, A., Booth Thomas, C. and Reed, W. (2005) Three Roads Back. *Time Magazine*, 21 March.

Boyle, M.V. (1997) Love the work hate the system: A qualitative study of emotionality, organisational culture and masculinity within an interactive service workplace. Unpublished doctoral thesis, University of Queensland, Brisbane, Australia.

Cannon, W. (1929) *Bodily Changes in Pain, Hunger, Fear and Rage.* New York: Appleton-Century.

Cannon-Bowers, J.A. and Salas, E. (eds.) (1998) *Making Decisions Under Stress. Implications for Individual and Team Training*. Washington, DC: American Psychological Association.

Cannon-Bowers, J.A., Salas, E. and Baker, C.V. (1991) Do you see what I see? Instructional strategies for tactical decision making teams. In *Proceedings of the 13th Annual Interservice/Industry Training Systems Conference.* Washington, DC: National Security Industrial Association.

Carver, C.S., Scheier, M.F. and Weintraub, J.K. (1989) Assessing coping strategies: A theoretically based approach. *Journal of Personality and Social Psychology*, 56, 267–283.

Charlton, D. (1992) Training and assessing submarine commanders on the Perishers' course. Paper presented at the First Offshore Installation Management Conference: Emergency Command Responsibilities. Robert Gordon University, Aberdeen.

Collyer, S.C. and Malecki, G.S. (1998) Tactical decision making under stress: History and overview. In J.A. Cannon-Bowers and E. Salas (eds.) *Making Decisions Under Stress. Implications for Individual and Team Training*. (pp. 3–16). Washington, DC: American Psychological Association.

Columbia Accident Investigation Board (2003) *Columbia Accident Investigation: Volume 1*. Washington, DC: U.S. Government Printing Office.

Cooper, C. and Clarke, S. (2003) *Managing the Risk of Workplace Stress: Health and Safety Hazards*. London: Routledge.

Cooper, C., Dewe, P. and Driscoll, M. (2001) *Organizational Stress; A Review and Critique of Theory, Research and Applications.* Thousand Oaks, CA: Sage.

Cooper, C.L. and Cartwright, S. (1994) Healthy and, healthy organisation: A proactive approach to occupational stress. *Human Relations*, 47, 455–471.

Cooper, C.L. and Cartwright, S. (2001) An overview of fatigue. In P.A. Hancock and P.A. Desmond (eds.) *Stress, Workload, and Fatigue* (pp. 235–248). Mahwah, NJ: Lawrence Erlbaum Associates.

Cooper, C.L., Cooper, R.D. and Eake, L.H. (1988) *Living with Stress*. London: Penguin Books.

Cooper, C.L. and Sadri, G. (1991) The impact of stress counselling in work. *Journal of Social Behaviour and Personality*, 6, 411–423.

Cox, T. (1993) *Stress Research and Stress Management: Putting Theory to Work.* Sudbury: HSE Books.

Cox, T. (1985) The nature and measurement of stress. *Ergonomics*, 28, 1155–1163.

Cox, T., Leather, P. and Cox, S. (1990) Stress, health and organisations. *Occupational Health Review*, 23, 13–18.

Cox, T., Randall, R. and Griffiths, A. (2002) *Interventions to Control Stress at Work in Hospital Staff.* Sudbury: HSE Books.

Cullen, D. (1990) *The Public Inquiry into the Piper Alpha Disaster: Volumes I and II*. London: HMSO.

DeLongis, A., Folkman, S. and Lazarus, R.S. (1988) The impact of daily stress on health and mood: psychological and social resources as mediators. *Journal of Personality and Social Psychology*, 54, 486–95.

Driskell, J.E., Salas, E. and Johnson, J. (2001) Stress management: Individual and team training. In E. Salas, C. Bowers and E. Edens (eds.) *Improving Teamwork in Organizations: Applications of Resource Management Training*. (pp. 55–72). Mahwah, NJ: Lawrence Erlbaum Associates.

European Foundation for the Improvement of Working Conditions (2000) *Third European Survey of Workers Conditions*. Dublin, Ireland: European Foundation for the Improvement of Working Conditions.

Everly, G.S. and Mitchell, J.T. (2000) The debriefing 'controversy' and crisis intervention: a review of lexical and substantive issues. *International Journal of Emergency Mental Health*, 2, 211–25.

Everly, G.S. and Mitchell, J.T. (1999) *Critical Incident Stress Management (Cism): A New Era and Standard of Care in Crisis Intervention.* Ellicott City, MD: Chevron Publishing.

Flannery, R.B. and Everly, G.S. (2000) Crisis intervention: a review. *International Journal of Emergency Mental Health*, 2, 119–25.

Flin, R. (1996a) *Sitting in the Hot Seat: Leaders and Teams for Critical Incident Management.* Chichester: John Wiley & Sons.

Flin, R. (1996b) Stress offshore. In R. Flin and G. Slaven (eds.) *Managing the Offshore Workforce* (pp. 65–81). Tulsa: PennWell.

Flin, R., Salas, E., Strub, M. and Martin, L. (eds.) (1997) *Decision Making Under Stress: Emerging Themes and Applications*. Aldershot: Ashgate Publishing.

Geipert, N. (2007) Don't be mad. More research links hostility to coronary risk. *Monitor on Psychology*, 38, 50–51.

Gunthert K.C., Cohen, L.H. and Armeli, S. (1999) The role of neuroticism in daily stress and coping. *Journal of Personality and Social Psychology*, 77, 1087–1100.

Hammond, J. and Brooks, J. (2001) The World Trade Center attack. Helping the helpers: the role of critical incident stress management. *Critical Care*, 5, 315–317.

Haynes, S.G., Levine, S., Scotch, S.N., Feinleib, M. and Kannel, W.B. (1978) The relationship of psychosocial factors to coronary heart disease in the Framingham study. I. Methods and risk factors. *American Journal of Epidemiology*, 107, 362–383.

Health and Safety Executive (2005a) *Tackling Stress: The Management Standards Approach.* Sudbury: HSE Books.

Health and Safety Executive (2005b) *Health and Safety Statistics Highlights 2004/2005*. Sudbury: HSE Books.

Health and Safety Executive (2005c) *Psychosocial Work Conditions in Great Britain in 2005*. Sudbury: HSE Books.

Health and Safety Executive (2000) *Management of Health and Safety at Work. Management of Health and Safety at Work Regulations 1999*. Sudbury: HSE Books.

Health and Safety Executive (1998) *Five Steps to Risk Assessment*. Sudbury: HSE Books.

Health and Safety Executive (1995) *Stress at Work: A Guide for Employers*. Sudbury: HSE Books.

Helmreich, R.L. and Merritt, A.C. (1998) *Culture at Work in Aviation and Medicine: National, Organizational and Professional Influences*. Aldershot: Ashgate.

Hodge, C.W., Castro, C.A., Messer, S.C., McGurk, D., Cotting, D.I., Koffman, R.L. (2004) Combat duty in Iraq and Afghanistan, mental health problems and barriers to care. *New England Medical Journal*, 351, 13-22.

Hodgkinson, P. and Stewart, M. (eds.) (1991) *Coping with Catastrophe*. Routledge: London.

Hodgson, J.T., Jones, J.R., Elliot, P.C. and Osman, J. (1993) *Self-reported Work-related Illness*. Sudbury: HSE Books.

Hogan, R., Curphy, G. and Hogan, J. (1994) What we know about leadership. *American Psychologist*, 49, 493–504.

Hokanson, M. and Worth, B. (2000) The critical incident stress debriefing process for the Los Angeles county fire department: automatic and effective. *International Journal of Emergency Mental Health*, 2000, 249–257.

Holmes, T.H. and Rahe, R.H. (1967) The social readjustment rating scale. *Journal of Psychosomatic Research*, 11, 213–218.

Houtman, I.L.D. (2005) *Work-related Stress*. Dublin, Ireland: European Foundation for the Improvement of Working Conditions.

Hytten, K. and Hasle, A. (1989) Fire fighters: a study of stress and coping. *Acta Psychiatrica Scandinavia, suppl., 355*, 80, 50–55.

Johnston, J.H. and Cannon-Bowers, J.A. (1996) Training for stress exposure. In J.E. Driskell and E. Salas (eds.) *Stress and Human Performance* (pp. 223–256). Mahawah, NJ: Erlbaum.

Kahn, R.L., Wolfe, D., Quinn, R.P., Snoek, J.D. and Rosenthal, R.A. (1964) *Organizational Stress: Studies in Role Conflict and Ambiguity*. New York: Wiley & Sons.

Karasek, A.R. and Theorell, T. (1990) *Healthy Work: Stress, Productivity, and the Reconstruction of Working Life*. New York: Basic Books.

Kivimaki, M., Elovainio, M. and Vahtera, J. (2000) Workplace bullying and sickness absence in hospital staff. *Occupational and Environmental Medicine*, 57, 656–60.

Kleinman, D.L. and Serfaty, D. (1989) Team performance assessment in distributed decision making. in The Interactive Networked Simulation for Training Conference. Orlando, FL.

Kobasa, S.C., Maddi, S.R. and Puccetti, M.C. (1982) Personality and exercise as buffers in the stress-illness relationship. *Journal of Behavioral Medicine*, 5, 391–404.

La Rocco, J.M. and Jones, A.D. (1978) Co-worker and leader support as moderators of the stress-strain relationship. *Journal of Applied Psychology*, 63, 629–631.

Lazarus, R.S. and Folkman, S. (1984) *Stress, Appraisal and Coping*. New York: Springer Publishing Company.

Li, C.-Y., Chen, K.-R., Wu, C.-H. and Sung, F.-C. (2001) Job stress and dissatisfaction in association with non-fatal injuries on the job in a cross-sectional sample of petrochemical workers. *Occupational Medicine*, 51, 50–55.

Little, L.F., Gaffney, I.C., Rosen, K.H. and Bender, M.M. (1990) Corporate instability in relation to airline pilots' stress symptoms. *Aviation, Space and Environmental Medicine*, 61, 977–982.

Maslach, C. (1978) The client role in burnout. *Journal of Social Issues*, 34, 111–124.

Morgan, B. and Bowers, C. (1995) Teamwork stress: implications for team decision making In R. Guzzo and E. Salas (eds.) *Team Effectiveness and Decision Making in Organizations* (pp. 262–290). San Francisco, CA: Jossey Bass.

Mumaw, R.J. (1994) *The Effects of Stress on NPP Operational Decision Making and Training Approaches to Reduce Stress Effects*. Washington, DC: US Nuclear Regulatory Commission.

Murphy, L.R. (1996) Stress management in work settings: a critical review of the health effects. *American Journal of Health Promotion*, 11, 112–135.

National Institute for Occupational Health and Safety (2004) *Worker Health Chartbook 2004*. Cincinnati, OH: National Institute for Occupational Health and Safety

National Institute for Occupational Health and Safety (1999) *Stress at Work*. Cincinnati, OH: National Institute for Occupational Health and Safety.

Noy, S. (1991) Combat Stress Reactions. In R. Gal and A.D. Mangelsorf (eds.) *International Handbook of Military Psychology*. London: John Wiley.

Okray, R. and Lubnau, T. (2004) *Crew Resource Management Training for the Fire Service*. Tulsa, OK: PennWell Corporation.

Orasanu, J. and Baker, P. (1996) *Stress and Military Performance*. In J. Driskell and E. Salas (eds.) *Stress and Performance* (pp. 89–126). Hillsdale, NJ: Lawrence Erlbaum.

Parker, K.T. and Sparkes, T.J. (1998) *Organizational Interventions to Reduce Work Stress: Are they Effective? A Review Of The Literature*. Sudbury: HSE Books.

Rick, J., Thompson, L., Briner, R.B., Oregan, S. and Daniels, D.K. (2002). *Review of Existing Supporting Scientific Knowledge to Underpin Standard of Good Practice for Key Work-related Stressors: Phase I*. Sudbury: HSE Books.

Ross, R.R. and Altmaier, E.M. (1994) *Intervention in Occupational Stress*. London: Sage Publications.

Salas, E., Driskell, J.E. and Hughes, S. (1996) Introduction: The study of stress and human performance. In J. Driskell and E. Salas (eds.) *Stress and Performance* (pp. 1–46). Hillsdale, NJ: Lawrence Erlbaum.

Saunders, T., Driskell, J.E., Johnston, J. and Salas, E. (1996) The effect of stress inoculation training on anxiety and performance. *Journal of Occupational Health Psychology*, 1, 170–186.

Sauter, S.L., Murphy, L.R. and Hurrell, J.J. (1990) Prevention of work-related psychological disorders. *American Psychologist*, 45, 1146–1158.

Schnal, P.L., Landsbergis, P.A. and Baker, D. (1994) Job strain and cardiovascular disease. *Annual Review of Public Health*, 15, 381–411.

Sefton, A. (1992) Introduction to the first offshore installation management conference: Emergency command. Robert Gordon University, Aberdeen, April.

Smith, A. (1991) A review of the non-auditory effects of noise on health. *Work and Stress*, 5, 49–62.

Smith, A., Johal, S., Wadsworth, E., Smith, G. and Peters, T. (2000) *The Scale of Occupational Stress: The Bristol Stress and Health at Work Study*. Sudbury: HSE Books.

Tversky, A. and Kahneman, D. (1974) Judgement under uncertainty: Heuristics and biases. *Science*, 185, 1124–1131.

Weaver, J.L., Bowers, C. and Salas, E. (2001) Stress and teams: Performance effects and interventions. In P.A. Hancock and P.A. Desmond (eds.) *Stress, Workload, and Fatigue* (pp. 83–106). Mahwah, NJ: Lawrence Erlbaum Associates.

Winkler, R. (1980) Occupational stress in Western Australian Ambulance Officers. Paper presented at The 14th Convention of Ambulance Authorities, Perth, Australia, November.

Wyler, A., Masuda, M. and Holmes, T. (1968) Seriousness of illness scale. *Journal of Psychosomatic Research*, 11, 363–375.

Chapter 8

Coping with Fatigue

Definition

Many high-risk industries necessitate 24 hours of operation, seven days a week, and consequently this can require long and unsocial hours of shift work. The machinery used by these industries (for example, engines, pumps, compressors) can continue for hours, days or months without requiring maintenance. However, there is one component within these systems that is not specifically designed for continuous operations – the human operator. Human fatigue is acknowledged to be a significant safety concern in high-risk industries (Rosekind et al., 1995). The experience of the feeling of heavy eyelids, head-nodding, grogginess, difficulty concentrating, low energy, and lack of *joie de vivre* has probably been experienced by everyone working in high-risk industries. However, researchers have found fatigue to be surprisingly difficult to define.

Åkerstedt (2000) says that fatigue is synonymous with drowsiness, sleepiness and tiredness. Cercarelli and Ryan (1996) state that fatigue involves a diminished capacity for work and possible decrements in attention, perception, decision-making and skilled performance. For the purpose of this chapter, Caldwell and Caldwell's (2003) definition will be used. They define fatigue 'as the state of tiredness that is associated with long hours of work, prolonged periods without sleep, or requirements to work at times that are "out of synch" with the body's biological or circadian rhythm' (p15). In many safety-critical environments, workers have to cope with fatigue, due to long working hours, difficult conditions, shift work or jet lag. The components of the skill category 'coping with fatigue' are shown in Table 8.1.

Table 8.1 Elements of coping with fatigue

Coping with fatigue	Identify causes of fatigue. Recognise effects of fatigue. Implement coping strategies.

This chapter summarises research on the significance of fatigue in accident causation, describes the state of sleep and how it is regulated, outlines techniques for optimising shift working, sleep regulation, and presents some measures that can reduce levels of fatigue or assist in coping with tiredness while at work. A list of resources on the science of sleep, fatigue and shift work is provided at the end of the chapter.

Fatigue and accidents

Fatigue has been implicated in major accidents in all industrial sectors (Coren, 1996; Maas, 1998), such as those in the nuclear power industry at Three Mile Island and Chernobyl (see also Box 8.1), and transportation accidents, e.g. the *Exxon Valdez* oil spill. Among a range of potential human factor failures that were considered to contribute to the space shuttle *Challenger* disaster were the possible effects of sleep loss, excessive duty shifts, circadian or daily rhythm effects, and the resulting fatigue on those who took the decision to launch the shuttle, despite concerns about safety (Hawkins, 1987).

Box 8.1 Asleep in the control room

In March 1987, the Nuclear Regulatory Commission (NRC) shut down the Peach Bottom nuclear power station in Pennsylvania, USA, after receiving information that control room operators had been observed sleeping while on duty in the control room and not effectively performing their duties.

At times during shifts, particularly from 11pm to 7am, one or more of the control room staff had slept or been otherwise inattentive to duties while on shift. Further, the shift supervisor and other senior management either knew and condoned these actions, or should have known about them and taken action to correct this situation (US NRC, 1987). Although no accidents resulted from the inattention of the control room operators, the Peach Bottom nuclear power station was shut down for a period of two years due to employees sleeping on shift.

Fatigue is the largest identifiable and preventable cause of accidents in transportation, surpassing that of drug- or alcohol-related incidents (Åkerstedt, 2000). In the aviation industry, it is estimated that fatigue may be involved in 4–7% of civilian aviation accidents (Kirsch, 1996) and anywhere between 4% and 25% of military aviation accidents (Caldwell et al., 2002; Ramsey and McGlohn, 1997). There are reports of commercial flights where both the pilot and co-pilot had fallen asleep (e.g. *Guardian*, 6 November 2007). In the maritime industry, Raby and Lee (2001) concluded that fatigue was a contributor to 16% of vessel accidents, and 35% of personnel injury accidents (see Box 8.2 for an example of the deadly effects of fatigue in the fishing industry).

Box 8.2 Fishermen warned to stay alert

Fishermen were issued with an urgent 'stay alert' warning by the UK Marine Accident Investigation Branch after an investigation into the sinking of the fishing vessel *FV Brothers* with two fatalities. On 1 June 2006, the vessel crashed at speed onto rocks off the Scottish island of Eilean Trodday and sank. The Marine Accident Investigation Branch report (2007) concluded that the vessel 'probably grounded due to one of the crew falling asleep in the wheelhouse, which allowed the vessel to sail past her intended fishing grounds and on to the shore. Both crew would have been suffering the effects of fatigue brought on by a number of long days at work, with only short, broken sleep periods. Both crew had also drunk some alcohol before the vessel left the harbour.'

On US highways, fatigue causes 100,000 crashes and 1,500 fatalities each year. In a survey of British car drivers, Maycock (1995) found that 15% of drivers who had been involved in an accident on a motorway cited tiredness as a contributing factor. In 2001, a man driving a Land Rover jeep on an English motorway fell asleep and the vehicle tumbled down an embankment onto a train line. It caused the crash of an express train, killing 10 passengers. In Australia, Howarth et al. (1988) estimate that 25–35% of truck crashes are due to fatigue. In an analysis of five studies of accidents reported in eight hours shift systems (morning, afternoon, and night), compared with the morning shift, the risk of injury is 15% higher in the afternoon and 28% higher on night shift (Spencer et al., 2006).

For health care workers, fatigue is recognised as a risk to patient safety. In the US, resident physicians commonly work on-call periods that last 24–36 hours, and some residents work 100–120 hours per week (Rall and Gaba, 2005). In a survey of junior doctors (known as residents in the US), 41% reported fatigue as the cause of their most serious mistake, with 31% of these errors resulting in a fatality (Wu et al., 1991). Similarly, 61% of anaesthetists and anaesthetic nurses surveyed, reported making an error in administering anaesthetic that they attributed to fatigue (Gravenstein et al., 1990). Medical duty hours have been reduced in Europe and the USA due to work legislation and this has been shown to decrease attentional failures (Lockley et al., 2004), but long shifts and on-call working can still leave doctors and other health care professionals in a state of fatigue while performing critical tasks.

Measuring fatigue

Early last century, Muscio (1921) stated that in order to define a phenomenon such as fatigue, we need a valid and reliable measurement instrument. However, nearly 100 years later, researchers still do not have such a tool. As there are no biochemical markers of fatigue, researchers have to infer levels of fatigue by using other subjective, behavioural, physiological or cognitive techniques.

Subjective methods include those methods in which people are asked to indicate how tired or sleepy they feel. Standardised tools such as the Epworth sleepiness scale (Johns, 1991) consist of items regarding the level of tiredness you feel when watching television or sitting for a few minutes at a traffic light. The Stanford sleepiness scale asks people how alert they feel at certain times of the day (Hoddes et al., 1972). These scales give a numerical measure of sleepiness. The advantage of subjective methods is that they are easy to administer. However, as discussed later, people are not necessarily good at judging their levels of fatigue, and so subjective measures may underestimate levels of sleepiness.

Behavioural methods involve looking for indicators of sleepiness such as yawning, microsleeps (brief periods in which sleep uncontrollably intrudes into wakefulness) (Caldwell and Caldwell, 2003), drooping eyelids or decreased social interaction. The difficulties with behavioural measures are that they are difficult to quantify and can be masked in a stimulating environment.

The multiple sleep latency test (MSLT) is the standard physiological method for assessing daytime sleepiness. This measure assesses how long it takes an individual to fall asleep in a sleep-inducing environment. Brain wave measurement is used to assess the onset of sleep. Ten to 20 minutes to fall asleep is normal, 5–10 minutes to fall asleep indicates mild to moderate sleepiness, and less than five minutes to fall asleep indicates severe sleepiness. Howard et al. (2002) gave the MSLT test to 11 anaesthesiology residents and found that during a daytime shift it took an average of 6.7 minutes to fall asleep, and it took 4.9 minutes to fall asleep immediately after a 24-hour work and in-house on-call period of work.

Researchers can also use standard cognitive tests to infer levels of fatigue. They compare the base line performance on standard cognitive tests of vigilance, arithmetic, verbal fluency and reaction time with the individual's performance when sleep deprived. However, it is worth stating that motivation has a large effect on the ability of individuals and teams to cope with fatigue. It has been suggested that the decrement in performance found in dull, lengthy tasks is due to a lack of interest in performing well. Johnson and Naitoh (1974) state that it is difficult to predict the effects of prolonged sleep loss of less than 60 hours on the performance of a highly motivated military force. Therefore, laboratory tests may not translate well into real-world performance.

Causes and effects of fatigue

The causes of fatigue include the obvious one of long hours of work, as well as a lack of sleep. Factors such as stress, temperature extremes, noise (>80 dB), physical work and vibration are all fatiguing. Also, the more boring the task, the more likely you are to suffer the effects of fatigue. In studies of fatigue in a simulated driving task, people were more likely to leave the road when driving on a straight stretch rather than on a corner (Caldwell and Caldwell, 2003). Fatigue has been shown to have detrimental effects on cognitive performance, motor skills, communication and social skills. Evidence for these detriments is given below, and summarised in Table 8.2.

Cognitive. After one night without sleep, cognitive performance may decrease by 25%, and after two nights without sleep, cognitive performance can fall to nearly 40% of baseline (Krueger, 1989). Bandaret et al. (1981) found that after 36 hours of constant work, soldiers became distracted from critical tasks and took longer on tasks that they had previously done more quickly. Samkoff and Jacques (1991) concluded that sleep-deprived junior doctors are able to perform effectively in crises or other novel situations. However, they may be more prone to errors on routine, repetitive tasks and tasks that require sustained vigilance.

Motor skills. Dawson and Reid (1997) compared the effects on performance of fatigue and alcohol intoxication using a computer-based tracking task. They demonstrated that one night of sleep deprivation produces performance impairment greater than is currently acceptable for alcohol intoxication. A loss of two hours' sleep produces a performance decrement on psychomotor tasks equivalent to drinking two or three beers (see Table 8.3).

Table 8.2 Summary of the effects of fatigue

Thinking (cognitive)

- adverse effect on innovative thinking and flexible decision-making
- reduced ability to cope with unforeseen rapid changes
- less able to adjust plans when new information becomes available
- tendency to adopt more rigid thinking and previous solutions
- lower standards of performance become acceptable

Motor skills

- less co-ordination
- poor timing

Communication

- difficulty in finding and delivering the correct word
- speech is less expressive

Social

- become withdrawn
- more acceptance of own errors
- less tolerant of others
- neglect smaller tasks
- less likely to converse
- increasingly irritable
- increasingly distracted by discomfort

Table 8.3 Comparison between sleep loss and alcohol consumption (Roehrs et al., 2003)

Sleep loss (hrs)	Equivalent US beers
8	10–11
6	7–8
4	5–6
2	2–3

Communication. Sleep deprivation has been shown to have a detrimental effect on communication. In a study of a four-man bomber crew it was found that, over a 36-hour exercise, there was a reduction in voice intonation and a slowing of speech (Whitmore and Fisher, 1996). Further, May and Klein (1987) found an impairment of verbal fluency and word retrieval in sleep-deprived military personnel.

Social. A lack of regard for normal social conventions, childishness, impatience, irritability and inappropriate interpersonal behaviour have all been described anecdotally by individuals participating in studies of sleep deprivation (Horne,

1993). In an analysis of 19 studies, Pilcher and Huffcutt (1996) found that mood is more affected by sleep deprivation than either cognitive or motor performance.

As with stress (see Chapter 7), people may not recognise when they are suffering from levels of fatigue that could lead to an accident. To illustrate, Helmreich and Merritt (1998) found that over 60% of a sample of medical doctors agreed with the statement: 'even when fatigued I perform effectively during critical operations.' The most obvious signs of fatigue in one's self or other team members is a degradation in task performance. For example, when driving, the frequency of risky overtaking manoeuvres increases with the number of hours driven (Brown et al., 1970). Nevertheless, people can be adept at increasing effort by concentrating harder on tasks to mitigate the effects of fatigue (Mackie and Miller, 1978). A recent study by Petrilli et al. (2006) studied long-haul flight crews' ability to perform decision-making tasks after a trans-Pacific flight, compared with rested crews. They found that the tired crews engaged in protective behaviours, such as communicating more, additional checking, and so on, but they also made more decision errors.

Recovering from fatigue: Sleep

The only way of recovering from fatigue is through sleep. A desire to sleep is a physiological drive just like thirst or hunger. A very sleep-deprived person will actually start to hallucinate (see Box 8.3). Further, rats that are forced to stay awake will eventually die. In a study of totally sleep-deprived rats, Everson et al. (1989) found no anatomical cause of death. All the rats showed a debilitated appearance, lesions on their tails and paws, and weight loss in spite of increased food intake. This study shows that sleep is essential. Individual requirements for sleep can range from 4–10 hours a night. Margaret Thatcher famously needed only four hours' sleep a night during her 11 years as British Prime Minister. However, on average, adults require seven to eight hours of sleep a day, and there is no research evidence suggesting that you can train yourself to get by on less sleep through constant sleep deprivation.

Box 8.3 Sleep deprivation record attempt

In January 1959 Peter Tripp, a 32-year-old New York disc jockey stayed awake for 200 hours (about eight days). Tripp had to fight to stay awake from the beginning. After 48 hours he began to have visual hallucinations (e.g. he reported finding cobwebs in his shoes). After 100 hours he had extreme difficulty with simple cognitive tests. His visual hallucinations also became worse: he thought a tweed suit worn by one of the scientists observing him was a suit of fuzzy worms.

By 170 hours Tripp was unsure as to who he was, and required frequent proof of his identity. Although he behaved as if he was awake, his brain wave patterns were as if he was asleep. After 200 hours, his hallucinations and reality had merged, and Tripp thought he was the victim of a conspiracy among the doctors and researchers observing him (Luce, 1966).

Although people may be able to function on less than their usual amount of sleep for a couple of nights, when sleep debt accumulates, there is a linear degradation in performance, especially on tasks for which people have a low motivation to perform (Gawron et al, 2001). However, after a period of sleep deprivation, people can return to normal levels of alertness after even one or two full nights of sleep.

Stages of sleep

There are different kinds of sleep, divided into five stages that are cycled through during an extended period of sleep, typically during one night (see Figure 8.1).

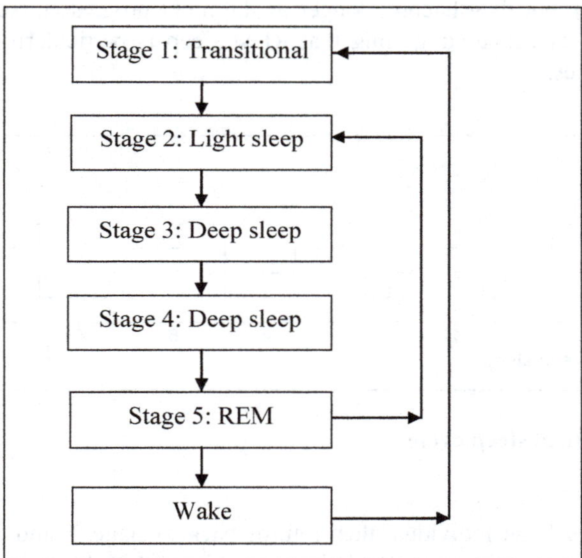

Figure 8.1 Stages of sleep

Stage 1: Transitional. This is the first stage of sleep and can last for five to 10 minutes. It is the transition between asleep and awake, and there may still be some awareness of activity in the surrounding environment. This state to and from consciousness to sleep is called hypnogogia and there can be an experience of images or mild hallucinations during this phase (Mavromatis, 1987). This is the stage that someone is in if they fall asleep at the wheel of a car, and is often described as microsleep. A microsleep can last from a few seconds to a minute in time.

Stage 2: Light sleep. This stage lasts 10–20 minutes. During stage 2, breathing and heart rate slows, and there is a slight decrease in body temperature. Brain waves become slower, interspersed with occasional bursts of rapid waves called sleep spindles.

Stages 3/4: Deep sleep. These are the stages of deep sleep and tend to last about 30 minutes. There is limited muscle activity, and the brain produces slow delta waves. Someone who is woken in these stages of sleep may be very groggy and could take a few minutes to wake properly. This grogginess (or sleep inertia) can last longer (up to 20–30 minutes). However, Rosekind et al. (1995) suggest that, in an emergency, the effects of adrenaline can rapidly overcome the negative effects of sleep inertia.

Stage 5: Rapid eye movement (REM) sleep. This is the stage in which dreams occur It is characterised by rapid eye movements behind the closed eyelids. Muscles relax, heart rate and brainwaves speed up, and breathing becomes rapid and shallow. There are several periods of REM sleep in a normal night's sleep. The first period is short (5–10 minutes), but they become longer as the night progresses (see Figure 8.2). There is some evidence suggesting that REM sleep is particularly important for storing memories.

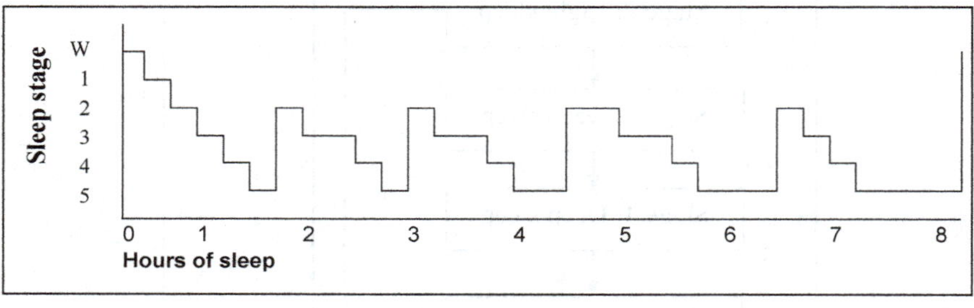

Figure 8.2 Typical sleep cycle

Following stage 5, an individual then returns back to stage 2, and cycles through stages 3, 4, before returning to REM sleep (see Figure 8.2). Most of the slow-wave sleep occurs during the beginning of a typical night's sleep, with more REM sleep in the latter half (except in day sleep most of the REM sleep occurs at the beginning of the sleep cycle and stages 3 and 4 in the second half of the sleep cycle). However, in a severely sleep-deprived individual, the REM sleep is pushed later in the period of sleep, with sleep at stages 3 and 4 occupying a greater proportion of the time asleep (Caldwell and Caldwell, 2003). As REM sleep occupies the period of time just prior to the end of a night's sleep, this is why many people remember their dreams.

Sleep regulation

Sleep homeostasis and circadian rhythms regulate when we feel alert and when we feel tired and fatigued. These two processes are at the core of many models addressing the regulation of fatigue and performance (Achermann, 2004). There are other factors that influence our level of fatigue (e.g. quality of sleep the previous night, level of activity, levels of interest in the task being performed, lighting).

Nevertheless, sleep homeostasis and circadian rhythms are the strongest factors influencing our level of alertness.

Sleep homeostasis

Sleep homeostasis is the amount of time awake since the last period of sleep. Obviously, workers are going to feel more tired if they have not slept in the last 16 hours than if they had just woken up from a nap an hour ago. A related factor is the cumulative effects of suboptimal periods of sleep over extended periods of time. Two nights with four hours of sleep each produce 'pathological' levels of sleepiness. Periods of recovery sleep are essential to allow a return to 'normal' levels of alertness. After seven days of only five hours of sleep each night, one full night of sleep followed by a day-time nap returned measures of sleepiness and fatigue to pre-restricted levels (Carsakadon and Dement, 1981).

Circadian rhythm

The circadian rhythm is the name given to the 'internal body clock' that regulates the approximately 24-hour cycle of biological processes in animals and plants. The body's physiological processes are not constant throughout the day, but are rhythmic and cyclical. There are circadian patterns for cognitive and psychomotor performance, physiological activities (e.g. digestion, immune function, temperature) (see Figure 8.3), alertness and mood. Even birth and death have circadian patterns that peak during the night (Kryger et al., 1994).

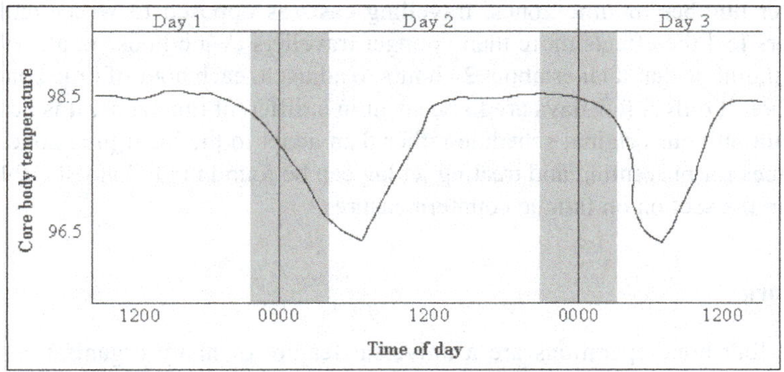

Figure 8.3 Typical core body temperature circadian cycle

Laboratory studies depriving volunteers of any time cues have shown that the circadian cycle actually runs slightly longer than 24 hours if allowed to 'free run' in the absence of external cues such as daylight (Czeisler et al., 1982). The circadian cycle is biphasic: our levels of alertness peak between 12:00 and 21:00 (usually around 16:00) and fall to a minimum between 03:00 and 06:00 hours. Evidence from traffic accidents and occupational accidents shows that a peak tends to occur in the early hours of the morning (between 1am and 6am), when performance is at its lowest. Further, particularly for older drivers, there is a secondary accident peak between 1pm and 5pm (Langlois et al., 1985).

There are individual differences in circadian rhythm. Questionnaires, such as the Morningness-Eveningness questionnaire, have been widely used to quantify individual propensity to be active and alert in the morning or in the evening (Horne and Ostberg, 1976). One in ten people is a morning person ('lark') who naturally likes to wake up early in the morning, and about two in ten are 'owls', who enjoy staying up long past midnight (Smolensky and Lamberg, 2000). The rest of us are somewhere between these two extremes. Thus, someone who is an owl may be more suited to night work than someone who is a lark.

Jet lag

Disruption to our circadian rhythm occurs when our internal circadian rhythms are not in tune with the external cues coming from the environment, the most influential of which is the light/dark cycle of day and night. This disruption can be caused by shift work (especially night shifts) or by crossing time zones (jet lag). The effects are we have trouble going to sleep, or wake up too early. As can be seen from Figure 8.3, if a traveller flies from London to Los Angeles (a minus eight-hour time difference), for the first few days they are likely to wake up very early, as our circadian cycle is peaking at around 11am London time, which is 3am in Los Angeles.

The circadian cycles are resistant to adjustment, and changes initiated by shift work or jet lag can take several days before the circadian cycles come into phase with a new sleep/wake cycle. Generally the effects of jet lag are increased by crossing a greater number of time zones, travelling east (as opposed to west), and older travellers feel the effects more than younger travellers (Waterhouse et al., 1997). A rule of thumb is that it takes about 24 hours to adjust to each hour of time zone shift. Therefore, if only a few days are to be spent in a different time zone, it is advisable to remain on your original schedule rather than adapt to the local time zone. Other techniques for preventing and treating jet lag can be found in O'Connell (1998) and below in the section on fatigue countermeasures.

Shiftwork

Twenty-four-hour operations are a universal feature of many organisations, such as transportation, manufacturing, production, health care and the military. Taylor et al. (1997) state that approximately 20–25% of employees in the manufacturing industry are shift workers. There are individual differences in the ability of people to adapt to shift work. People with a history of sleep disorders, who are over 50 years of age, have a history of gastrointestinal complaints, or a preference for rising early, are associated with difficulties in adjusting to night work (Monk, 1990). The ability to sleep at unusual times and overcome drowsiness are associated with shift work tolerance (Costa et al., 1989). Knauth and Hornberger (2003) identify a number of measures that can be taken to optimise the well-being of shift workers and mitigate the effects of shift work on performance and alertness. Each of these preventative and compensatory measures for shift workers is discussed below.

Ergonomic shift system design

There is no ideal shift system. No system is without drawbacks, so it comes down to selecting the 'least worst' shift system. It is possible to make a number of recommendations regarding the direction and speed of shift rotation, shift duration, when shift changeover should occur, and time off between shifts:

- *Direction and speed of shift rotation.* A forward rotation (morning to evening to night) produces less fatigue than a reverse rotation (night to evening to morning). As described above, the internal push from the circadian cycle to run slightly longer than 24 hours would favour the forward rotation.

  Speed of rotation can range from none (fixed shifts), to very slow (three to four weeks), to rapid (every two to three days). The worst possible schedule is to work between four and seven nights and then switch to day shift. This results in the circadian rhythm being disrupted, and then beginning to adapt just prior to being disrupted again. Having a permanent night shift team may seem to be the ideal solution by having people who are fully adapted to working at night. However, working a permanent night shift may be detrimental to the mental well-being of night-shift workers as it minimises the contact with friends and family, and only those individuals who are 'night owls' are likely to fully adapt. Spencer et al. (2006) reviewed six studies that examined adjustment of the circadian rhythm of permanent night-shift workers. They concluded that only 7% of workers showed evidence of 'good' adjustment. However, 29% of workers showed evidence of at least some adjustment that would benefit performance on night shift.

  Wedderburn (1991) proposes that the ideal schedule is to work as few night shifts as possible in a row (ideally one) so that the disruption to the circadian rhythm is minimised. However, a slow rotation has the advantage of allowing a worker's circadian rhythm to adapt to the schedule. Whether a fast or slow rotation is best depends on the type of work being carried out. A fast rotation means that a night shift is staffed by unadapted individuals who are suffering from partial sleep loss. Therefore, in this situation it would be desirable to avoid high-risk, demanding tasks during the night shift. Alternatively, if a high work tempo during the night shift cannot be avoided, a slow shift rotation may be more desirable in terms of safety and productivity.

- *Time of shift changeover.* Twenty-four-hour operations are usually divided into two or three shifts. In a three-shift structure, day shift starts around 5am–8am, and ends around 2pm–6pm. Evening shift starts around 2pm–6pm, and ends around 10pm–2am. Night shift starts around 10pm–2am and ends around 5am–8am (Rosa and Colligan, 1997). Knauth and Hornberger (2003) recommends that the morning shift should not start too early (6am is better than 5am) to maximise hours of sleep. The evening shift should not end too late (10pm is better than 11pm) and should end earlier at the weekend to increase social contact and maximise hours of sleep. Finally, the night shift should end as early as possible to increase the hours of sleep during night-time. It can be seen that it is impossible to improve the timing of one shift without having a negative effect on the other shifts.

- *Time off between shifts.* Knauth and Hornberger (2003) recommend a maximum of five to seven days of work, with adequate time off (at least 11 hours) between shifts. Research has found that the shorter the time between shifts, the less time people sleep (Kurumatani et al., 1994). Fatigue has also been shown to be a factor in accidents in the offshore oil industry due to the long periods of work (14–21 days). Miles (1999) describes research carried out with offshore accident rates by Connolly (1997) in which the incidence of serious injury in comparison to all injuries was found to increase with increasing days working offshore time. Knauth (1995) also outlined several studies that showed an increase in accidents and errors over consecutive night shifts. Spencer et al. (2006) examined seven studies in which accidents and injuries were reported over at least four successive night shifts. They found that on average, when compared with the first night, the risk of injury was 6% higher on the second night, 17% higher on the third night, and 36% higher on the fourth night.
- *Shift duration.* There is no strong evidence on the length of shift that is the best in terms of safety and productivity. Knauth and Hornberger (2003) state that many studies have been published comparing eight-hour with nine-hour, or 10-hour with 12+ hours shift lengths, with contradictory results. Dembe et al. (2005) found that working at least 12 hours per day was associated with a 37% increased hazard rate and working at least 60 hours per week was associated with a 23% increased hazard rate. Tucker et al. (1996) compared two groups of chemical workers: one group working a 12-hour shift, and one group working an eight-hour shift. The two groups did not differ on most outcome measures, although the differences that did exist suggested advantages for the 12-hour shift workers over the eight-hour shift workers, except for levels of alertness at certain times of day. Thus, it would appear that the other factors associated with shift design are of greater importance than the duration of the shift.

A summary of recommendations for ergonomic shift design is shown in Table 8.4.

Table 8.4 Summary of ergonomic shift design recommendations

Direction of rotation	Forward rotation
Speed of rotation	Either fast, or slow, dependent upon the type of activity.
Time of shift change	Morning shift – should not start too early. Evening shift – should not end too late. Night shift – should end as early as possible.
Time off between shifts	A maximum of 5–7 days continuous work, with at least 11 hours between shifts.
Shift duration	No clear recommendations.

Shift worker participation

Knauth and Hornberger (2003) recommend that the design of a shift schedule should be a compromise between the goals of the organisation and the wishes of the workers. There is a need to have worker participation, information and communication about a new shift schedule, champions of change, effective management of the change, evaluation of the effects of the new schedule on the company and the workers, and time for people to adapt to the new system. Duchon et al. (1994) found that metal mine workers supported a 12-hour shift schedule with a compressed work schedule as compared with an eight-hour schedule. A positive effect was found on absenteeism, morale, health, stress, eating habits, family life, and sleep quantity and quality.

Shift working conditions, personal health-related and behavioural resources

Rosa (1995) states that attention should be given to the levels of staffing and workloads of individuals working in shifts. Other factors such as lighting, breaks, exercise, naps, type of food and frequency of eating are discussed in detail in the section concerned with fatigue countermeasures. Long-term shift work has been associated with cardiovascular, psychiatric, chronic fatigue and gastrointestinal problems (Monk and Folkard, 1992; Smith et al., 2003). There is increasing evidence to suggest that shift work, and particularly night work, may present special risks to women. Specific health outcomes for pregnant workers linked to shift work include increased risk of spontaneous abortion, low birth weight and prematurity (Harrington, 2001). In a review of the link between breast cancer and shift work, Swerdlow (2003) states that the evidence of a link is appreciable, but not definitive. So while there is consistent evidence that workers can suffer health problems related to shift working, not all studies, e.g. Nicholson and D'Auria (1999), have found a statistical increase in the morbidity (incidence of disease) in shift workers.

Health care management

As described above, some authors believe there may be a link between shift working and long-term health effects. Knauth and Hornberger (2003) suggest that to address these issues companies should monitor the health of shift workers to allow the early identification of any problems. Further, a company's medical staff should be knowledgeable about current research on sleep disorders and shift maladaption syndrome (insomnia during the main sleep period or excessive sleepiness during the major awake period associated with night shift work or frequently changing shift work; American Psychiatric Association, 2000).

Commuting

Obviously a worker will be tired after a 12-hour night shift, and if they are driving home after work, fatigue could affect their ability to concentrate. Knauth and Hornberger (2003) summarise a number of measures that could be considered to stop shift workers having a driving accident on the way home due to fatigue. They suggest that the employer have a company car pool to take workers home or offer a place to take a nap before driving home. They also advise that shift workers should keep their car cool, turn on the radio, vary their route, use public transport, and live close to their place of work.

Sleep at home

Night shift workers report a lower quality and quantity of sleep than other shift workers. The sleep problems that more commonly occur for night shift workers include difficulty falling asleep, disturbed sleep and spontaneous early awakenings. Evidence suggests that night shift workers may obtain up to 25% less sleep during their main sleep period than they do when working during the day. Unless you live alone, another obvious problem with sleeping at home during the day is that other family members or flatmates, and the majority of the world, are not on night shift, and therefore sleep may be disrupted. Measures should be taken to ensure good sleep hygiene (see the fatigue countermeasures section).

Family and social support

One of the main difficulties with shift work is the lack of social contact with family and friends when working evening, night or weekend shifts. Knauth and Hornberger (2003) recommend using a family planning calendar so that everyone is aware of the plans of other family members, hold family meetings, create time for being alone with your partner, and meeting other shift worker families.

Fatigue countermeasures

Fatigue countermeasures are techniques designed to ensure adequate sleep for all workers and optimise circadian adaptation for travellers or shift workers. Eight fatigue countermeasures for improving the quality of sleep and/or maintaining alertness while at work are discussed below. For more detailed advice on how to improve the quality of sleep, as well as fatigue countermeasures, see Horne (2006),

Maas (1998) or the Eurocontrol Fatigue and Sleep Management Brochure designed for air traffic controllers (website at end of chapter).

Education

The US Bureau of Labor states that of the over 27 million workers who are in managerial and professional work, only 1.3% work the night shift and only 1.7% work the evening shift. Therefore, those individuals in managerial positions have little experience of working shifts, and so are not likely to have an understanding of the difficulties associated with shift working. There is a need to educate both managers and shift workers about the effects of shift work on performance and alertness, the advantages of different shift cycles, the effects of sleep deprivation, circadian disruption and possible solutions to help mitigate the effects of fatigue.

Sleep hygiene

Sleep hygiene refers to behaviours that promote improved length and quality of sleep. It is important that there is an environment conducive to sleep. The room should be quiet, dark and cool. Also, it is important that the bed is comfortable. People should attempt to relax before going to sleep and, in the few hours prior to going to bed, avoid activities such as heavy exercise, large amounts of food, caffeine, alcohol and napping

Rest breaks

Vernon (1920) studied munitions workers during World War I and found that they would regularly work more than 60 hours a week. When the number of hours was reduced to between 50 and 55 hours a week, there was an increase in productivity, and with the introduction of 10-minute breaks, output further increased. These findings have been replicated in a number of other studies. Therefore, rest breaks would seem to be an effective mechanism for reducing the effects of fatigue. Other techniques that could be considered are rotating duties, talking to team members and moving around to attempt to remain engaged in the job and to prevent boredom. Changing posture can also be helpful. For example, work standing up rather than sitting for long periods of time, or a short period of light exercise can be performed to attempt to maintain alertness.

Napping

Napping is an effective strategy for reducing fatigue. Even a short nap of 10 minutes can improve functioning. The US Navy recommends that aviation commands should encourage, and at times mandate, napping during sustained operations (Naval Strike and Air Warfare Center, 2000), although in many occupations this is not permitted while on duty. Caldwell and Caldwell (2003) give a number of recommendations for effective napping:

1. Time naps in accordance with the circadian rhythm. The best time to have a nap is during the natural dips associated with the circadian rhythm to promote the natural instigation of sleep. Therefore, the best time for a nap during a night shift is between 1am and 6am, and during the daytime is between 2pm and 4pm.
2. Time naps close to the beginning of long periods of duty. Obviously the closer the nap to when you start work, the greater benefit you will receive of that period of rest when working.
3. Place naps early in a period of sleep deprivation. When in situations in which sleep loss is inevitable, naps should be used to prevent fatigue, rather than restore performance after it has already deteriorated. Therefore, a nap should be taken before you feel the extreme effects of fatigue.
4. Make naps as long as possible. People who have seven or eight hours of sleep at night perform better than when they have had three hours of sleep. The same is true of napping, with naps of 40 minutes to two hours recommended. However, obviously 10 minutes of sleep is better than nothing.
5. Consider the effects of sleep inertia. As described in the section on sleep, sleep inertia is the post-sleep grogginess that is sometimes experienced following a period of sleep. The recommendation of naps of 40 minutes to two hours is so that the stages of deep sleep have either not been entered, or have been cycled through. It is also advisable not to perform safety-critical tasks immediately upon waking from a nap.

Diet

It is recommended for fatigue management that everyone, especially those on shift working, should eat regular meals with a balanced diet. Greasy food should be avoided, particularly during the night as it is harder to digest. Shift workers should also consider having their largest meal after the main sleep period, and having a light meal in the middle of night shift. If you are feeling hungry or thirsty before going to sleep, it is suggested that a light snack and/or a small drink is preferable to a large meal.

The use of alcohol to aid sleep is also not advised. Although it promotes sleep by aiding relaxation, it also results in a lighter sleep, and can lead to dehydration and a headache the following morning.

Caffeine can be as effective as a chemical stimulant in maintaining performance when you are fatigued. Caffeine is most effective for people who do not normally consume large quantities on a daily basis. To increase alertness, 200 milligrams of caffeine (one small cup) is recommended for consumption every two hours up to five hours before the next sleep break. However, it is important to note that there are large individual differences in our ability to metabolise caffeine. To illustrate, Blanchard and Sawyer (1983) found the half-life, or time it takes for the amount of caffeine in the blood to decrease by 50%, ranges from 2.7 to 9.9 hours in 10 healthy male volunteers. Caldwell and Caldwell (2003) recommend that sleep-deprived people should drink a minimum of 200mg of caffeine every two hours (up to five hours before the next sleep break).

Medication

In military aviation, stimulants, or 'go pills', such as dexedrine (dextroamphetamine) are used to increase alertness and maintain performance. However, their use is carefully monitored by medical personnel and is generally only authorised during combat or exceptional circumstances of operational necessity. A newer medicine that is receiving some attention for attenuating the performance effects of fatigue is modafinil. Originally used to treat narcolepsy, modafinil has been shown to enhance cognitive performance (particularly during the circadian troughs in performance and alertness) in normal, sleep-deprived adults. However, Wesensten et al. (2002) found no significant differences in performance and alertness between caffeine and modafinil in sleep-deprived individuals.

Melatonin is another potential aid to facilitate changes in the circadian cycle. Melatonin is a naturally occurring hormone secreted by the pineal gland. Melatonin levels rise during the night and decline at dawn. Research has shown that oral administration of doses of melatonin from 0.1mg to 240mg can help night shift workers obtain improved sleep during the day. However, studies of melatonin administered to promote night-time sleep have produced mostly negative results (Sharkey et al., 2001). Melatonin is also beneficial to counteract the effects of jet lag, particularly if flying east through multiple time zones.

Bright light

Bright light has been proposed as a countermeasure to increase the speed at which people adapt to different shifts. It has been shown to be superior to helping people adapt to a different shift than melatonin (Dawson et al., 1995). When changing from day shift to night shift, bright light exposure late in the evening (starting around 8pm) for two to four hours helps to adapt to night shift (Boivin and James, 2004). However, night workers should also limit light exposure on the way home (e.g. by wearing sunglasses). Night shift workers who are going back to day shift should do the reverse: have bright light exposure in the morning and limit exposure to light in the evening.

It is not clear how much bright light is required to have an effect on the circadian cycle. Even bright indoor lighting (1,000–2,000 lux) has been shown to be beneficial (Phipps-Nelson et al., 2003). Other studies have shown benefits in pulsing periods of light of 5,000 lux (40 minutes) with normal room lighting (<500 lux) (Baehr et al., 1999). However, it would seem that light at around 10,000 lux (equivalent to light exposure in a cloudy day) would be preferable. Similarly, for travellers, if you are trying to adjust to a new time zone it is important that you expose yourself to daylight on arrival to speed up the synchronisation of the body's circadian rhythms to the new time zone.

Planning for fatigue

Just as managers plan for the number of personnel, equipment required and cost of operations, fatigue should be another factor that should be taken into account and managed during planning. Attempts should be made to avoid intricate or risky activities between 03:00 and 06:00. Researchers have developed algorithms and scheduling software to aid project managers and schedulers to plan for the effects of fatigue. Two examples of fatigue modelling tools that are currently available are described below.

Fatigue Avoidance Scheduling Tool (FAST™). The FAST™ tool was developed by US Department of Defence laboratories. This model uses the Sleep, Activity, Fatigue and Task Effectiveness (SAFTE) algorithm to integrate information about circadian rhythms, cognitive performance recovery rates associated with wakefulness, and cognitive performance decay rates associated with sleep inertia to produce a three-process model of human cognitive performance. The SAFTE provides a measure of 'task effectiveness' measured on a scale from 0% to 100% effective. The SAFTE was developed for use in both military and commercial settings, with current users including the US Air Force and the US Federal Railroad Administration (Mallis et al., 2004).

The FAST™ scheduling tool is a software interface that provides the schedule input predictions in graphical and tabular form, and parameter tables for adjusting the model (Mallis et al., 2004). It can be used to provide planners with a method for evaluating the outcome of different schedules on the effectiveness of personnel. To illustrate, Figure 8.4 shows the percentage effectiveness predicted by the tool during the first three night shifts.

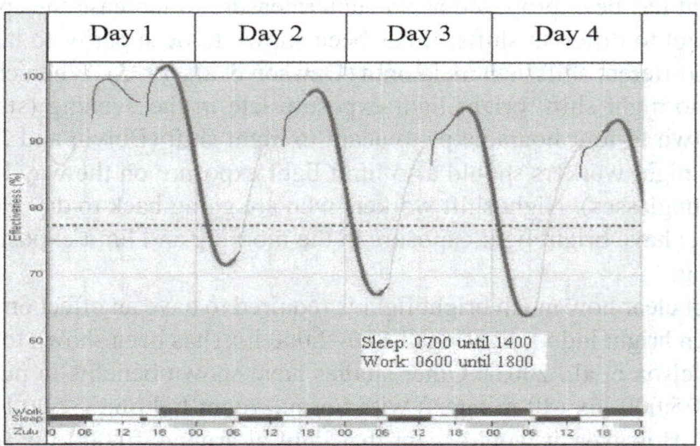

Figure 8.4 Percentage effectiveness during the first three nightshifts as predicted by the FAST™ scheduling tool

Optimal schedules may be selected based on average effectiveness for proposed work periods or critical events. Further, the tool can take input from an actigraph. An actigraph is a small accelerometer that is attached to the wrist and systematically records the movement of the wearer, both when awake and asleep. Therefore, real-life records of a worker's amount of sleep can be fed into the model to allow predictions to be made about levels of fatigue.

The tool can also be used for retrospective analysis of fatigue-related factors that may have contributed to an accident, error or safety-related incident. Information about the work and sleep schedules of individuals involved in an accident can be entered into the tool to provide an estimate of the effects of fatigue on the performance of those involved in an accident.

Fatigue and risk index calculator. The UK Health and Safety Executive (HSE) has also commissioned the development of a tool to allow managers to assess the levels of risk arising from fatigue associated with different work patterns. The tool uses a spreadsheet to provide the user with a measure of fatigue, and the relative risk of the occurrence of an incident on a particular shift (see Spencer et al., 2006, for more details). The fatigue and risk indices are calculated based upon five variables (time of day, shift duration, rest periods, breaks and cumulative fatigue). The fatigue and risk index calculator allows a comparison to be made between the impact of different working patterns on the levels of fatigue and the risk of injury. The fatigue and risk index calculator can be downloaded from the HSE website (www.hse.gov.uk).

Conclusion

Working while fatigued or when out of phase with your circadian cycle is common in high-reliability industries, as well as military operations. Fatigue has been implicated in many accidents and is recognised to have a detrimental effect on safety and performance. This chapter has outlined the effects of fatigue on human performance and identified strategies that can be taken by organisations, and individuals, to reduce or cope with levels of fatigue and decrease the time taken to synchronise your circadian rhythm with a new work schedule.

Key points

- Fatigue is difficult to define and quantify.
- Fatigue has implications for both the safety and productivity of workers.
- Fatigue has detrimental effects on cognitive performance, motor skills, communication and social skills.
- There are five stages of sleep that are cycled through during a normal period of sleep.
- Shift working schedules can be designed to improve the performance of workers and decrease the amount of time it takes to adapt to a different shift.
- Measures can be taken at organisational and individual levels to reduce fatigue and decrease the time taken for workers to adapt to a different work schedule.

Suggestions for further reading

Caldwell, J.A. and Caldwell, J.L. (2003) *Fatigue in Aviation*. Burlington, VT: Ashgate.

Coren, S. (1996) *Sleep Thieves*. New York: Free Press.

Horne, J. (2006) *Sleepfaring. A Journey through the Science of Sleep*. Oxford: Oxford University Press.

Maas, J. (1998) *Power Sleep*. New York: Villard.

Rosa, R.R. and Colligan, M.J. (1997) *Plain Language about Shiftwork*. Cincinatti, Ohio: US Department of Health and Human Services.

Websites

BBC Science: www.bbc.co.uk/science/humanbody/sleep

Eurocontrol Fatigue and Sleep Management Brochure: www.eurocontrol.int/humanfactors/public/subsite_homepage/homepage.html

Loughborough Sleep Research Centre: www.lboro.ac.uk/departments/hu/groups/sleep

NASA Ames fatigue countermeasures group: /human-factors.arc.nasa.gov/zteam/

National Sleep Foundation (USA): www.sleepfoundation.org

University of Southern Australia Centre for Sleep Research: www.unisa.edu.au/sleep/

References

Achermann, P. (2004) The two-process model of sleep regulation revisited. *Aviation, Space, and Environmental Medicine*, 75, A37–A43.

Åkerstedt, T. (2000) Consensus statement: Fatigue and accidents in transportation operations. *Journal of Sleep Research*, 9, 395.

American Psychiatric Association (2000) *Diagnostic and Statistical Manual of Mental Disorders DSM-IV-TR*. Washington, DC: *American Psychiatric Association*.

Baehr, E.K., Fogg, L.F. and Eastman, C.I. (1999) Intermittent bright light and exercise to entrain human circadian rhythms to night work. *American Journal of Physiology*, 277, R1598–R1604.

Bandaret, L.E., Stokes, J.W., Franscesconi, R., Kowal, D.M. and Naithow, P. (1981) Artillery teams in simulated sustained combat: Performance and other measures. In L.C. Johnson, D.I. Tepas and W.P. Colquhoun (eds.) *Biological Rhythms, Sleep and Shift Work* (pp. 455–497). New York: Pergamon Press.

Blanchard, J. and Sawyer, S.J.A. (1983) The absolute bioavailability of caffeine in man, *European Journal of Clinical Pharmacology*, 24, 93–98.

Boivin, D.B. and James, F.O. (2004) Light treatment and circadian adaptation to shift work. *Industrial Health*, 43, 34–48.

Brown, I.D., Tickner, A.H. and Simmonds, C.V. (1970) Effects of prolonged driving on overtaking criteria. *Ergonomics*, 13, 239–242.

Caldwell, J.A. and Caldwell, J.L. (2003) *Fatigue in Aviation.* Burlington, VT: Ashgate.

Caldwell, J.A., Gilreath, S.R. and Erickson, B.S. (2002) A survey of aircrew fatigue in a sample of army aviation personnel. *Aviation, Space, and Environmental Medicine*, 73, 472–480.

Carsakadon, M.A. and Dement, W.C. (1981) Nocturnal determinants of daytime sleepiness. *Sleep*, 5, S73–S81.

Cercarelli, L.R. and Ryan, G.A. (1996) Long distance driving behaviour of Western Australian drivers. In L.R. Hartley (ed.) *Proceedings of the Second International Conference on Fatigue and Transportation* (pp. 35–45). Caning Bridge, Australia: Promaco.

Connolly, S. (1997) Analysis of time factors and experience in relation to incidents and accidents over the period 1989/90 to 1995/96. London: HSE. Unpublished report.

Coren, S. (1996) *Sleep Thieves.* New York: Free Press.

Costa, G., Lievore, F., Casaletti, G., Gauffuri, E. and Folkard, S. (1989) Circadian characteristics influencing inter-individual differences in tolerance and adjustement to shift work. *Ergonomics*, 32, 373–385.

Czeisler, C.A., Moore-Ede, M.C., and Coleman, R.M. (1982) Rotating shift work schedules that disrupt sleep are improved by applying circadian principles, *Science*, 217, 460–463.

Dawson, D., Encel, N. and Lushington, K. (1995) Improving adaptation to simulated night shifts: timed exposure to bright light versus daytime melatonin administration. *Sleep*, 18, 11–21.

Dawson, D. and Reid, K. (1997) Fatigue, alcohol and performance impairment. *Nature*, 388, 23.

Dembe, A.E., Erickson, J.B, Delbos, R.G. and Banks, S.M. (2005) The impact of overtime and long work hours on occupational injuries and illnesses: new evidence from the United States. *Occupational and Environmental Medicine*, 62, 588–597.

Duchon, J.C., Keran, C.M. and Smith, T.J. (1994) Extended workdays in an underground mine: A work performance analysis. *Human Factors*, 36, 258–268.

Everson, C.A., Bergmann, B.M. and Rechtschaffen, A. (1989) Sleep deprivation in the rat: III. Total sleep deprivation, *Sleep*, 12, 13–21.

Gawron, V.J., French, J. and Funke, D. (2001) An overview of fatigue. In P.A. Hancock and P.A. Desmond (eds.) *Stress, Workload, and Fatigue* (pp. 581–595). Mahwah, NJ: Lawrence Erlbaum Associates.

Gravenstein, J.S., Cooper, J.B. and Orkin, F.K. (1990) Work and rest cycles in anaesthesia practice, *Anaesthesiology*, 72, 456–461.

Harrington, J.M. (2001) Health effects of shift work and extended hours of work. *Occupational and Environmental Medicine*, 58, 98–72.

Hawkins, F.H. (1987) *Human Factors in Flight.* Aldershot: Gower Technical Press.

Helmreich, R.L. and Merritt, A.C. (1998) *Culture at Work in Aviation and Medicine.* Aldershot: Ashgate.

Hoddes, E., Dement, W.C. and Zarcone, V. (1972) The development and use of the Stanford sleepiness scale (SSS). *Psychophysiology*, 9, 150.

Horne, J. (2006) *Sleepfaring. A Journey through the Science of Sleep.* Oxford: Oxford University Press.

Horne, J.A. (1993) Human sleep, sleep loss and behaviour: implications for the prefrontal cortex and psychiatric disorder. *British Journal of Psychiatry*, 162, 413–419.

Horne, J.A. and Ostberg, O. (1976) A self-assessment questionnaire to determine morningness-eveningness in human circadian rhythm. *International Journal of Chronobiology*, 4, 97–110.

Howard, S.K. Gaba, D.M., Rosekind, M.R. and Zarcone, V.P. (2002) The risks and implications of excessive daytime sleepiness in resident physicians. *Academic Medicine*, 77, 1019–1025.

Howarth, N.L., Triggs, T.J. and Grey, E.M. (1988) *Driver Fatigue: Concepts, Measurement and Crash Countermeasures*. Canberra, Australia: Federal Office of Road Safety.

Johns, M.W. (1991) A new method for measuring daytime sleepiness: The Epworth sleepiness scale. *Sleep*, 14, 540–545.

Johnson, L. and Naitoh, P. (1974) *The Operational Consequences of Sleep Deprivation and Sleep Deficit.* (AGARD AG-193). Neuilly-sur-Seine: NARO AGARD.

Kirsch, A.D. (1996) Report on the statistical methods employed by the U.S. FAA and its cost benefit analysis of the proposed 'flight crewmember duty period limitations, flight time limitations and rest requirements'. *Comments of the Air Transport Association of America to FAA notice 95-18, FAA docket number 28081, Appendix D*, 1–36.

Knauth, P. (1995) Speed and direction of shift rotation *Journal of Sleep Research*, 4, 41–46.

Knauth, P. (1998) Innovative worktime arrangements. *Scandinavian Journal of Work and Environmental Health. Journal of sleep research*, 4, 13–17.

Knauth, P. and Hornberger, S. (2003) Preventative and compensatory measures for shift workers. *Occupational Medicine*, 53, 109–116.

Kurumatani, N., Koda, S., Nakagiri, S., Hisashige, A., Sakai, K., Saito, Y., Aoyama, H., Dejima, M. and Moriyama, T. (1994) The effects of frequently rotating shiftwork on sleep and the family life of hospital nurses. *Ergonomics*, 37, 995–1007.

Krueger, G.P. (1989) Sustained work, fatigue, sleep loss and performance: A review of the issues. *Work and Stress*, 3, 121–141.

Kryger M.H., Roth T. and Carskadon, M. (1994) Circadian rhythms in humans: An overview. In M.H. Kryger, T. Roth, and W.C. Dement (eds.) *Principles and Practice of Sleep Medicine* (pp. 301–308), Philadelphia: W.B. Sauders Co.

Langlois, P.H., Smolensky, M.H., His, B. and Weir, F.W. (1985) Temporal patterns of reported single-vehicle car and truck accidents in Texas, USA, during 1980–1983. *Chronobiology International*, 2, 131–146.

Lockley, S. et al. (2004) Effects of reducing interns' weekly work hours on sleep and attentional failures. *New England Journal of Medicine*, 351, 1829–1848.

Luce, G.G. (1966) *Current Research on Sleep and Dreams*. Washington, DC: Public Health Service.

Maas, J. (1998) *Power Sleep.* New York: Villard.

Mackie, R.R. and Miller, J.C. (1978) *Effects of hours of service, regularity of schedules, and cargo loading on truck and bus driver fatigue* (Rep. No. 1765-F). Goleta, CA: Human Factors Research.

Mallis, M.M., Mejdal, S., Nguyen, T.T. and Dinges, D.F. (2004) Summary of the key features of seven biomathematical models of human fatigue and performance. *Aviation, Space and Environmental Medicine*, 75, A4–A14.

Marine Accident Investigation Branch (2007) *Report on the Investigation of the Grounding of FV Brothers With the Loss of Two Lives off Eilean Trodday On 1 June 2006*. Southampton, UK: Author.

Mavromatis, A. (1987) *Hypnogogia.* London: Routledge.

May, J. and Klein, P. (1987) Measuring the effects upon cognitive abilities of sleep loss during continuous operations. *British Journal of Psychiatry*, 78, 443–455.

Maycock, G. (1995) *Driver Sleepiness as a Factor in Car and HGV Accidents.* Crowthorne, England: Transportation Research Laboratories.

Miles, R. (1999 March) Progress in the understanding of shift working offshore. Paper presented at the I.ChemE/Ergonomics Society Conference. London.

Monk, T.H. (1990) The relationship of chronobiology to sleep schedule and performance demands. *Work and Stress*, 4, 227–236.

Monk T.H. and Folkard S. (1992) *Making Shiftwork Tolerable*. London: Taylor & Francis.

Muscio, B. (1921) Is a fatigue test possible? *British Journal of Psychology*, 12, 31–46.

Naval Strike and Air Warfare Center (2000) *Performance Maintenance During Continuous Flight Operations: A Guide for Flight Surgeons*. Fallon, Nevada: Author.

Nicholson, P.J. and D'Auria, D.A. (1999) Shift work, health, the working time regulations and health assessments. *Occupational Medicine*, 49, 127–137.

O'Connell, D. (1998) *Jetlag. How to Beat It*. London: Ascendant.

Petrilli, R.M., Thomas, M.J.W., Dawson, D. and Roach, G.D. (2006) *The Decision-making of Commercial Airline Crews Following an International Pattern.* Paper presented at the 7th International Australian Aviation Psychology Symposium, Sydney, November.

Phipps-Nelson, J., Redman, J.R., Dijk, D.J. and Rajaratnam, S.M. (2003) Daytime exposure to bright light, as compared to dim light, decrease sleepiness and improves psychomotor vigilance performance. *Sleep*, 26, 695–700.

Pilcher, J.J. and Hufcutt, A.I (1996) Effects of sleep deprivation on performance: A meta-analysis. *Sleep*, 19, 318–326.

Raby, M. and Lee, J.D. (2001) Fatigue and workload in the maritime industry. In P.A. Hancock and P.A. Desmond (eds.) *Stress, Workload, and Fatigue* (pp. 566–580). Mahwah, NJ: Lawrence Erlbaum Associates.

Rall, M. and Gaba, D.M. (2005) Human performance and patient safety. In R.D. Miller (ed.) *Miller's Anaesthesia* (pp. 3021–3072). Philadelphia: Elsevier.

Ramsey, C.S. and McGlohn, S.E. (1997) Zolpidem as a fatigue counter measure. *Aviation, Space, and Environmental Medicine*, 68, 926–931.

Roehrs, T., Burduvali, E., Bonahoom, A., Drake, C. and Roth, T. (2003) Ethanol and sleep loss: A 'dose' comparison of impairing effects. *Sleep*, 26, 981–985.

Rosa, R.R. (1995) Extended workshifts and excessive fatigue. *Journal of Sleep Research*, 4, 51–56.

Rosa, R.R. and Colligan, M.J. (1997) *Plain Language about Shiftwork*. Cincinnati, OH: US Department of Health and Human Services.

Rosekind, M.R., Smith, R.M., Miller, D.L., Co, E.L., Gregory, K.B., Webbon, L.L., Gander, P.H. and Lebaqcz, J.V. (1995) Alertness management: Strategic naps in operational settings. *Journal of Sleep Research*, 4, 62–66.

Samkoff, J. and Jacques, C. (1991) A review of studies concerning effects of sleep deprivation and fatigue on residents performance. *Academic Medicine*, 66, 687–693.

Sharkey, K.M., Fogg, L.F. and Eastman, C.I. (2001) Effects of melatonin administration on daytime sleep after simulated night shift work. *Journal of Sleep Research*, 10, 181–192.

Smith, C., Folkard, S. and Fuller, J. (2003) Shiftwork and working hours. In J. Quick and L. Tetrick (eds.) *Handbook of Occupational Health Psychology.* Washington: APA Books.

Smolensky, M. and Lamberg, L. (2000) *The Body Clock Guide to Better Health: How to Use your Body's Natural Clock to Fight Illness and Achieve Maximum Health*. New York: Henry Holt and Co.

Spencer, M.B., Robertson, K.A. and Folkard, S. (2006) *The Development of a Fatigue/Risk Index for Shiftworkers*. London: HSE Books.

Swerdlow, A. (2003) *Shiftwork and Breast Cancer: A Critical Review of the Epidemiological Evidence*. London: HSE Books.

Taylor, E., Briner, R.B. and Folkard, S. (1997) Models of shiftwork and health: an examination of the influence of stress on shiftwork theory. *Human Factors*, 39, 67–82.

Tucker, P., Barton, J. and Folkard, S. (1996) Comparison of eight and 12 hour shifts: impacts on health, wellbeing, and alertness during the shift. *Occupational and Environmental Medicine*, 53, 767–772.

US Nuclear Regulatory Commission (1987) *Shutdown Order Issues Because Licensed Operators Asleep While on Duty*. Information notice 87-21. Washington, DC: Office of the Nuclear Regulatory Commission.

Vernon, H.M. (1920) Industrial efficiency and fatigue. In E.L. Collis (ed.) *The Industrial Clinic* (pp. 51–74). London: John Bale and Sons.

Waterhouse, J., Reilly, T. and Atkinson, G. (1997) Jet-lag. *Lancet*, 350, 1609–1614.

Wedderburn, A. (1991) Guidelines for shiftworkers. *Bulletin of European Shiftwork Topics, No. 3*. Dublin: European Foundation for the improvement of living and working conditions.

Wesensten, N., Belenky, G., Kautz, M.A., Thorne, D.R., Reichardt, R.M. and Balkin, T.J. (2002) Maintaining alertness and performance during sleep deprivation: modafinil versus caffeine. *Psychopharmacology*, 159, 238–247.

Whitmore, J. and Fisher, S. (1996) Speech during sustained operations. *Speech Communication*, 20, 55–70.

Wu, A.W., Folkman, S. McPhee, S.J. and Lo, B. (1991) Do house officers learn from their mistakes? *Journal of the American Medical Association*, 26, 2089–2094.

Chapter 9

Identifying Non-Technical Skills

Introduction

In order to train or assess non-technical skills, the specific skills for a given occupation and work setting must be determined. The question then, is, 'Where do we start?' In aviation, the basic skill set, usually called CRM skills, is fairly well established, although this may be customised for particular aircraft or types of operation. There have been research programmes (Wiener et al., 1993), the larger airlines have conducted their own analyses (Flin and Martin, 2001) and the regulators and other bodies have produced recommended skill lists and training syllabi (for example, the CAA (2006) have produced guidance for CRM training, see Chapter 10). For other occupations, there may not be non-technical skills lists and a number of methods are described here that can be used to identify the relevant non-technical skills, such as interviews, questionnaires, observations and reviews of accident/incident reports. These diagnostic tools can be used individually but results can be enhanced by using more than one data collection technique (e.g. triangulation) (Cohen et al. 2007), for example, combining a questionnaire with interviews and observations. Given existing expertise, advice is taken from the aviation sector about how to go about determining a non-technical skills set (Seamster and Keampf, 2001). There are also other sources of relevant advice and techniques from the occupational literature on devising competence frameworks (Lucia and Lepsinger, 1999; Whiddett, 2003) and conducting job and task analyses (Wilson and Corlett, 2005).

This chapter specifically addresses the main diagnostic tools used to gather the data to identify the skills that underpin efficient and safe performance. No standard method of task analysis exists that is appropriate for all situations and domains, so the method to be used depends upon the needs of the analysis (Zachary et al., 1998). In situations where tasks have a high cognitive component, a set of techniques that have been designed to specifically target the cognitive aspects of tasks (i.e. cognitive task analysis (CTA)) have become popular (Crandall et al., 2006; Schraagen et al., 2000; Strater, 2005). Cognitive aspects of tasks draw on the worker's knowledge base, enabling experts to make inferences to fill in pieces of the situation or link it to their knowledge base to select appropriate actions, or other types of processing (e.g. interpretation, goal development, judgement or prediction). This is not to say that other, more traditional methods of task analysis, such as hierarchical task analysis (Annett, 2004), or task decomposition (Kirwan and Ainsworth, 1992), and CTA are mutually exclusive. Indeed, much can be gained when these two methods are used in parallel, such that both the observable and cognitive aspects of a task can be analysed.

As more non-technical skill taxonomies are developed, it is apparent that many of the basic skill categories are generic, especially across higher-risk occupations (for example, decision-making, situation awareness, communication, team co-ordination, stress/fatigue management) (see Chapter 11). These broad skill categories were outlined in the previous chapters. However, when a given occupation is examined, the elements and specific behaviours for each category look very different and clearly vary from one technical setting to another. This is why it is generally inadvisable to use a non-technical skills taxonomy or behavioural marker system devised for one domain (e.g. aviation) in a different work setting (e.g. health care) without clearly examining the similarities and differences within the domain, the task and the proposed use of the taxonomy, behavioural marker system or non-technical skills training programme.

In essence, a two-stage process is employed to develop a taxonomy of skills:

1. identify the skills and related behaviours deemed to influence safe and efficient performance, and
2. refine the resulting list and to organise it into a concise, hierarchical structure or taxonomy (see Chapter 11).

The level of detail and scope of the analysis will depend on the purpose for which the taxonomy or training programme is being developed. A set of behaviours identified for a research tool could be more complex and comprehensive than one being developed for practitioners to use as the basis for training (as described in Chapter 10) or assessment (Chapter 11). The focus here, as in previous chapters, is again on individual skills rather than on team skills, although the individuals will normally be working in a team setting.

Obviously the first place to look for information is in the published research literature for the job in question. There are now thousands of studies examining behaviour in a range of different workplaces, and a wealth of information exists not only about typical behaviours but noting which of these contribute to safe or unsafe performance (e.g. Barling and Frone, 2004; Glendon et al., 2006). Nowadays searching the published literature is a relatively easy task due to the advent of electronic databases and powerful search engines (e.g. Google: www.google.com, Google Scholar: http://scholar.google.com, Web of Science: http://wos.mimas.ac.uk). Organisations may also have documentary analyses that cover non-technical skills for a particular occupation including job assessments, task analyses, competence frameworks, training programmes, assessment and appraisal systems. If sufficient information cannot be found in these sources, then the data collection methods described below can be used.

A number of different techniques used to identify non-technical skills are listed in Table 9.1 and discussed in the following sections. Most of these have been used for safety-critical occupations in aviation, energy sector, rail, marine, military and acute medicine. These techniques can be grouped into three different approaches: analysis of events from accident/incident reports, questioning and observation.

Table 9.1 Techniques to identify non-technical skills

Type	Techniques
Event-based analyses: examining accident or near-miss reports to identify patterns of behaviours	• accident/near-miss analysis • confidential reporting systems analysis (including reports of safety concerns, as well as actual events)
Questioning techniques: soliciting information directly from role-holder/s from the job under investigation	• interview (structured, unstructured, semi-structured) • focus groups • questionnaires and surveys
Observational techniques: watching individuals or teams carrying out one or more tasks	• direct • participant • remote (e.g. from video recording)

Event-based analyses

Event-based analyses include both accident/near-miss analysis and confidential reporting systems. Each of these two sources of data will be described.

Accident/near-miss analysis

Most organisations have systems for recording safety incidents, accidents and sometimes also near-miss events. These can be analysed to provide data from actual events and can include an examination of the human factors, or non-technical skills, inherent in the accident or near-miss (see Box 9.1). Reports are submitted by employees to the organisation and the more serious accidents will normally have to also be reported to a regulatory authority. Reports of incidents and accidents are usually coded for causal factors by the reporter or by a supervisor or by a specialist in the safety department. In some work settings, the human factors coding of accidents has tended to be minimal. Gordon et al. (2005) argued that many accident reporting systems used by the offshore oil industry in the UK lacked a firm theoretical framework for identifying human factor, or non-technical skill, causes of accidents. In sectors such as health care this recently has begun to change with much more effort devoted to gathering, recording and analysing adverse events (Holden and Karsh, 2007; Johnson, 2006). Accident investigations are conducted for more severe incidents to establish what happened, and to prevent a similar event happening again. Although only one diagnostic source, the analysis of accident data, is essential for improving workplace safety (Dismukes et al., 2007; Kayten, 1993; Wiegmann and Shappell, 2003). A post-incident inquiry begins with a negative outcome and considers how and when the defences built into the system failed (Hollnagel, 2004).

An accident investigation method based on robust human factors models allows a broader interpretation of accident records, thus potentially reducing the likelihood of future accidents (see Stanton et al. (2005) for tools focusing on human error). Examples of accident analysis tools that include both human and organisational factors are:

1. Safety Through Organisational Learning (SOL) (Fahlbruch and Wilpert, 1997)
2. Human Factors Accident Classification System (HFACS) (Shappell et al., 2007)
3. Technique for Retrospective Analysis of Cognitive Errors (TRACEr) (Shorrock and Kirwan, 2002)
4. TapRooT (Paradies et al., 1996)
5. Tripod (BETA and DELTA) (Hudson et al., 1994)
6. Human Factors Investigation Tool (HFIT) (Gordon et al., 2005).

Box 9.1 Example of accident analysis in aviation

Accident data were used by Orasanu and Fischer (1997) to examine non-technical skills in a study identifying decision-making strategies in the cockpit. Analyses of aircraft accidents were conducted in the US by the National Transportation Safety Board (NTSB). These analyses (based on crew conversations that are recorded by the cockpit voice recorder (from the 'black box'), physical evidence, aircraft systems data and interviews with survivors) provide contextual factors that contributed to the pilots' decision-making, including sources of difficulty, types and sources of error, and decision-making strategies. This information was also combined with data from actual observations in a high-fidelity simulator, along with data from the Aviation Safety Reporting System (ASRS), which is a confidential reporting system maintained by the National Aeronautics and Space Administration (NASA). The use of these three data sources allowed decision situations and decision strategies to be identified, and revealed that unsafe flight conditions were related to failures in pilots' non-technical skills rather than lack of technical knowledge, flying ability or aircraft malfunction.

Accident analyses are often constrained, as the accident analysis system may be vulnerable to underreporting, have incomplete recordings and may present only part of the overall picture of the event (Hollnagel, 2004; Stoop, 1997). To illustrate, in an examination of US Navy diving accident reports, the largest proportion (70%) of the diving mishaps were attributed to unknown causes; only 23% were attributed to human factors (O'Connor et al., 2007), a rate far below the 80% generally attributed to this source in high-reliability industries.

Confidential reporting systems

Confidential reporting systems gather data from individuals about their mistakes made in the workplace, or safety concerns that they feel unable to report through normal channels. These differ from standard incident reporting systems in that the reporter enters their name but there is a guarantee (under most conditions) that their identities will not be revealed, especially to their employers or to regulatory authorities. These systems are sometimes administered by an independent agency rather than the employer. The results of the analysis of these data can then be fed back to employers and employees to lead to safety improvements in the workplace. This can be another source of information on non-technical skills required for a given occupation.

Reason (1997) comments that it is not an easy task to persuade people to file critical incident and near-miss reports, especially if this requires divulging their own errors. Disincentives to participating in a reporting system include extra work, scepticism (that management will not act upon the information gathered in the system), and a lack of trust and fear of reprisals. However, confidential reporting systems can and have been successfully instituted in organisations, such as the NASA Aviation Safety Reporting System (ASRS), British Airways Safety Information System (BASIS), Confidential Human Factors Incident Reporting Programme (CHIRP), and the UK railway's Confidential Incident Reporting and Analysis System (CIRAS).

Five factors are important in the success of a confidential reporting system (Reason, 1997): (i) indemnity against disciplinary proceedings (as far as possible); (ii) confidentiality or de-identification; (iii) separation of the agency or department collecting and analysing the reports from those bodies with the authority to institute disciplinary proceedings and impose sanctions; (iv) rapid, useful, accessible and intelligible feedback to the reporting community; (v) ease of making the report.

The purpose of confidential reporting systems is to provide valid feedback on the factors that promote errors and incidents, and to achieve a useful outcome in terms of providing meaningful information to the domain. Following a series of Royal Navy diving accidents, the Royal Navy Superintendent of Diving instigated a confidential reporting system. Feedback is provided to divers on any recent diving accidents and near-misses by way of a quarterly newsletter (O'Connor, 2007). However, reports must be confidential to protect reporters and their colleagues from disciplinary action as much as possible. Reports also need to be carefully analysed to identify the key human factor, or non-technical skills, issues associated with the incident or near-miss. Hale et al. (1998) describe the importance to organisations of learning from accidents, in that the initial response after an incident is to look for someone to blame, next to want to understand the incidents, and finally to learn from the incident so that it can be avoided in the future.

Summary of event-based analyses

Event-based analyses, integrating the non-technical skills aspects of performance, are increasingly being introduced in industries. An analysis of accident data, which includes an analysis of non-technical skills, helps to improve workplace safety by identifying any recurring shortfalls in non-technical skills performance, which can then be used to identify training needs. Moreover, these data can be used in the design and development of safe systems of work to minimise the recurrence of subsequent incidents.

Questioning techniques

Questioning techniques can be subdivided into individual interview methods, focus groups and questionnaires, often administered as organisational surveys. In all cases, the job-holders (usually experts) are asked a number of questions about their work. This may be to solicit their knowledge of the job, key tasks and their relative importance, and the cognitive, social and other skills needed to complement their technical expertise. In addition, interviews and surveys can be used to ask about attitudes, safety concerns, typical workplace behaviours of self and others and aspects of organisational life that influence work behaviour.

Interview methods

Interviews are frequently used to examine system usability, reactions, attitudes and job analysis (Stanton et al., 2005). Interviews can be structured, semi-structured or unstructured (Gillham, 2005; Kvale, 2004), as described below:

- *Structured interviews* use pre-determined questions, either open or closed, dependent on the data requirements to elicit information from the interviewee. Benefits of structured interviews include ease of administration and collection of focused data. A further benefit of a structured interview is that it facilitates comparison of responses from different respondents to specific questions. However, this type of interview demands that the interviewer has relevant background knowledge of the topic under examination, and requires the interviewer to access static knowledge that the interviewee must verbalise in some form.
- *Unstructured interviews* comprise open-ended questions, encouraging substantial discussion and interaction between the interviewee and interviewer, and does not pre-suppose any previous knowledge of the topic by the interviewer (Heiman, 1995). Unstructured interviews provide a useful introduction to the domain for the interviewer, as well as increasing familiarity with the appropriate terminology.
- *Semi-structured interviews*, on the other hand, generally use pre-determined, but open-ended, questions, which can be adapted to individual requirements, and allow a certain degree of flexibility in exploring a wide range of issues that may arise during the interview session (Breakwell et al., 2006). The questions are designed to obtain information about how an individual performs the sub-tasks of the main task under investigation, thus is very useful for collecting task knowledge and information about certain types of skills, such as procedural skills. Benefits of a semi-structured interviews include the efficient collection of declarative knowledge, further exploration of issues raised by the interviewee, and a structure for a set of consistent questions such that the interviewer can react to the information given and seek clarification if required. Furthermore, unanticipated topics can be discussed (Redding and Seamster, 1994), and the interviewer requires relatively little prior knowledge of the domain beyond necessary terminology. One of the main disadvantages

of semi-structured interviews is that, similar to structured interviews, the interviewer generally requires some understanding of the topic area to know when to follow up particular points.

Each interview type has its advantages and disadvantages (see Table 9.2), and the selection of interview type depends on the purpose of the data being gathered.

Table 9.2 Advantages and disadvantages of types of interview

Type of interview	Advantages	Disadvantages
Structured	• ease of administration • focused data • ability to compare responses	• requirement for relevant background knowledge by interviewer • lack of flexibility to follow up topics raised
Semi-structured	• flexibility to follow up topics raised	• some previous relevant background knowledge required
Unstructured	• flexibility • no previous relevant background knowledge required • minimal preparation	• digress from purpose of interview

Semi-structured interviewing can also include techniques under the heading of diagramming. Diagramming techniques, e.g. laddered grid (Burton and Shadbolt, 1987), refer to the construction of a diagram representing relationships between particular tasks or job concepts, within a domain involving the relationship between elements, or subsystems. For example, experts can be provided with cards and asked to arrange the cards to represent the relationships between them (Hoffman et al., 1995). In addition, acquired data can be used to validate results from scaling and rating tasks, and can also be used to direct a structured interview. Disadvantages are that it relies on key points from the task or job being known beforehand, and that it is more suited to rule-based or procedural domains.

Interviewing technique example: critical decision method

Cognitive interviews refer to questions and probes being used to elicit information about mental processes, for example critical decision method (CD method) (Klein et al., 1989), a knowledge elicitation strategy, based on Flanagan's critical incident technique (1954). It can be used to elicit information about the complexities involved in making decisions, i.e. use of resources, teamwork, communication and stress management, as well as technical understanding and application of knowledge. How this is affected by external and internal factors such as organisational and time pressures can also be incorporated into the analysis. The CD method is particularly useful for examining the cognitive component of proficient performance in naturalistic environments. However, it may also reveal aspects of other required skills, such as stress management, leadership or team co-ordination.

The CD method involves experienced personnel describing critical incidents that they remember, and targets key judgements and decisions during non-routine incidents, where expert skills made a difference and a novice might have faltered (Hoffman et al., 1998; Crandall et al., 2006). It consists of a number of phases summarised below, which follow on from an initial period as the interviewer gathers demographic information and establishes rapport with the interviewee.

Sweep 1: Selecting an incident: candidate incidents are identified and an appropriate incident is selected for deepening. The interviewee provides an account of the story from beginning to end.

Sweep 2: During this phase, a clear, refined and verified overview of the incident is obtained. Key events and segments are identified, which then act as a framework for the remainder of the interview. Greater detail of the incident is recalled by the interviewee and the initial brief description of the incident is expanded, while creating a timeline.

Sweep 3: Deepening: The interviewer attempts to identify cognitive aspects, such as the interviewee's perceptions, expectations, goals, judgements, confusions and uncertainties, as well as concerns. A deeper understanding of the event is thus built up.

Sweep 4: 'What if' queries: The interviewer's insight into the interviewee's experience, skill and knowledge is heightened by addressing hypotheticals around the incident. This might be about the incident as a whole or parts of the incident.

Non-routine incidents are usually targeted as these are considered to provide the richest source of data from skilled personnel. By asking participants to repeatedly describe an event, the interviewer is able to identify a timeline for the incident, as well as identifying situational cues and decision points. The decision requirements of the task can then be identified, and the decision-making process can be modelled. Decision requirements include the decisions made; the assessments made by decision-makers; the cues attended to; factors affecting how cues are interpreted; the cognitive processes that decision-makers invoke to make assessments; and, the differences between experienced and less experienced decision-makers (Miller, 2001; Seamster et al., 1997).

By employing focused probe questions, the CD method allows more varied, and specific, information about cognitive tasks to be acquired than verbal reports. The interviewee is asked to think back to a previous experience or incident and, while being guided by the interviewer, provides information within the context of the specific incident. For example, information can be gleaned about decision strategies, perceptual discriminations, pattern recognition, expectancies, cues and errors (Seamster et al., 1997). During the description of the incident, the interviewee is encouraged to consider what cues were influential when changing an assessment of a situation, or when selecting a particular course of action, what other options could have been possible, but why they chose the one they did. The CD method is particularly effective at identifying distinctions between experts and novices and a further benefit is that unplanned themes generated during discussion can emerge from the data.

However, the technique does have some limitations. Klein et al. (1986) suggest that it is preferable to use two interviewers, rather than one, to avoid breakdowns in attention and suffer the consequent loss or misunderstanding of information, but this is particularly resource intensive, and may also be more intimidating for the interviewee. Taping the interview, with the interviewee's permission, is a useful compensatory technique. Additional disadvantages include that in some instances there may be no real experts, or that parts of the task are distributed over time, space or personnel. Finally, interviewees may be unable to generate incidents that can be probed, for example where events become blurred due to high workload or pressure, as may arise during combat conditions (Crandall et al., 2006).

Box 9.2 Examples of interviewing methods in offshore oil and gas industry

Decision-making by offshore installation managers (OIMs) in the offshore oil and gas industry was researched by Flin et al. (1996). Interviews were conducted with 16 OIMs who had managed an offshore emergency to examine the factors influencing their decision-making. Adapting the critical decision method, the managers were asked to talk through the event with probes at critical points requiring decisions. In a pilot study, a verbal protocol analysis was used (see below). This was based on the 'withheld information' approach (Marshall et al., 1981) with a paper-based scenario of an emergency. Experienced OIMs were asked to think aloud and to describe their thoughts and anticipated decisions. In this case, the researchers were interested not only in the expressed thoughts about the situation but also in which of the missing information the OIMs sought to make decisions regarding the management of an emergency.

The CD method has been used in medicine (Crandall and Getchell-Reiter, 1993; Fletcher et al., 2004), fire ground command (Klein et al., 1989; Burke and Hendry, 1997), naval command and control (Kaempf et al., 1992; Thordsen et al., 1992), as well as programmers and systems analysts (Crandall and Klein, 1988). To study decision-making during emergencies in the offshore oil industry, two interview studies based on the CD method were carried out based on real and simulated events (see Box 9.2). This method provides a wealth of qualitative data that can be analysed in terms of decision points, decision strategies and time. However, it must not be overlooked that the CD method relies upon subject introspection, retrospective recall and the ability to verbally express thoughts and experiences. In addition, the coding employed for analysis must be reliable and valid.

The CD method, described above, was combined with a cued recall technique to develop a human factors review interview protocol (HFRIP) (Omodei et al., 2004; 2005) to identify the human factors that operate to influence wildland firefighters' behaviours. The interview protocol was based on classifications based on the human factors and analysis classification system (HFACS) (Wiegeman and Shappell, 2001) and the technique for the retrospective and predictive analysis of cognitive errors (TRACEr) (Shorrock and Kirwan, 2002), and was designed to maximise the quality and quantity of information obtained in post-incident interviews. The interviewee was encouraged to maintain an 'insider' perspective, i.e. to discover what is in and on someone's mind. The interviewee described how they experienced the situation. The interviewer adopts a stance of being courteous, attentive, with interested curiosity. The interviewer then proceeds by using leading questions to facilitate recall, i.e. using open-ended questions, encouraging statements and paraphrasing. The interviews are then analysed to identify the human factors related to the individual (predisposing physiological states, predisposing mental states, perception, memory, decision-making, communication and action execution), as well as to the context including small-group factors (the effects on the individual of their immediate colleagues) and large-group factors (the effects on the individual of the wider organisation or community).

Psychological scaling methods

Psychological scaling methods can also be used to gather information about how tasks are performed in the workplace, and include sorting and rating methods whereby individuals sort, rate or rank concepts, principles, rules or problems, albeit that these methods are not specific to analysing cognitive skills. A diagram representing the inter-relationships of concepts can be drawn from statistical analysis of the data, indicating how aspects are grouped. The example in Figure 9.1 illustrates how different types of decisions (shown as 'd1' etc) made by members of a nuclear emergency response organisation can be clustered. The groupings of the decisions are shown in the figure, and the dimensions are identified, which can be used to explain the groupings. Concurrence about the identification of the dimensions was obtained from four subject matter experts (Crichton et al., 2005).

Scaling methods include, as previously stated, sorting, rating (including repertory grids), and ranking tasks (Rugg and McGeorge, 1997). These methods are particularly useful for testing predictions about expertise or subjective judgement, for example identifying the key dimensions influencing decision-making, but are also appropriate for use in the conceptualisation and refinement phases of knowledge acquisition (Hoffman et al., 1995; McGraw and Harbison-Briggs, 1989). Use of such techniques allows the structure of a task or knowledge domain to be determined, in particular how key concepts are organised in memory. Examples of the use of scaling methods in aviation and nuclear power production are described in Box 9.3.

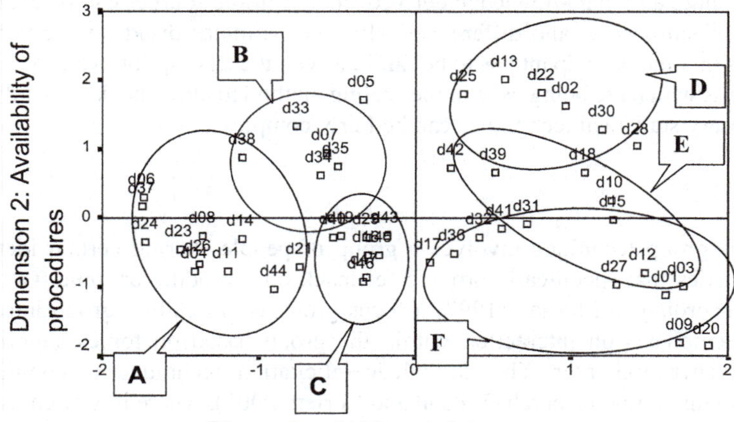

Legend: A = Make up of teams; B = Access and deployment; C = Personnel and resources; D = levels of contamination and temperature; E = Hazardous environmental issues; F = Team tasks

Figure 9.1 Example of analysis of card sorting data (Crichton et al., 2005)

Box 9.3 Examples of psychological scaling methods in aviation and in nuclear power production

Fischer et al. (1995) used free and directed card-sorting tasks, analysed using hierarchical clustering and multidimensional scaling, to investigate situational aspects used by expert pilots in making decisions and to validate aspects of the aviation decision process model (see Chapter 2). The cards contained a brief description of a scenario, and participants had to infer the appropriate decision for each incident, and then group similar decisions together. Differences relating to salient factors (e.g. risk, time, situational complexity) could thus be identified across the three groups of participants (captains, flight engineers and first officers).

More recently, decision-making by on-scene incident commanders in a nuclear emergency response organisation (ERO) has been studied using card-sorting tasks (Crichton et al., 2005). The card-sorting task in this study used types of decisions that had been obtained previously from interviews conducted with on-scene incident commanders in a nuclear ERO – examples of the types of decisions shown on the card are: *'Deployment and make up of teams as required'*, *'The procurement and use of station equipment needed to accomplish a task'*. The results of the card-sorting task identified that decision-making was influenced by four main factors: availability of procedures; uncertainty; typicality of the decision; and advice from others.

Advantages of such methods include that participants receive feedback during the knowledge elicitation session, that the knowledge engineer can extend their domain understanding, and that knowledge derived from multiple sources can be compared in terms of similarities and differences. However, a major disadvantage with this method is that the participant has to be familiar with the descriptions (e.g. decisions) on the cards which are being used in the scaling method to elicit the data. In addition, the necessary statistical techniques can be fairly complex.

Focus groups

The focus group technique involves a group of people sharing certain identified characteristics that specifically provide feedback on a product or topic (Coolican, 2004). According to Morgan (1997) a focus group is basically a group interview, with the emphasis on interaction within the group based on topics supplied by the researcher/moderator. This knowledge-elicitation technique is common in qualitative marketing research (Wilson and Corlett, 2005), where it is often used for idea generation.

Box 9.4 Example of focus groups in medicine

A project to develop behavioural markers for teamwork in neonatal resuscitation (Thomas et al., 2004) used focus groups of neonatal care providers combined with surveys and observation of resuscitations. Focus groups were used to gather information from the providers' perspective, i.e. their impressions of working together and what influenced their work.

A total of seven focus groups, each with three to seven participants, were conducted using a series of open-ended questions. For five of the focus groups, the participants consisted of different providers, e.g. transport nurses, staff nurses, residents, fellows, attending physicians, with two groups comprising multiple providers. Transcripts from the focus groups were then analysed using qualitative methods to identify common themes, namely providers, workplace factors and group influences. Descriptive elements were then associated with the themes:

1. Providers: provider personalities, reputations, and egos
2. Workplace factors: staffing, and organisation of care processes
3. Group influences: communication styles, relationships, team functioning.

The focus group data emphasised the unique complexity of this particular work environment and the multiplicity of factors that influence how neonatal providers work together. When the focus group results were combined with the results of the survey and observations, 10 behavioural markers for neonatal resuscitation were identified: information-sharing, inquiry, assertion, intentions shared, teaching, evaluation of plans, workload management, vigilance/environmental awareness, teamwork overall and leadership.

Focus groups can range from being the principal source of data, a supplementary source of data (often combined with surveys), or integrated into multimethod studies that combine two or more data-gathering methods, such as observation and individual interviews. An example of the use of focus groups, combined with surveys and observation, is described in Box 9.4. One of the major strengths of the technique is that concentrated amounts of data on the exact topic of interest can be acquired, especially on topics that may not be directly observable. On the other hand, this very factor can lead to uncertainty about the accuracy of the data obtained, as the researcher/moderator also guides the group and may influence the data by their very visibility. In addition, the reliance on interaction within the group can also have a simultaneously positive and negative impact on the method (Morgan and Krueger, 1993). Participants may compare experiences and opinions, providing a valuable source of insights into behaviours and motivations, but the group interaction may also influence individual contributions through, for example, a tendency to conformity.

Questionnaires

Questionnaires are frequently used to collect information on respondents' attitudes and beliefs regarding behaviour in the workplace and can also be used to gather self-reports about the respondent's own or others' behaviour at work. They are an economical and efficient method of gathering this kind of data, especially from large numbers of personnel; however, there are limitations in the type and quality of information that can be collected. They can either be specifically designed or there are standard questionnaires available that measure particular aspects of work or working life. Useful sources for designing questionnaires and conducting surveys are Fink (2003) or Oppenheim (1992).

Standard questionnaires can be employed in the workplace to examine particular non-technical skills, such as leadership (Multifactor Leadership Questionnaire, Bass, 1985), team functioning (Team Climate Inventory, Anderson and West, 1994), team roles (Belbin, 1993), cognitive failures (Wallace and Chen, 2005) or situation awareness (Sneddon et al., under review). There is no one specific questionnaire available to capture information on all the non-technical skills or related behaviours. Questionnaires have been developed to measure many aspects of working life, such as organisational commitment (Balfour et al., 1996; Mowday et al., 1979) or job satisfaction (Warr et al., 1979). The instruments that have most typically been used in the identification of non-technical skills tend to fall into the category of attitude questionnaires.

Attitude questionnaires

An 'attitude' is a generic term to include beliefs, opinions, values and preferences (Schuman and Presser, 1996). Oppenheim (1992) defined it as a 'state of readiness, a tendency to respond in a certain manner when confronted with certain stimuli' (p174), that is reinforced by beliefs, feelings, and which can lead to specific behaviours or action tendencies. An attitude statement, then, is 'a single sentence that expresses a point of view, a belief, a preference, a judgement, an emotional feeling, a position for or against something.' (p174). A well-designed questionnaire should have carefully constructed attitude statements that result in respondents making full use of the range of response categories (e.g. from strongly agree through to strongly disagree).

The reason attitude surveys are of interest to the identification of non-technical skills is that our attitudes influence our behaviour, although the converse is also true, i.e. that our attitudes are shaped by the way we see ourselves behaving in a given situation. Questionnaires have been developed to measure pilots' attitudes to non-technical skills, such as communication, stress management and crew co-ordination. Gregorich et al. (1990) describe how instilling desired attitudes in flight crew members is a focus of training to improve the quality of crew co-ordination. The best known of these questionnaires is the FMAQ (flight management attitudes questionnaire (see Helmreich and Merritt, 1998)). This has been used to assess whether training in crew resource management (CRM) has produced any attitude change (see Chapter 10). There are also attitude questionnaires designed to measure safety climate in industry (Flin et al., 2000; HSE, 2001) and in health care (Flin et al., 2006; Madsen et al., 2007). These often ask questions about non-technical skills, such as risk-taking behaviours, willingness to speak up about safety or to challenge a colleague's unsafe act, and support of colleagues. Attitude questionnaires are more appropriate for measuring attitudes to team and leadership behaviours – the social skills – as opposed to the cognitive skills of decision-making and situation awareness.

Summary of questioning techniques

These techniques are used to gather the information from individuals and teams that can then be used to identify the key non-technical skills associated with different roles and within different settings. Data are gathered using these techniques either individually or, preferably, in combination, and are then further analysed to help to develop a taxonomy of non-technical skills.

Observation

Observations can be used to collect information about a workplace setting or a problem. An expert may be observed to record how a task is performed or a problem solved. Therefore observations can be used to identify communication skills, study leadership behaviours, to chart the information required for a task, and to verify an expert's description of the task. Different methods of conducting an observational study exist, namely: direct observation, participant observation and remote (e.g. from videotape) (Stanton et al., 2005). An example of the use of observation in nuclear power production is shown in Box 9.5.

1. Direct observation refers to the situation where normally an observer, as unobtrusively as possible, watches how tasks are performed to record people's behaviour in their workplace.
2. Participant observation means that the participant observer takes part in the events being observed, but this is less frequently used by researchers, who tend to lack the required technical skills.

Box 9.5 Example of observation methods in nuclear power production

Much of previous research into behaviour of staff on nuclear power installations has focused on control room personnel. For example, Roth et al. (1994) used observation to investigate operator performance in cognitively demanding emergencies during simulator-based scenarios. The aim was to identify situations where higher-level cognitive activities (e.g. situation assessment and response planning) were used to deal with situations not fully handled by procedures, and to document the behaviours employed to handle the situations.

Performances by 22 crews (each with five members), using simulators at two separate sites, were videotaped and analysed, as well as partial transcripts made of their dialogue. Types of observable behaviours were developed from a model of higher-level cognitive activity and provided the framework for analysis. Only extra-procedural activities (i.e. those not directed by the specific step in the procedures) were analysed. The results showed that situation assessment and response planning were important for successful performance, even when emergency operating procedures (EOPs) were used. Performance differences were found between the crews, based on situation assessment and team interaction skills. One limitation in the study is that only observable behaviours were analysed, and higher-level cognitive activities were inferred, but this raises problems with non-observable activities and observer bias. The authors concluded that, as was found in the offshore environment, an overreliance on procedures may neglect the necessity for more immediate decision-making based on recognition built up through experience.

Observation recording techniques include *in vivo* observation, videotaping and audiotaping (Seamster et al., 1997). When recording observations *in vivo*, that is, watching the individual(s) directly while they are performing their tasks, these can be recorded in note form, on a paper rating scale or increasingly using a hand-held computer. In higher-risk work settings, where the observer could be unsafe or interfere with the tasks, cameras may be installed to provide video recordings (i.e. remote observation) – see Box 9.6. There are a considerable number of practical and legal issues involved in routinely audio or video recording workers' behaviour for research purposes (Mackenzie and Xiao, 2007).

Observations can be made in an unstructured fashion, which is that the researcher makes notes of what is happening or key tasks or points of particular interest. More typically, the records are made in a structured format using some kind of rating scale or coding system, e.g. an observational checklist, to capture the tasks under investigation. Obviously for structured observations, the researchers need to have some understanding of the behaviours they expect to see in order to devise or select a behaviour rating tool. Generic tools for rating observed behaviours have been developed and behaviour rating systems have been designed for particular occupations, such as NOTECHS for pilots or ANTS for anaesthetists – these are discussed in Chapter 11, which deals with the assessment of non-technical skills. However, a downside of this is that the observer may be less sensitive to what is going on (Crandall et al., 2006).

Box 9.6 Example of observation methods in medicine

One of the best examples of behaviour observational research is at the R. Cowley Adams Shock Trauma Center at the University of Maryland in Baltimore. Since the early 1990s there have been cameras installed in parts of this unit allowing researchers to examine the videotapes in order to analyse the behaviour of teams diagnosing and treating trauma patients. Their research has shown how failures in communication and decision-making can hamper the execution of critical tasks, such as an emergency intubation (Mackenzie et al., 1994; Mackenzie and Xiao, 2007).

Observation and analysis of behaviours in operating theatre simulators have been employed by researchers studying anaesthetists (Botney et al., 1993; Gaba et al., 1994) to identify factors influencing performance, especially in critical situations. De Keyser and Nyssen (2001) conducted observations in a simulator to assist the development of a model of problem-solving strategies identified in novice and more experienced trainee anaesthetists. Their data were collected from a total of 200 observation hours, as well as verbalisations and interviews with different members of the operating team. More recently, psychologists and surgeons have been conducting observations of behaviour in operating theatres during cases to identify factors that complicate cognitive and collaborative performance (Carthey, 2003; Catchpole et al., 2006; Healey et al., 2004; Roth et al., 2004).

A typical procedure for conducting observations includes eight steps (Stanton et al., 2005):

- Step 1: Define the objective of the analysis.
- Step 2: Define the scenario(s) to be observed.
- Step 3: Observation plan (i.e. what is to be observed, what is being observed, and how it is to be observed).
- Step 4: Pilot observation to assess any problems with collecting the data.
- Step 5: Conduct observation.
- Step 6: Analyse data (e.g. frequency of tasks, verbal interactions, sequence of tasks).
- Step 7: Further analysis using additional techniques as appropriate.
- Step 8: Participant feedback.

A disadvantage of observation is that the observation itself may not be feasible, in that the observation may be risky to the observers, or the observers may impede the tasks being undertaken. In addition, the events being captured may be atypical, and the observers may need to be highly skilled in that setting to adequately capture the activities (Crandall et al., 2006). Nevertheless, observations can provide a valuable insight into the work demands, the strategies used by personnel to cope with the task, and communication and co-ordination requirements.

Verbal report methods

Verbal protocol analysis is a method commonly used to analyse an expert actually solving domain problems (Rasmussen, 1985; Shadbolt, 2005; Simon, 1955) and this can accompany observations of the task being performed (as described in Box 9.7). The expert's performance is recorded, either by video or audio, or by written notes, from which protocols can be developed, and meaningful rules can be extracted. The self-report technique is when an expert describes aloud their own actions, whereas shadowing refers to another expert providing a running commentary of what the expert is doing.

Disadvantages with this technique is that the very act of thinking aloud – self-report – may cause distortions in performance, and that subjects may be unable to introspect about their own performance (Ericsson and Smith, 1991). Also, interviewees may not be able to explicitly describe exactly why they did things (Nisbett and Wilson, 1977). Similarly, retrospective analysis of an expert performance by another expert may be based purely on observed behaviours, with unobservable behaviours being inferred. Nevertheless, verbal protocols are common laboratory-based methods to study expert/novice differences (Chi et al., 1988).

Team communication analysis involves the analysis of verbal communication between team members from observations or recordings (Seamster et al., 1997). Although useful for identifying the content and quality of communication, it relies on inference, particularly of individual skills, to understand team performance. As its name implies, this technique is more suited to a team, although individual skills for working in a team may be revealed.

Box 9.7 Studying conversations in aviation

In the aviation industry, pilots are accustomed to having their workplace conversations stored on the cockpit voice recorders, i.e. the black box. These are tape recorded on a continuous loop storing the last 30 minutes of flight deck conversation, as well as communication on the radio to air traffic control or other ground centres. Governance of these tapes is very strictly controlled and they are normally only accessed after an incident. The analysis of these tape recordings have been very revealing of pilots' non-technical skills, especially when coping with unanticipated conditions (Orasanu and Fischer, 1997; see also Box 9.1).

Computer-based observational recording

Sometimes observations are recorded and archived on computers, especially for individuals performing high-risk work, either in real-life settings or high-fidelity simulators. There are special software packages for recording and coding observations; these offer frame marking, coding systems, etc. Some are available commercially, such as The Observer (Noldus The Observer XT 6.1, 2006) and others are under development, such as RATE (Guerlain et al., 2005), and a mobile evaluation system for distributed team training (MESDTT) (Hiemstra et al., 2003).

The RATE tool (a multi-track, synchronised, digital audio-visual recording system) has been developed for use in the hospital operating theatre to study team behaviour. The system (software and hardware) digitally records, scores, annotates and analyses performance for a team of up to eight people. The system can be used to support surgical team debriefing and training, especially surgical team situational awareness, and team communication and co-ordination strategies. It is also proposed that the system could be adapted to collect and analyse team behaviour in other domains, including distributed teams (Guerlain et al., 2005).

The MESDTT is a measurement tool that is installed on a hand-held computer. The quality of observations is improved due to the use of a standardised format for scoring behaviours. An observer can quickly score targeted behaviours within a team and between distributed teams. Team performance dimensions include communication, information exchange, initiative/leadership and supporting behaviour (Smith-Jentsch et al., 1998). Additional behaviours for distributed teams include pre-task preparation and convergence. The system includes digital links, drop-down windows, checklists, and also incorporates a debrief organisation screen that can be used by an observer to prepare an after-action review to facilitate team learning.

Summary of observation

Conducting observations of individuals or teams performing their tasks helps to gather data about the physical and verbal aspects of performance, such as errors that may be made, communication (e.g. between team members), and the use of technology in a task. As with all other techniques, observations have advantages and disadvantages. One of the key benefits from observation of tasks is that a variety of data can be collected, e.g. verbal interactions, task times, errors. Disadvantages include that being observed may actually inhibit an individual or team's performance of the task under observation, observations and subsequent analysis are time-consuming, and cognitive aspects may not be captured. Observation can be used as an initial means through which to gather information, but the data obtained can then be subjected to further analysis to confirm or expand on the results obtained through alternative methods. Once a prototype taxonomy of non-technical skills has been developed, observations can also be conducted to evaluate the applicability and usability of the taxonomy as an observational tool, whether any glaring omissions were noted and whether the skill descriptors were appropriate.

Combined techniques

Examples of research using combinations of the techniques described above to identify non-technical skills critical for safe and efficient performance are given below from aviation, nuclear and medicine.

Aviation

In the USA, several centres, including NASA Ames and Helmreich's group at the University of Texas at Austin, led the field in identifying pilots' cognitive and social skills (see Wiener et al., 1993, and Chapter 11) and today there is also significant aviation psychology research from other countries (Goeters, 1998; Edkins and Pfister, 2003). These teams study the behaviour of pilots by conducting systematic observations from flight decks and in simulators, as well as using questionnaires, interviews and accident reports.

Box 9.8 Development of a behaviour rating system for pilots from interviews, observations and accident analyses (LMQ, see CAA, 2006: appendix 11)

Interviews were conducted by an aviation consultancy company, with several hundred training captains to identify the qualities of a professional pilot, as well as those of someone who would be considered as being a liability. Qualities included being *open and honest, having control of the situation and can see the big picture*, and *lets others know what they are thinking and planning*. In contrast, liabilities included: *is insensitive, pompous and aggressive, has little respect for others*, and *has a closed mind and rigid views*.

The results of the interviews were combined with observations of exercises undertaken by crews when practising CRM skills, and accident analyses, to develop CRM standards. Examples of the observable actions for each of the five skill categories include:

Communication	Know when, what, how much and to whom they need to communicate. Pass messages and information clearly, accurately, timely and adequately.
Teamworking	Agree and are clear on the team's objectives and members' roles. Demonstrate respect and tolerance for other people.
Workload management	Prioritise and schedule tasks effectively. Offer and accept assistance, and delegate when necessary.
Situation	Are aware of what the aircraft and its systems are doing. Are able to recognise what is likely to happen, to plan and stay ahead of the game.
Problem-solving and decision making	Identify and verify why things have gone wrong and do not jump to conclusions or make assumptions. Use and agree the most effective decision-making process.

This research plus interviews, observations and accident analyses have been used for the development of behaviour rating systems (see Chapter 11). Helmreich's work has now evolved into a major international programme based on in-flight observations, called LOSA, which is described in Chapter 11. Another behaviour rating method from aviation is the NOTECHS system devised as the basis for an assessment system for European pilots' non-technical skills as described in Appendix 1. A third example from aviation is shown in Box 9.8.

Nuclear industry

The nuclear industry also produced a list of non-technical skills for control room operators, which appears to have been based on the earlier University of Texas behavioural markers for pilots. Team skills for nuclear power plant operating staff have also been identified by O'Connor et al. (2001), and include six categories: shared situation awareness, team-focused decision-making, communication, co-ordination, influence and team climate. This taxonomy of team skills has been designed for training and feedback purposes. Moreover, a list of non-technical skills required by members of nuclear emergency response teams has been identified (Crichton and Flin, 2004), and describes the five critical skills as being decision-making, situation assessment, communication, teamwork and stress management.

Medicine

Using a number of the methods described above, such as critical incident interviews, observations, questionnaires followed by asking panels of experts to refine and organise these, sets of non-technical skills that can be observed and rated have been developed. These include ANTS for anaesthetists (Fletcher et al., 2003), NOTSS for surgeons (Yule et al., 2006), OTAS for surgical teams (Healey et al., 2004), and UTNR for neonatal resuscitation teams (Thomas et al., 2004).

Conclusion

This chapter has described a range of different diagnostic tools that can be used to identify the non-technical skills inherent in safe and effective performance. We have seen here that diagnostic tools can include self-report questionnaires, interviews, observation (by trained observers), accident/near-miss analysis and confidential reporting systems. No single tool on its own provides a full picture of the relevant non-technical skills, and it is recommended that combinations of two or more tools are used. Although core non-technical skills can be identified across domains (e.g. decision-making, situation awareness, communication, leadership, teamwork), behaviours, or elements, emerge specific to each domain. Once the non-technical skills appropriate to a setting have been identified and rigorously analysed, specific training objectives, training evaluation and feedback can be designed and introduced.

Key points

- Identification of the specific skills relative to an occupation or task set is the first step in designing training or assessment tools for non-technical skills.
- A variety of diagnostic tools exist, which can be used individually, but acquired data are enhanced by using more than one technique. Questioning, observation and event-based techniques are used to identify the non-technical skills. These data are then used to develop a taxonomy that comprises categories of the relevant skills and elements or behaviours that describe those skills.
- Regardless of the technique, or techniques used, careful analyses are required to identify the key non-technical skills.
- Cognitive task analysis (CTA) techniques relate to a particular application of the above techniques when they are used to identify the cognitive skills that support task performance. These techniques are particularly relevant in dynamic tasks that demand high cognitive complexity, and has been used to assess non-technical skills in high-hazard industries, such as the offshore oil and gas industry, nuclear production installations, aviation and anaesthetics.
- An accident reporting tool provides data that begin with a negative outcome and considers how and when the defences built into the system failed. Accident analyses are only useful for non-technical skills identification if the presence or absence of relevant behaviours have been recorded.
- Confidential reporting systems gather data from individuals about errors or safety concerns that can be fed back into the system to improve safety.
- On the basis of the data collected from these techniques, a preliminary taxonomy of non-technical skills can be developed, validated and evaluated, which can then be used to guide a non-technical skills training programme.

Suggestions for further reading

Websites

Anaesthetists Non-Technical Skills (ANTS) behavioural marker system: http://www.abdn.ac.uk/iprc/ants

BASIS (British Airways Safety Information System): www.winbasis.com

Civil Aviation Authority (CAA): http://www.caa.co.uk

Confidential Human Factors Incident Reporting Programme (CHIRP): www.chirp.co.uk

Confidential Incident Reporting and Analysis System (CIRAS): www.ciras.org.uk

Line Operations Safety Audit (LOSA) website: http://homepage.psy.utexas.edu/homepage/group/HelmreichLAB/Aviation/LOSA/LOSA.html

NASA Aviation Safety Reporting System (ASRS) website: http://asrs.arc.nasa.gov/main_nf.htm

Non-Technical Skills for Surgeons (NOTSS): http://www.abdn.ac.uk/iprc/notss

TNO Mobile Evaluation System for Distributed Team Training: http://www.tno.nl/defensie_en_veiligheid/projecten/mobile_evaluation_system/index.xml

Key texts

Carayon, P. (ed.) (2007) *Handbook of Human Factors and Ergonomics in Health Care and Patient Safety*. Mahwah, NJ: LEA.

Crandall, B., Klein, G. and Hoffman, R. (2006) *Working Minds: A Practitioner's Guide to Cognitive Task Analysis*. Cambridge, MA: Bradford.

Hale, A., Wilpert, B. and Freitag, M. (eds.) (1998) *After the Event: From Accident to Organisational Learning*. Oxford: Elsevier Science.

Kvale, S. (2005) *Interviews. An Introduction to Qualitative Research Interviewing*. Thousand Oaks, CA: Sage Publications.

Fink, A. (2003) *The Survey Kit (2nd ed)*. London: Sage.

Seamster, T.L., Redding, R.E. and Kaempf, G.L. (1997) *Applied Cognitive Task Analysis in Aviation*. Aldershot: Ashgate.

References

Anderson, N. and West, M.A. (1994). *The Team Climate Inventory: Manual and User's Guide.* Windsor, UK: ASE Press.

Annett, J. (2004) Hierarchical task analysis. In N.A. Stanton, A. Hedge, K. Brookhuis, E. Salas and H. Hendrick (eds.) *Handbook of Human Factors and Ergonomics Methods.* Boca Raton, FL: CRC Press.

Balfour, D.L. and Wechsler, B. (1996) Organisational commitment: Antecedents and outcomes in public organisations. *Public Productivity and Management Review*, 29, 256–277.

Barling, J. and Frone, M.R. (2004) *The Psychology of Workplace Safety*. Washington, DC: American Psychological Association.

Bass, B.M. (1985) *Leadership and Performance: Beyond Expectations*. New York: Free Press.

Belbin, R.M. (1993) *Team Roles at Work: A Strategy for Human Resource Management*. Oxford: Butterworth-Heinemann.

Botney, R., Gaba, D. and Howard, S. (1993) Anesthesiologist performance during a simulated loss of pipeline oxygen. *Anesthesiology*, 79, A1118.

Breakwell, G., Fife-Schaw, C., Smith, J.A. and Hammond, S. (2006) *Research Methods in Psychology* (3rd ed). London: Sage.

Burke, E. and Hendry, C. (1997) Decision making on the London incident ground. An exploratory study. *Journal of Managerial Psychology*, 12, 40–47.

Burton, M., and Shadbolt, N. (1987) Knowledge engineering. In N. Williams and P. Holt (eds.) *Expert Systems for Users*. London: McGraw-Hill.

CAA (2006) *Crew Resource Management (CRM). Training. CAP737. Guidance for Flight Crew, CRM Instructors (CRMIs). and CRM Instructor-Examiners (CRMIEs).* Hounslow, Middlesex: Civil Aviation Authority.

Carthey, J. (2003) The role of structured observational research in health care. *Quality and Safety in Health Care*, 12, *Suppl II,* ii13–16.

Catchpole, K., Giddings, A., de Leval, M. et al. (2006) Identification of systems failures in successful paediatric cardiac surgery. *Ergonomics*, 49, 567–57.

Chi, M.T.H., Glaser, R. and Farr, M.J. (1988) *The Nature of Expertise*. Hillsdale, NJ: Lawrence Erlbaum Associates.

Cohen, L., Manion, L. and Morrison, K. (2007) *Research Methods in Education* (6th ed). London: Routledge.

Coolican, H. (2004) *Research Methods and Statistics in Psychology* (4th ed). London: Hodder Arnold.

Crandall, B. and Getchell-Reiter, K. (1993) Critical decision method: A technique for eliciting concrete assessment indicators from the intuition of NICU nurses. *Advances in Nursing Science*, 16, 42–51.

Crandall, B. and Klein, G.A. (1988) *Key Components of MIS Performance*. Dayton, OH: Klein Associates Inc.

Crandall, B., Klein, G.A. and Hoffman, R.R. (2006) *Working Minds. A Practitioner's Guide to Cognitive Task Analysis*. Cambridge, MA: MIT Press.

Crichton, M. and Flin, R. (2004). Identifying and training non-technical skills of nuclear emergency response teams. *Annals of Nuclear Energy*, 31, 1317–1330.

Crichton, M., Flin, R. and McGeorge, P. (2005). Decision making by on-scene incident commanders in nuclear emergencies. *Cognition, Technology and Work*, 7, 156–166.

De Keyser, V. and Nyssen, A.-S. (2001) The management of temporal constraints in naturalistic decision making: The case of anaesthesia. In E. Salas and G. Klein (eds.) *Linking Expertise and Naturalistic Decision Making.* Mahwah, NJ: Lawrence Erlbaum.

Dismukes, K., Berman, B. and Loukopoulous, L. (2007) *The Limits of Expertise. Rethinking Pilot Error and the Causes of Airline Accidents.* Aldershot: Ashgate.

Edkins, G. and Pfister, P. (eds.) (2003) *Innovation and Consolidation in Aviation.* Aldershot: Ashgate.

Ericsson, K.A. and Smith, J. (1991) Prospects and limits of the empirical study of expertise: An introduction. In K.A. Ericsson and J. Smith (eds.) *Toward a General Theory of Expertise: Prospects and Limits*. New York: Cambridge University Press.

Fahlbruch, B. and Wilpert, B. (1997) Event analysis as problem solving process. In A. Hale, B. Wilpert and M. Freitag (eds.) *After the Event: From Accident to Organisational Learning*, 113–129.Oxford: Elsevier Science.

Fink, A. (2003) *The Survey Kit* (2nd ed). London: Sage.

Fischer, U., Orasanu, J. and Wich, M. (1995) Expert pilots' perception of problem situations. Paper presented at the Eighth International Symposium on Aviation Psychology, April 24–28, Columbus, Ohio.

Flanagan, J. (1954) The critical incident technique. *Psychological Bulletin*, 51, 327–358.

Fletcher, G., Flin, R., McGeorge, P., Glavin, R., Maran, N. and Patey, R. (2003) Anaesthetists' non-technical skills (ANTS): Evaluation of a behavioural marker system. *British Journal of Anaesthesia*, 90, 580–588.

Fletcher, G., Flin, R., McGeorge, P., Glavin, R., Maran, N. and Patey, R. (2004) Rating non-technical skills: Developing a behavioural marker system for use in anaesthesia. *Cognition, Technology and Work*, 6, 165–171.

Flin, R., Burns, C., Mearns, K., Yule, S. and Robertson, E. (2006) Measuring safety climate in health care. *Quality and Safety in Health Care*, 15, 109–115.

Flin, R. and Martin, L. (2001) Behavioural markers for Crew Resource Management: A review of current practice. *International Journal of Aviation Psychology*, 11, 95–118.

Flin, R., Mearns, K., O'Connor, P. and Bryden, R. (2000) Measuring safety climate: identifying the common features. *Safety Science*, 34, 177–192.

Flin, R., Slaven, G. and Stewart, K. (1996) Emergency decision making in the offshore oil and gas industry. *Human Factors*, 38, 262–277.

Gaba, D.M., Fish, K. and Howard, S. (1994) *Crisis Management in Anesthesia*. New York: Churchill-Livingstone.

Gillham, W. (2005) *Research Interviewing: The Range of Techniques*. Milton Keynes: Open University Press.

Glendon, I., Clarke, S. and McKenna, E. (206) *Human Safety and Risk Management* (2nd ed). London: Taylor & Francis.

Goeters, K.M. (ed.) (1998) *Aviation Psychology. A Science and a Profession*. Aldershot: Ashgate.

Gordon, R., Flin, R. and Mearns, K. (2005) Designing and evaluating a human factors investigation tool (HFIT) for accident analysis. *Safety Science*, 43, 147–171.

Gregorich, S., Helmreich, R. and Wilhelm, J. (1990). The structure of cockpit management attitudes. *Journal of Applied Psychology*, 75, 682–690.

Guerlain, S., Adams, R., Turrentine, B., Shin, T., Guo, H., Collins, S. and Calland, F. (2005) Assessing team performance in the operating room: Development and use of a 'Black-Box' recorder and other tools for the intraoperative environment. *Journal of American College of Surgeons*, 200, 29–37.

Healey, A., Undre, S. and Vincent, C. (2004) Developing observational measures of performance in surgical teams. *Quality and Safety in Healthcare, 13 (Suppl 1),* i33–i40.

Heiman, G.A. (1995) *Research Methods in Psychology*. Boston: Houghton Mifflin Co.

Helmreich, R. and Merritt, A.C. (1998) *Culture at Work in Aviation and Medicine. National, Organizational and Professional Influences*. Aldershot: Ashgate.

Hiemstra, A.M.F., van Berlo, M.P.W. and Hoekstra, W. (2003) The development of MOPED – A mobile tool for performance measurement and evaluation during distributed team training. In Proceedings of the Human Factors of Decision Making in Complex Systems Conference, 8–11 September, 2003. Dunblane, UK.

Hoffman, R.R., Crandall, B. and Shadbolt, N. (1998) Use of the critical decision method to elicit expert knowledge: A case study in the methodology of cognitive task analysis. *Human Factors*, 40, 254–276.

Hoffman, R.R., Shadbolt, N.R., Burton, A.M. and Klein, G.A. (1995) Eliciting knowledge from experts: A methodological analysis. *Organizational Behavior and Human Decision Making*, 62, 129–158.

Holden, R.J. and Karsh, B. (2007) A review of medical error reporting system design considerations and a proposed cross-level systems research framework. *Human Factors*, 49, 257–276.

Hollnagel, E. (2004) *Barriers and Accident Prevention*. Aldershot: Ashgate.

HSE (2001) *Health and Safety Climate Survey Tool*. London: HSE Books.

Hudson, P., Primrose, M.J. and Edwards, C. (1994) Implementing tripod-DELTA in a major contractor. (SPE 27302) In *Proceedings of the SPE International Conference on Health, Safety and Environment, Jakarta, Indonesia.* Richardson, TX: Society of Petroleum Engineers.

Johnson, C. (2006) Incident Analysis in Health Care. In P. Carayon (ed.) *Handbook of Human Factors and Ergonomics in Health Care and Patient Safety*. Mahwah, NJ: LEA.

Kayten, P. (1993) The accident investigator's perspective. In E. Wiener, B. Kanki and R. Helmreich (eds.) *Cockpit Resource Management.* San Diego: Academic Press.

Kaempf, G.L., Wolf, S.P., Thorsden, M.L. and Klein, G.A. (1992) *Decision making in the AEGIS Combat Information Center* (Technical report No 1; Contract No N66001-90-C-6023). San Diego, CA: Naval Command, Control and Ocean Surveillance Center.

Kirwan, B. and Ainsworth, L.K. (1992). *A Guide to Task Analysis.* London: Taylor & Francis.

Klein, G., Calderwood, R. and Clinton-Cirocco, A. (1986). Rapid decision making on the fire ground. Paper presented at the Human Factors Society 30th Annual Meeting, San Diego, CA.

Klein, G., Calderwood, R. and Macgregor, D. (1989) Critical decision method for eliciting knowledge. *IEEE Transactions on Systems, Man and Cybernetics*, 19, 462–472.

Kvale, S. (2004) *Interviews. An Introduction to Qualitative Research Interviewing* (2nd ed). Thousand Oaks, CA: Sage Publications.

Lucia, A. and Lepsinger, R. (1999) *The Art and Science of Competency Models.* New York: Wiley.

Mackenzie, C., Craig, C., Parr, F., Horst, R. and the LOTAS group. (1994) Video analysis of two emergency tracheal intubations identifies flawed decision making. *Anesthesiology*, 81, 763–771.

Mackenzie, C. and Xiao, Y. (2007) Video analysis in health care. In P. Carayon (ed.) *Handbook of Human Factors and Ergonomics in Health Care and Patient Safety.* Mahwah, NJ: Lawrence Erlbaum Associates.

Madsen, M., Anderson, H. and Itoh, K. (2007) Assessing safety culture and climate in healthcare. In P. Carayon (ed.) *Handbook of Human Factors and Ergonomics in Healthcare and Patient Safety.* Mahwah, NJ: Lawrence Erlbaum.

Marshall, E., Duncan, K. and Baker, S. (1981). The role of withheld information in the training of process plant fault diagnosis. *Ergonomics*, 24, 711–724.

McGraw, K.L. and Harbison-Briggs, K. (1989) *Knowledge Acquisition: Principles and Guidelines*. Englewood Cliffs, NJ: Prentice-Hall International.

Miller, T.E. (2001) A cognitive approach to developing tools to support planning. In E. Salas and G. Klein (eds.) *Linking Expertise and Naturalistic Decision Making* (pp. 95–112). Mahwah, NJ: Lawrence Erlbaum.

Morgan, D.L. (1997) *Focus Groups as Qualitative Research* (2nd ed). Thousand Oaks, CA: Sage.

Morgan, D.L. and Krueger, R.A. (1993) When to use focus groups and why. In D.L. Morgan (ed.) *Successful Focus Groups: Advancing the State of the Art* (pp. 3–19). Newbury Park, CA: Sage.

Mowday, R., Steers, R. and Porter, L. (1979) The measurement of organisational commitment. *Journal of Vocational Behaviour*, 14, 224–247.

NASA ASRS website (2005) http://asrs.arc.nasa.gov/main_nf.htm

Nisbett, R.E. and Wilson, T.D. (1977) Telling more than we can know: Verbal reports on mental processes. *Psychological Review*, 84, 231–259.

O'Connor, P. (2007) The nontechnical causes of diving accidents: Can U.S. Navy divers learn from other industries? *Journal of the Undersea and Hyperbaric Medical Society*, 34, 51–59.

O' Connor, P., Flin, R. and O'Dea, A. (2001) Team skills for nuclear power plant operating staff. Technical Report to Nuclear Industry management Committee. Industrial Psychology Group, University of Aberdeen.

O'Connor, P., O'Dea, A. and Melton, J. (2007) The development of a taxonomy to classify the nontechnical causes of U.S. navy diving accidents. *Human Factors*, 49, 214–226.

Omodei, M., McLennan, J. and Reynolds, C. (2005) Understanding the reasons even good firefighters make unsafe decisions: A human factors interview protocol. Paper presented at the Seventh Conference on Naturalistic Decision Making, Amsterdam, NL, June.

Omodei, M., McLennan, J. and Wearing, A.J. (2004) How expertise is applied in real-world dynamic environments: Head-mounted video and cued-recall as a methodology for studying routines of decision making. In T. Besch and S. Haberstrohe (eds.) *The Routines of Decision Making*. Mahwah, NJ: LEA.

Oppenheim, A.N. (1992) *Questionnaire Design, Interviewing and Attitude Measurement*. London: Pinter.

Orasanu, J. and Fischer, U. (1997) Finding decisions in natural environments: The view from the cockpit. In C. Zsambok and G. Klein (eds.) *Naturalistic Decision Making* (pp. 343–358). Mahwah, NJ: LEA.

Paradies, M., Unger, L. and Busch, D. (1996) *TapRooT. Root Cause Tree Users Manual.* Knoxville, TS: System Improvements Inc.

Rasmussen, J. (1985) The role of hierarchical knowledge representation in decision making and system management. *IEEE Transactions on Systems, Man and Cybernetics*, 15, 234–243.

Reason, J. (1997) *Managing the Risks of Organisational Accidents*. Aldershot: Ashgate.

Redding, R.E. and Seamster, T.L. (1994) Cognitive task analysis in air traffic controller and aviation crew training. In N. Johnston, N. McDonald and R. Fuller (eds.) *Aviation Psychology in Practice* (pp. 190–222). Brookfield, VT: Ashgate.

Roth, E.M., Mumaw, R.J. and Lewis, P.M. (1994) *An Empirical Investigation of Operator Performance in Cognitively Demanding Simulated Emergencies* (NUREGCR-6208). Washington, DC: Division of Systems Research, Office of Nuclear Regulatory Research.

Roth, E., Christian, C., Gustafon, M., et al. (2004) Using field observations as a tool for discovery: analysing cognitive and collaborative demands in the operating room. *Cognition, Techncology and Work*, 6, 148–157.

Rugg, G. and McGeorge, P. (1997) The sorting techniques: A tutorial paper on card sorts, picture sorts and item sorts. *Expert Systems*, 14, 80–94.

Schuman, J. and Presser, S. (1996) *Questions and Answers in Attitude Surveys. Experiments on Question Form, Wording, and Context*. Thousand Oaks: Sage Publications.

Seamster, T. and Keampf, G. (2001) Identifying resource management skills for airline pilots. In E. Salas, C. Bowers and J. Cannon-Bowers (eds.) *Improving Teamwork in Organisations. Applications of Resource Management Training*. Mahwah, NJ: Lawrence Erlbaum.

Seamster, T.L., Redding, R.E. and Kaempf, G.L. (1997) *Applied Cognitive Task Analysis in Aviation*. Aldershot: Ashgate.

Shadbolt, N. (2005) Eliciting expertise. In J.R. Wilson and N. Corlett (eds.) *Evaluation of Human Work* (3rd ed). London: Taylor & Francis.

Shappell, S., Detwiler, C., Holcomb, K., Hackworth, C., Boquet, A. and Wiegmann, D. (2007) Human error and commercial aviation accidents: An analysis using the Human Factors Analysis and Classification System. *Human Factors*, 49, 227–242.

Shorrock, S.T. and Kirwan, B. (2002). Development and application of a human error identification tool for air traffic control. *Applied Ergonomics*, 4, 319–336.

Simon, J.A. (1955) A behavioral model of rational choice. *Quarterly Journal of Economics*, 69, 99–118.

Smith-Jentsch, K.A., Johnston, J.H. and Payne, S. (1998) Measuring team-related expertise in complex environments. In J.A. Cannon-Bowers and E. Salas (eds.) *Making Decisions Under Stress. Implications for Individual and Team Training*. Washington, DC: American Psychological Association.

Sneddon, A., Mearns, K. and Flin, R. (under review) Workplace Situation Awareness.

Stanton, N.A., Salmon, P.M., Walker, G.H., Baber, C. and Jenkins, D.P. (2005) *Human Factors Methods. A Practical Guide for Engineering and Design*. Aldershot: Ashgate.

Stoop, J. (1997) Accident scenarios as a tool for safety enhancement strategies in transportation systems. In A. Hale, B. Wilpert and M. Freitag (eds.) *After the Event: From Accident to Organisational Learning*. Oxford: Elsevier Science, 77–93.

Strater, O. (2005) *Cognition and Safety: An Integrated Approach to Systems Design and Performance Assessment*. Aldershot: Ashgate.

Thomas, E.J., Sexton, J.B. and Helmreich, R.L. (2004) Translating teamwork behaviours from aviation to healthcare: Development of behavioural markers for neonatal resuscitation. *Quality and Safety in Health Care*, 13, 57–64.

Thordsen, J., Militello, L.G. and Klein, G.A. (1992) Cognitive task analysis of critical team decision making during multi-ship engagements. Dayton, OH: Klein Associates.

Wallace, J. and Chen, G. (2005) Development and validation of a work-specific measure of cognitive failure: implications for occupational safety. *Journal of Occupational and Organizational Psychology*, 78, 615–632.

Warr, P., Cook, J. and Wall, T. (1979) Scales for the measurement of some work attitudes and aspects of psychological well being. *Journal of Occupational and Organizational Psychology*, 52, 129–148.

Whiddett, S. (2003) *A Practical Guide to Competencies*. London: Chartered Institute for Personnel and Development.

Wiegeman, D.A. and Shappell, S.A. (2001) Human error analysis of commercial aviation accidents: Application of the human factors analysis and classification system (HFACS). *Aviation, Space, and Environmental Medicine*, 72, 1006–1017.

Wiegmann, D.A. and Shappell, S.A. (2003) *A Human Error Approach to Aviation Accident Analysis: The Human Factors Analysis and Classification System*. Aldershot: Avebury Aviation.

Wiener, E., Kanki, B. and Helmreich, R. (eds.) (1993) *Cockpit Resource Management*. San Diego: Academic Press.

Wilson, J.R. and Corlett, N. (eds.) (2005) *Evaluation of Human Work* (3rd ed). London: Taylor & Francis.

Yule, S., Flin, R., Paterson-Brown, S., Maran, N. and Rowley, D. (2006) Developing a behavioural rating system to assess surgeons' non-technical skills NOTSS. *Medical Education*, 40, 1098–1104.

Zachary, W., Ryder, J. and Hicinbothom, J. (1998). Cognitive task analysis and modeling of decision making in complex environments. In J.Cannon-Bowers and E. Salas (eds.) *Making Decisions Under Stress. Implications for Individual and Team Training*. Washington, DC: American Psychological Association.

Chapter 10

Training Methods for Non-Technical Skills

For individuals and teams to perform effectively in high-risk environments, they must be proficient in the non-technical skills discussed in the earlier chapters. However, should these non-technical skills need to be introduced or enhanced, what are the most effective training techniques to achieve this? Further, following the implementation of a training course, how do you know the non-technical skills have actually improved? Organisations are limited in terms of the amount of time and resources that can be spent on training. Therefore, the training time available must be used efficiently.

The purpose of this chapter is to provide guidance on developing training designed to improve the non-technical skills of individuals working in risky environments. The most widely used strategy to train non-technical skills used in high-risk industries is crew resource management (CRM) training (resources specific to CRM training are listed at the end of the chapter). A general framework for training development will be presented, potential training methods and strategies identified, and methods of evaluating the effectiveness of the training will be described. Although not specific to the training of non-technical skills, there are a number of useful basic texts on designing, running and evaluating training courses that are worth reading prior to developing a training course (e.g. Goldstein and Ford, 2002; Truelove, 1997; and see the recommendations for further reading at end of the chapter).

Framework of training development

Table 10.1 Method for designing and delivering effective training (adapted from Goldstein and Ford, 2002)

Phase		Steps
1.	Needs assessment	1A. Training needs assessment 1B. Define training objectives
2.	Training and development	2A. Select and design training programme 2B. Devise training strategy
3.	Training evaluation	3A. Design assessment measures 3B. Training evaluation goals

Classical models of training start with an identification of the training needs, defining the objectives of the training, development of the training, and evaluation of the training. This is typical of the majority of the training models in the literature (e.g. Goldstein and Ford, 2002; Lambert, 1993). Table 10.1 provides an overview of a method for designing and delivering training: Each of the steps will be described in detail below.

Phase 1: Needs assessment

Conducting a training needs assessment is crucial. In the context of non-technical skills, a training needs assessment is the identification of skills that need to be trained. It is an easy stage to either ignore, or not carry out adequately. However, a good training needs assessment will pay dividends when it comes to designing the training. There is no point in developing a training course that does not address the needs of a given job and ultimately the needs of the organisation, because it was based upon a poor training needs assessment. This is a waste of time, resources and money. Further, the opportunity for developing a worthwhile training course is lost.

1A: Training needs assessment Techniques for identifying non-technical skills required for a particular task or profession are discussed in detail in Chapter 9. A training needs assessment is designed to identify the gaps between present levels of skill and the required levels. In practice, much of the non-technical skills training outside aviation is run at an introductory level. For more advanced and targeted skills training, the gap analysis may be based on regular performance evaluations, safety auditing or organisational data, such as incident analysis.

1B: Training objectives The content of a non-technical skills training course is dependent on the identified training needs, and the ability to design methods to address these training needs in an effective manner. For each non-technical skill that was identified in the training needs assessment, training objectives should be written. These objectives should be recorded in such a way that they can be empirically evaluated to determine whether or not they were accomplished. The training objectives then guide the development of the content of the course. Training objectives are crucial as these can be empirically evaluated to assess whether or not they were achieved through the training (Goldstein and Ford, 2002).

Phase 2: Training and development

2A: Select and design training programme The methods of training are the tools and techniques used to deliver the training to the team. Salas and Cannon-Bowers (1997) distinguish between three different types of training delivery methods: information-based, demonstration-based and practice-based.

Information-based: Information-based techniques are the most widely used method in training. These are passive lecture-type training for conveying information, which can be complemented with reading material and web-based information. Advantages of this type of method include its ease of delivery to large groups and cost. It also works well when trainees are being introduced to unfamiliar concepts and topics. Disadvantages include that exposure to information is no guarantee of learning. It is not known if the participants are actually taking in, and processing, the information presented.

Demonstration-based: Demonstration-based methods allow the participants to observe the required behaviours, actions or strategies. They may be presented with effective (or ineffective) examples. While actors can be employed, the most commonly used technique is to use video clips. For example, the airlines use films of pilots' decision-making in emergencies with re-enactments based on cockpit voice recordings. Engineers are also able to recreate the technical setting for the last minutes prior to an accident from the black box recording. A graphic of the aircraft's flight path and attitude, the position of the flight controls, the readings of key instruments, and the voice recording provide a detailed view of a flight crew's activities prior to the crash. These types of recordings are used to great effect in the US Navy CRM programme. Watching these accident video clips allows participants to discuss where errors occurred and what could have been done to prevent them. Participants can also be asked to observe and rate non-technical skills demonstrated by the actors in these videos.

The aviation accident videos (several have been produced as television documentaries) can be used to good effect with other industries, although participants pay most attention to video re-enactments of events from their own work domain. Some companies make their own videos by filming simulated accidents at their own worksites. There are also a number of training videos for health care that illustrate the path of fatal error trajectories, e.g. a useful film about the errors relating to the drug *Vincristine* is available from the UK Department of Health website (www.dh.gov.uk/Home/fs/en). The advantages and disadvantages of demonstration-based methods are similar to information-based delivery methods. However, demonstration-based methods have the additional benefit of engaging the audience more by providing examples of situations to which the audience can relate.

Practice-based: Practice-based methods are arguably the most effective method of non-technical skills training. However, to be genuinely useful, these methods need to be supported by activities such as cueing, feedback or coaching to help the participants to understand, organise and assimilate the learning objectives (Dismukes and Smith, 2000; Salas and Cannon-Bowers, 1997). Particular examples of practice-based methods are:

- Small syndicate exercises, typically used to discuss a scenario or case study. The class is broken into smaller groups of between three and five people to discuss a particular event. Each group reports their findings to the whole class.
- Role play involves acting a part in a make-believe situation. The aim is for the course participants to gain an understanding of a differing point of view or to practise a cognitive skill or a social interaction.

- Desktop exercises give participants a written scenario in which they are asked to indicate how they would respond. This type of exercise also includes tactical decision games (TDGs). A TDG is a scenario-based 'what if' facilitated simulation designed to provide the opportunity to make decisions, to review the consequences of these decisions, and to examine the rationale underlying the reason for the decision (Crichton et al., 2000). These 'games' can be used for groups of equivalent role-holders, or for teams. Facilitated debriefing encourages participants to enhance non-technical skills and to self-critique (see Chapter 3 for a more detailed discussion of TDGs).
- Many industries and health care organisations have simulators that can be used to model both normal and emergency work situations, or they can simulate using existing facilities on the worksite. Computer-based simulation can also be employed, for example the Vector Command system (www.vectorcommand.com) used to train fire officers and other incident commanders. Scenarios can be developed that emphasise particular non-technical skills, e.g. decision-making or team co-ordination. The events modelled can be based upon real incidents or on accident reports. See Riley (in press) or Rall and Gaba (2005) for information on simulation training in health care. For flight-deck crews, non-technical skills are often practised and assessed in flight simulator sessions known as line-oriented flight training (LOFT). LOFT refers to 'aircrew training which involves a full mission simulation of situations which are representative of line operations, with special emphasis on situations which involve communications, management, and leadership' (ch5-1: Civil Aviation Authority, CAA, 2002). It is important to indicate that the purpose of LOFT training is not to keep adding more problems for the team to deal with until they ultimately fail. Rather, the difficulties and emergencies included in LOFT training should be realistic (see CAA, 2002, for a detailed discussion of designing effective LOFT training scenarios).
- Full-scale exercises in which team members participate in a particular simulated event, such as a site emergency, can also be used for training. The advantage of this practice method is that it is very powerful for putting the learning objectives across to the participants. However, using a full-scale simulation can be costly and time-consuming to both develop and stage.

2B: Design training strategy The training methods combine with training objectives to shape the development of specific training strategies. The strategy chosen should be theoretically based and use sound instructional principles (Paris et al., 1999).

The main method of training non-technical skills is in some form of crew resource management course, although these may be labelled human performance, human factors, non-technical skills, team resource management, crisis resource management or safety skills courses. The background to CRM training was presented in Chapter 1 and is covered in detail in other sources such as Wiener et al. (1993). There have been a number of recent publications on CRM training (e.g. CAA, 2006a; Salas et al., 2006b; 2006c), there are also specialist texts on CRM training, especially for aviation (e.g. Jensen, 1995: Macleod, 2005: McAllister, 1997; Walters, 2002), but also for other occupations, such as the fire service (Okray and Lubnau, 2004) and sources on CRM for health care professions (Bleakley et al., 2004; Baker et al., 2007; Flin et al., 2007; Howard et al., 1992). In addition, there are websites that include background material for CRM developers (e.g. the Royal Aeronautical Society). These resources are listed at the end of this chapter. Therefore, only a brief description of the background to CRM training is included here.

Crew resource management (CRM) training

Lauber (1984) defined CRM as 'using all the available resources – information, equipment, and people – to achieve safe and efficient flight operations' (p20). CRM training was initially used to train civilian pilots. However, CRM training is now being used by many other high-reliability industries. Those who adopted it first were, unsurprisingly, involved in aviation: aviation maintenance, cabin crew and air traffic control. However, CRM training has now begun to be used in organisations unrelated to aviation: nuclear power generation, health care, the fire service, the maritime and rail industries, and offshore oil and gas production (see Flin et al., 2002; Salas et al., 2006a, for examples).

Approaches to CRM training have evolved since it was first introduced by United Airlines in the early 1980s. Many of the early courses were not popular with the pilots, who saw them as 'charm schools' or attempts to manipulate their personalities. CRM training is now used by virtually all the large international airlines and is recommended by the European (Joint Aviation Authorities; JAA.[1] 2006) and US (Federal Aviation Administration; FAA, 2004) civil aviation regulators. Helmreich et al. (1999) proposed that, in aviation, CRM training was in its fifth generation by the end of the 1990s. The assumption underlying the fifth generation of CRM is that human errors are inevitable. For CRM to be accepted as a system for protecting against human limitations, there must be an organisational recognition, and acceptance, of the certainty of human error – but not violations (Reason, 1990). The fifth generation of CRM proposes that it can be seen as a set of three barriers to the occurrence of errors:

a. prevent errors occurring in the first instance, but
b. if errors do occur, then it provides crews with techniques of identifying, and trapping, errors before they have an operational effect, and
c. mitigate the consequences of errors (Helmreich et al., 1999).

Helmreich and his group at the University of Texas have proposed a model of threat and error management that can be used to train operators in high-risk organisations to understand how threat and error, and their management, interact to affect safety (Helmreich, 2000; see the end of the chapter for resources on the threat and error management model).

1 The Joint Aviation Authorities (JAA) is transitioning to the European Aviation Safety Agency (EASA).

Topics covered in CRM training 'are designed to target knowledge, skills, and abilities as well as mental attitudes and motives related to cognitive processes and interpersonal relationships' (p173; Gregorich and Wilhelm, 1993). The most effective CRM training should involve at least three distinct phases (CAA, 2002).

1. Awareness phase. This is generally the classroom-based portion of CRM training in which the participants are introduced to the theoretical background of non-technical skills. This phase provides team members with a common frame of reference and language for discussing, and thinking about, non-technical skills. An introductory CRM course generally is conducted in a classroom for two or three days. Teaching methods include lectures, practical exercises, role-playing, case studies and films of accident re-enactments. The syllabi recommended for flight crew in Europe (JAA) and the US (FAA) are shown in Table 10.2. These syllabi are provided for guidance only. As stated in the section on needs assessment, the training should be specifically designed to address the specific skill requirements of operators within individual organisations.

Table 10.2 JAA and FAA CRM curricula recommendations for flight crew

JAA (2006)	FAA (2004)

• human error and reliability, error chain, error prevention and detection
• company safety culture, standard operating procedures, organisational factors
• stress, stress management, fatigue and vigilance
• information acquisition and processing, situation awareness, and workload management
• decision-making
• communication and co-ordination inside and outside the cockpit
• leadership and team behaviour synergy
• automation (for type of aircraft)
• specific type-related differences
• case-based studies
	1. Communications processes and decision behaviour:
	• briefings
	• safety, security
	• inquiry/advocacy/assertion
	• crew self-critique (decisions and actions)
	• conflict resolution
	• communication and decision-making
	
	2. Team-building and maintenance:
	• leadership/followership/ concern for task
	• interpersonal relationships/ group climate
	• workload management and situation awareness
	  • preparation/planning/vigilance
	  • workload distribution/ distraction avoidance
	• individual factors/stress reduction

2. Practice and feedback phase. During the second phase of training, the participants are provided with the opportunity to practise the non-technical skills discussed during the awareness phase. In aviation, this phase generally consists of a LOFT flight.

However, other industries also make use of simulators to practise the non-technical skills discussed in the awareness phase (see Box 10.1). If simulator training is not possible, other practice-based methods could be used.

Box 10.1 Scottish Clinical Simulation Centre: Crisis Avoidance and Resource Management for Anaesthetists (CARMA)

Course background. The course was developed as a result of research into the role of human factors in medicine and a growing recognition that 70% of adverse events in operating theatres are caused by human error.

Course development and objectives. The course (www.scsc.scot.nhs.uk) was designed to raise awareness of the factors associated with critical incidents in anaesthesia. The objectives of the course are:

- to reduce the incidence of errors committed by anaesthetists
- to reduce the effects of errors committed by anaesthetists
- to improve the ability of anaesthetists to deal with such crises as can and do arise.

Course content. The course is delivered over two days, and the main topics include: situation awareness, communication and leadership, decision-making, stress and fatigue, and the role of senior and junior anaesthetists. A formal presentation is given on each topic, as well as exercises, and group work. Participants practise the skills by participating in realistic emergency scenarios in the anaesthetic simulator.

Training evaluation. The course is evaluated using a self-assessment questionnaire tool. Assessments are made immediately after day one, day two and each scenario. In addition to self-assessment, informal subjective observation by supervisor, co-worker or other person is used to assess the impact of training in terms of participant satisfaction, evaluation of usefulness of scenarios, impact of session on clinical practice.

3. Continual reinforcement phase. It is unrealistic to think that a single exposure to a training course will have a lasting effect. Therefore, no matter how effective the earlier two phases are, the organisation must continue to reinforce the concepts covered in the training. Refresher training, normally a half- or whole-day course focusing on a specific CRM topic, is recommended. The reason for this is because in the absence of recurrent training and reinforcement, attitudes and practices decay (Klinect et al., 1999: see the section on training evaluation).

Despite the widespread use of CRM training, the literature contains comparatively few studies in which the effects of the training have been evaluated. Reviews of the studies evaluating the effects of CRM training have shown it to produce:

1. positive reactions
2. enhanced learning
3. desired behavioural changes in a simulated or real environment (O'Connor et al., 2002; Salas et al., 2001; 2006a).

Methods for evaluating the effectiveness of non-technical skills training are described later in the chapter.

Other techniques for training non-technical skills

Although CRM training is the most commonly used method for training non-technical skills, it is not the only type of training that can be employed. Other types of training techniques that can be used to improve the non-technical skills of team members are: cross-training, team self-correction training, event-based training, and team facilitation training. These strategies are discussed in detail in Chapter 5. Therefore only a brief description will be included here:

- Cross-training is a training strategy in which each team member is trained in the duties of his or her team-mates. Salas and Cannon-Bowers (1997) recommend this type of training as being particularly useful where there is a high turnover of personnel.
- Team self-correction training works on the premise that effective teams review events, correct errors, discuss strategies and plan future events (Salas and Cannon-Bowers, 2000). Therefore, the training simply provides direction on processes that typically occur.
- 'Event based training is an instructional approach that systematically structures training in an efficient manner by tightly linking learning objectives, exercise design, performance measurement and feedback' (Dwyer et al., 1999: p191).
- Team facilitation training is designed to help team leaders stimulate learning by creating an effective learning environment, supporting more formal training experiences, and facilitating and encouraging team discussions (Tannenbaum et al., 1998).

Which training strategy should be used?

The most effective strategy is dependent upon the issues that need to be addressed, the resources available and the make-up of the team to be trained. It is possible to make a number of recommendations about when the training strategies described above are likely to be most effective (see Table 10.3).

However, these training techniques are not mutually exclusive, and a combination of strategies may be the most appropriate and effective (see Box 10.2 for a description of a non-technical skills training course for the offshore oil and gas drilling industry).

Table 10.3 Recommendations for the use of particular types of training strategies

Type of training	Specific recommendations
Team co-ordination training	• Effective even with teams that do not have a fixed set of personnel. • The training addresses a particular set of non-technical skills.
Cross-training	• Team has high levels of interdependence between members. • There is a lack of knowledge about the roles of other team members. • High staff turnover.
Team self-correction training	• Team has high levels of interdependence between members. • Low staff turnover.
Event-based training	• Useful when there are problems with a particular subset of tasks, and the tasks can be simulated.
Team facilitation training	• There are limitations in training resources.

Box 10.2 Example of non-technical skills training from the offshore oil and gas drilling industry

Oil and gas drilling industry project teams are brought together to complete a particular operation. The team members are typically from different disciplines, may be of different nationalities, have different levels of experience, and may be geographically distributed across different locations across the world. For example, a drilling and completions team consists of a vertically aligned organisation including the offshore-based well site leader as well as onshore-based roles of drilling engineer, operations superintendent, wells manager and field manager. An analysis of drilling and completions team functioning identified the critical areas as being communication, co-ordination and teamwork (Crichton and Flin, 2003). The knowledge, skills and abilities required of team members as individuals, and across the team as a whole, were identified based on the goals and objectives of the team, and whether the team was stable or whether changes to team members occur frequently. The training needs identified, and the interventions designed to address these needs, are summarised in Table 10.4.

Table 10.4 Training needs and identified interventions for oil and gas drilling teams

Training need	Training strategy
Introduction to non-technical skills	• CRM-based course as introduction to non-technical skills
Situation awareness	• event-based training • tactical decision games
Decision-making	• event-based training • tactical decision games
Teamwork training	• team process and facilitation training • role and responsibility review
Clarification of roles when managing unexpected events	• role and responsibility review • competency identification • event-based training • tactical decision games • command and control training course
Communication	• tactical decision games • communication exercises

Phase 3: Training evaluation

It is vital that a training course is evaluated to determine whether objectives have been achieved. The FAA (2004: p12) states that for CRM training 'it is vital that each training programme be assessed to determine if CRM training is achieving its goals. Each organisation should have a systematic assessment process. Assessment should track the effects of the training programme so that critical topics for recurrent training may be identified and continuous improvements may be made in all other respects.' The recommended approach to training evaluation is one that is multifaceted and considers several separate methods of assessment.

3A: Design assessment measures The evaluation methods can be categorised into what is described by training researchers as different levels of training effects – ranging from individual to organisational indicators. Kirkpatrick's (1976; 1998) hierarchy is a popular model for guiding training evaluation. It provides a useful framework to assess the effects of a training intervention on an organisation by considering training evaluations at different levels of evaluation: (1) reactions, (2) learning, (3) behaviour, and (4) organisation. These represent a sequence of methods to evaluate a training programme. Each level builds on the previous one, with the process becoming more difficult and time-consuming to perform at each higher level, but also providing more valuable information. Each level of evaluation is discussed in more detail below.

Level 1: Reactions. Reactions are concerned with how the participants react to the training. A paper-based questionnaire is sufficient to gain feedback on the participants' reactions to the training. These are sometimes called 'happy sheets'. A brief questionnaire should be administered after each module of the course has been completed. Generally, after two or more days of training, it may be difficult for participants to remember the comment they wished to make on the early modules.

It is suggested that the questionnaire consists of closed statements to which the participants respond on a five-point Likert scale from 1 (very poor) to 5 (excellent), and with open-ended questions allowing them to write their comments. As shown in Table 10.5, the statements should focus on whether the participants found the course content interesting and useful. It may also be beneficial to collect information on the quality of the trainer's performance, or the suitability of the facilities.

It is important to indicate that positive reaction does not ensure learning – although a negative reaction almost certainly reduces the likelihood that this has taken place (Kirkpatrick, 1998).

Table 10.5 An example of a reaction questionnaire for assessing a particular topic

Workpackage 2: Decision-making					
	Very Poor	Poor	Satisfactory	Good	Excellent
How interesting did you find this section of the course?	1	2	3	4	5
What did you think about the presentation of the teaching?	1	2	3	4	5
What did you think about the structure of the teaching?	1	2	3	4	5
What did you think of the exercises?	1	2	3	4	5
What did you think about the standard of the course materials (handouts, etc)?	1	2	3	4	5
What did you think of the relevance of this topic to your job?	1	2	3	4	5
Other comments:					

Level 2: Learning. Learning is the second level in the hierarchy, and refers to 'the principles, facts, and skills which were understood and absorbed by the participants' (Kirkpatrick, 1976: p11). This level is concerned with whether the participants have acquired knowledge, or have modified their attitudes or beliefs as a result of attending the training course. It is important to measure learning, as no change in

behaviour can be expected if no new knowledge or change in attitudes has occurred. At the learning level, attitude change and knowledge can be assessed.

Attitudes. The attitude assessment can be carried out using a paper-based attitude questionnaire that should, at the minimum, be based on an established instrument such as the Cockpit Management Attitudes Questionnaire (CMAQ; Gregorich and Wilhelm, 1993). The CMAQ is a well-established training, evaluation and research tool developed to assess the effects of CRM training for flight crew. It comprises 25 items chosen to measure a set of attitudes that are either conceptually or empirically related to CRM. The statement topics cover 'communication and co-ordination', 'command responsibility', and 'recognition of stressor effects'. The 'communication and co-ordination' sub-scale encompasses communication of intent and plans, delegation of tasks and assignment of responsibilities, and the monitoring of crew members. 'Command responsibility' includes the notion of appropriate leadership and its implications for the delegation of tasks and responsibilities. Disagreement with items on this sub-scale suggests a belief in the captain's autocracy. 'Recognition of stressor effects' emphasises the consideration of – and possible compensation for – stressors. Disagreement with items on this sub-scale suggests a belief in one's own imperviousness to stressors (Chidester et al., 1991; Gregorich et al., 1990). For each statement in the questionnaire, the degree to which the participants agree is assessed using a five-point Likert scale ranging from 1 (strongly disagree) to 5 (strongly agree; see Table 10.6). Where this has been applied before and after CRM training, it allows an assessment of the changes in the attitudes in the individual trainee (e.g. Awad et al., 2005).

Table 10.6 Example items from an adaptation of the CMAQ for US Navy divers (O'Connor, 2005)

Please answer the following items by using the following scale by writing your response beside each item.				
A	B	C	D	E
Disagree Strongly	Disagree Slightly	Neutral	Agree Slightly	Agree Strongly
____1. Junior divers should not question the Master Diver's decisions in emergencies.				
____2. Even when fatigued, I perform effectively during critical times in a dive.				
____3. Divers should be aware of, and sensitive to, the personal problems of other team members.				
____4. Divers should not question actions of the Dive Supervisor, except when they threaten the safety of the dive.				
____5. I let other divers know when my workload is becoming (or is about to become) excessive.				

The CMAQ provides, at a minimum, a good starting point for the design of an attitude assessment questionnaire. It has been used as the basis of Aircrew Coordination Attitudes Questionnaire (ACAQ) designed for military pilots, the Control Room Operations Attitude Questionnaire designed for nuclear control room personnel (Harrington and Kello, 1992), the Maintenance Resource Management/Technical Operations Questionnaire (MRM/TOQ) designed for aviation maintenance personnel (Taylor, 1998), the Air Traffic Control Safety Questionnaire (ATCSQ) (Woldring and Isaac, 1999), and US Navy divers (O'Connor, 2005) and Operating Room Management Attitudes Questionnaire (ORMAQ) (Helmreich and Merritt, 1998; see also Flin et al., 2006). However, depending on the training, it is possible that the CMAQ will not measure the range of attitudes that should have changed. The CMAQ was developed solely to assess pilots' attitudes regarding 'interpersonal components' of the flight crew's job performance and to link these attitudes to behaviour (Gregorich et al., 1990; Helmreich, 1984). It does not address attitudes to the *cognitive* aspects of the role of the flight crew such as situation awareness, decision-making and workload management.

Knowledge. Knowledge assessment is a useful method for evaluating the effects of training. The suggested method is a paper-based test. This is a fairly quick and simple way of receiving feedback on knowledge acquisition. However, the questions and answers (both correct and false if using a multi-choice test) must be designed very carefully to avoid either a floor (too difficult) or ceiling effect (too easy).

Another method for assessing learning of non-technical skills, rather than have an explicit test of participants' knowledge of the curriculum, is to present participants with accident vignettes to identify human factors. Kerlinger (1996) considers vignettes to be a one type of unobtrusive measure. He defines them as '... brief concrete descriptions of realistic situations so constructed that responses to them will yield measures of variables' (p475). Other advantages outlined by Gliner et al. (1997) are that they include standardised information, are easy to repeat, and if properly constructed can hold the participant's interest while approximating realistic psychological and social situations. Disadvantages are that they represent an artificial situation and responses are not necessarily the same as they would be in the actual situation. It is crucial that if vignettes are to be used, they must be realistic, have sufficient detail, be simple and easy to understand (Njaa, 2000).

Level 3: Behaviour. An evaluation at the behaviour level is the assessment of whether knowledge learned in training actually transfers to behaviours on the job, or a similar simulated environment. Kirkpatrick (1998) outlines the danger of only carrying out an evaluation at this level of the hierarchy. If no behavioural change was found, an obvious conclusion is that the training was ineffective. Nevertheless, reactions may have been favourable and the learning objectives could have been met. He suggests that a number of preconditions must be present to make the jump from a positive evaluation at levels 1 or 2, to positive evaluations at levels 3 and 4:

- The participants must have a desire to change.
- The participants must know what and how to make the change.

- The organisational climate must be conducive to allowing for the change.
- The participant must be rewarded for changing (e.g. positive feedback).

A widely used technique for assessing non-technical skills in flight crew is by instructing training captains to use an observational system based on ratings of behaviour to assess flight-crew performance. Similar behaviour rating systems are employed to evaluate non-technical skills in air traffic control, anaesthesia, surgery and maritime settings. These are discussed in detail in Chapter 11.

Level 4: Organisation. This is the highest level of evaluation in Kirkpatrick's (1976) hierarchy. The ultimate aim of any training programme is to produce tangible evidence at an organisational level, such as an improvement in safety and productivity. The problems with the evaluation of training at this level are that it can be both difficult to establish discernible indicators, and be able to attribute these to the effects of a single training course.

The Royal Aeronautical Society (1999) identifies a number of potential measures of assessing the effects of CRM training. These include: fuel management, punctuality, job satisfaction, insurance costs and damage to aircraft. Another possible assessment method is to use an organisational climate tool. Boehm-Davis et al. (2001) suggest that this is a useful process to carry out prior to the introduction of CRM training in an organisation to help identify training requirements. However, it is important that the results of such an evaluation method are examined with caution because other changes in the organisation (e.g. downsizing or structural changes) could also affect the responses to a survey.

3B: Training evaluation goals Once the evaluation measures have been identified, there needs to be a decision as to the most effective design for applying these measures to the training evaluation. A design is the specific manner in which the study will be conducted (Heiman, 1995). The design selected is based upon the evaluation question that is to be asked. Goldstein and Ford (2002) differentiate between four different potential goals for the design of the training evaluation. These goals will be discussed in detail below.

Did the trainees learn during the training? When asking this question, Goldstein and Ford (2002) are only referring to learning. However, as described in the section on evaluation, learning is unlikely to occur unless the reactions to the training course are positive, and there is a change in attitudes to the concepts covered in the training. Therefore, to assess whether learning has occurred, evaluation at both levels 1 (reactions) and 2 (knowledge and attitude) of Kirkpatrick's evaluation hierarchy should be carried out. Obviously, reactions to a training course can only be examined after the training has been completed. Therefore, Figure 10.1 (design 1) represents the most suitable way of assessing the reactions of trainees to training.

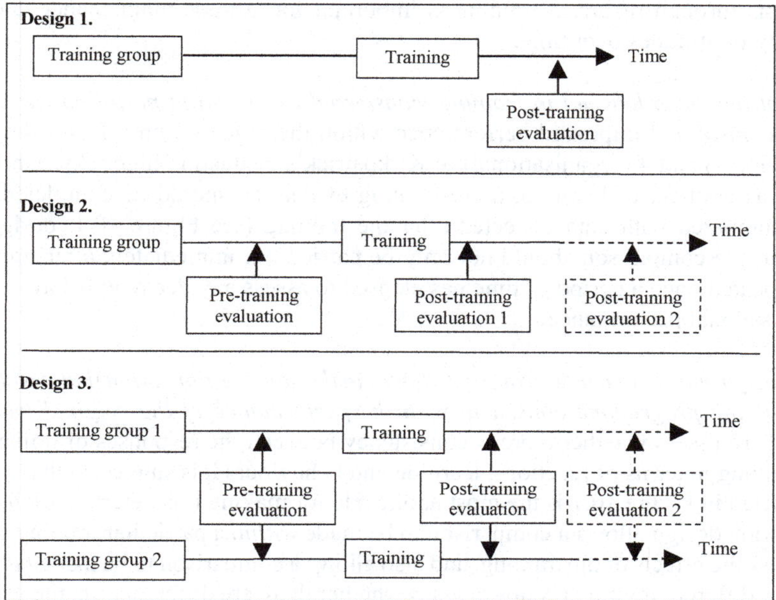

Figure 10.1 Training evaluation designs

To assess the effects of training on learning, a pre-training evaluation of knowledge and attitudes should be completed, and then these responses should be compared to data collected after the training (see Figure 10.1, design 2). It should be stated that non-technical skills training does not necessarily lead to a positive attitude change. To illustrate, there has always been a small subset of people, known as 'boomerangs', 'cowboys' or 'drongos', who have rejected the concept of CRM (Helmreich and Wilhelm, 1991). Efforts at remedial training for these individuals have not been found to be effective. Irwin (1991) found that 8% of a sample of 5,830 civilian flight crew in seven different organisations failed to respond to CRM training, and suggested that this may need to be addressed through avenues other than training.

It is also suggested that training should not only be assessed immediately after training, but also after a period of time has passed since the training (see Figure 10.1, design 2). For example, Irwin (1991) found a decay in pilots' positive attitudes to CRM over time (as measured using the CMAQ). Attitudes were measured at five different time intervals: a baseline (two years prior to initial CRM training), immediately prior to initial CRM training, immediately after initial CRM training, immediately prior to recurrent CRM training (one year later), and immediately after recurrent CRM training. It was found that there was an overall decline in attitudes to CRM training during the intervals between training interventions. The recurrent training resulted in another positive shift in attitudes back to the level found after initial training. Therefore, Irwin (1991) concluded that the reinforcement of CRM concepts through recurrent training is important for attitude maintenance and the stability of attitudes over time.

Is what has been learned in training transferred as improved performance within the organisation? Improved performance within the organisation refers to levels 3 (behaviours) and 4 (organisational) of Kirkpatrick's evaluation hierarchy. Similarly to the assessment of learning, a pre-training evaluation should be completed, and then compared with data collected after the training (see Figure 10.1, design 2). Further, the comparison should not only be carried out immediately after training, but repeated once a period of time has elapsed to assess any decay in behaviours or organisational performance.

Is the performance of a new group of trainees in the same organisation that developed the training programme consistent with the performance of the original training group? To assess whether there is consistency between the responses of trainees to the training in terms of reactions, learning and behaviours it is suggested that design 3 depicted in Figure 10.1 is the most applicable for making this determination. This evaluation design allows a comparison to be made *within* a particular training group to assess the effects of the training, and also allows a comparison to be made *between* the two different training groups to asses whether there are differences in the groups' responses to the training.

Has an attempt been made to determine whether a training programme validated in one organisation can be used successfully in another organisation? As the principal causes of human errors for teams in high-reliability industries do not differ, there may be a temptation for non-technical skill training developers to simply use exactly the same training across different organisations. For example, if aviation CRM is to be adapted for an industrial audience, the training materials must be customised for the particular industry, as explained in Chapter 9. For the training to be effective, it is imperative that the skills that are required have been identified through a training needs analysis. The relevant psychological concepts must be translated into the language of the participants receiving the training. Lastly, relevant practical examples and case studies should be used to illustrate the concepts. The training is not likely to be effective unless examples poignant to the particular industry are used.

One of the main criticisms of participants of the early aviation CRM courses was that there was too much psychological theory and not enough relevance to aviation (Helmreich et al., 1999). 'I am not suggesting the mindless import of existing programmes; rather, aviation experience should be used as a template for developing data driven actions reflecting the unique situation of each organisation' (p784). Thus there has been a need for other industries to tailor their CRM training so that it specifically addresses the issues of their personnel.

Other issues to consider

Multi-level training analysis. Ideally, training should be assessed at more than one level of Kirkpatrick's hierarchy. Using a multi-level analysis approach allows the return on investment (ROI) of the training to be calculated. Taylor (2000) proposes the following equation for calculating the ROI of CRM training:

$$\text{ROI} = \left(\frac{[\text{net CRM benefits}] \times \text{causal operator}}{\text{CRM costs}} \right) \times 100$$

The *net CRM benefits* are the benefits of the training minus the cost of the development of the training (*CRM costs*). These are financially calculable savings such as reduction in accidents, reduction in damage to aircraft, etc. The *causal operator* is a corrective calculation that is used to take account of the many different factors that may act to improve the behaviours or outcomes. The value of the causal operator is the square of the correlation between the training outcome (i.e. knowledge gained, improvement in attitudes or behaviours) and subsequent safety results. The square of a given correlation coefficient is equal to the proportion of the variance in one of the two measures explained by the other and is known as the coefficient of determination (Howell, 1992). This provides a 'conservative, quantitative estimates of "credit" to be allocated to organisational effectiveness interventions' (Taylor, 2000: p4).

Taylor (2000) provides the following example to illustrate how this equation is used:

> The development of a maintenance resource management (MRM) course for aviation maintenance technicians and delivery to 1,600 employees costs \$251,660. A post-training survey showed a significant improvement in attitudes, and in the two years following the training there was a decrease in lost time injuries of 80% (an injury resulting in four or more days off work). A correlation of –0.24 was found between lost time injuries and attitudes towards participative leadership and assertiveness in a post-training survey. Therefore, the coefficient of determination is 0.24^2 or 0.0576. Taylor (2000) estimated the cost of an injury as \$13,465. Thus, a reduction of 80% on 91 incidents per year is a saving of \$1,314,150. Therefore, the ROI calculation is as follows:

$$\text{ROI} = \left(\frac{[\$1{,}314{,}150 - \$251{,}660] \times .0576}{\$\,251{,}660} \right) \times 100 = 24.3\%$$

It follows that even with the conservative estimate of 5.76% lost time injuries benefit accounted for by the MRM training, the training paid for itself plus an additional 24% return in two years.

Establish appropriate participants. Obviously the training audience is at the centre of the learning environment, thus its needs must be clearly identified. Each operational crew or team is composed of unique characteristics. When designing a CRM training course for the offshore oil production industry, it was necessary to experiment with a variety of course participants (e.g. participants from two different platforms, supervisory staff, members of a single crew) (O'Connor and Flin, 2003). It was found that for this industry, a cross-section of staff from the most senior offshore manager to the most junior offshore worker from a single shift was optimal.

Individual, task and team characteristics play a significant role in the manner in which a learning environment is constructed. For example, at a team level, these may include the degree to which team members hold a shared understanding of the tasks and their team-mates; the team's organisational structure; and the physical proximity of team members to each other.

Identify appropriate instructors. As important as the content of the training course are the skills and credibility of the instructor. An enthusiast from the workforce at a local level who has experience of working 'in the field' is recommended for the course to be successful (Drury, 1998; Robertson, 2001). These individuals are necessary to help in the construction of the course and can also help to present the courses to give credibility and help answer any very technical questions. The UK Civil Aviation Authority (CAA) has provided a set of standards criteria that a pilot must meet to be designated as a CRM instructor and examiner (see CAA, 2006b). US Navy/Marine Corps aviation has also recognised the need for the credibility of CRM instructors and insisted on the requirement of naval aviators to be CRM trainers, rather than contracting the task out to civilian companies. A suitable enthusiast is an individual who has many years of experience and commands the respect of his or her peers. He or she must also have an understanding of, and enthusiasm for, human factors (Drury, 1998).

Establish the appropriate teaching style. The majority of personnel who work in high-risk industries will be familiar with technical training in which the teacher stands in front of the class and describes a piece of equipment, or a procedure for carrying out a particular task. This teaching style is known as instruction. Instruction is useful if the goal of the training is only to impart knowledge. However, the aim of non-technical skills training is to reinforce the desirable non-technical skills of personnel who are already technically very well trained. A facilitation style of training is more appropriate when attempting to change behaviours.

The goal of facilitation is for the teacher to guide the students towards recognising the appropriate behaviour, rather than simply telling them the correct way to behave. For instructors, and students, used to an instructional style of teaching, facilitation may take some practice (for more information on facilitating non-technical skills, see CAA, 2006a). Nevertheless, a facilitation teaching style is less likely to cause participants to object to being told how to behave or think – as may be the case if an instructional style of teaching is used. Therefore, facilitation is particularly useful when the trainer is less experienced or junior to the members of the class. A facilitation style of instruction is also appropriate when debriefing after a simulator exercise (see Chapter 11).

Another issue to consider is that many organisations are moving away from classroom training and replacing it with computer-based training. If well designed, computer-based training may be an efficient method for imparting knowledge, but is arguably less appropriate for training designed to change behaviours. Further, organisations should be wary of the impression that computer-based training can give employees about the training material. When one of the authors asked a junior US Navy sailor about the computer-based safety training he had just completed, he was told that 'the subject can't be very important, otherwise they would have had a real person deliver the training'.

Ensure management support. It is important that senior management at a corporate level show their commitment to any training programme (Drury, 1998). If the management do not demonstrate that they believe the training is important, then it is unlikely that the workforce will show much enthusiasm.

Refresher training and reinforcement of training. A one-time training course is unlikely to have a lasting impact with reinforcement. In the aviation industry refresher CRM training is given to flight crew members. This is normally a half- or whole-day course focusing on a specific CRM topic. For pilots, CRM skills are then practised and can be assessed on the job or in flight simulator sessions.

Conclusion

Training design is not an art (Salas and Cannon-Bowers, 2000), but rather should be a systematic process that allows for the creation of an environment that enables dissemination of the non-technical skills required to improve job performance. This chapter has outlined the steps that should be taken when designing a non-technical skills training course. It is recognised that limitations in terms of personnel, time resource and access may mean that it is not possible to carry out the steps exactly as described above. Nevertheless, the efforts made to complete a systematic approach to training development will pay off in terms of the quality and relevance of the training, and as a mechanism for enhancing individual, team and organisational effectiveness.

Key points

- A training needs assessment is an easy step to ignore, or 'pay lip service to', but it is crucial for the design of an effective non-technical skills training course. The needs assessment should be based upon a training requirements assessment and used to formulate training objectives.
- The training method and strategy should be based upon the type of skills to be trained, time and resources available. Further, it is likely that a combination of different methods and strategies will be the most appropriate.
- Training evaluation is crucial to ensure that the training course is meeting its objectives. A training evaluation should be carried out at as many levels of Kirkpatrick's hierarchy as is practicable, and an evaluation design appropriate to meet the evaluation issues that need to be addressed.

Suggestions for further reading

Training

Cannon-Bowers, J. and Salas, E. (eds.) (1998) *Making Decisions Under Stress: Implications for Individual and Team Training*. Washington, DC: American Psychological Association.

Goldstein, I.L. and Ford, K.J. (2002) *Training in Organizations: Needs Assessment, Development, and Evaluation*. Belmont, CA: Wadsworth.

Kirkpatrick, D.L. (1998) *Evaluation Training Programs*. San Francisco: Berrett-Koehler.

Truelove, S. (1997) *Training in Practice*. Oxford, UK: Blackwell.

Crew Resource Management (also see references in Chapter 1)

CAA (2006) *Crew Resource Management (CRM) Training*. London: Author. (This report can be downloaded from www.caa.co.uk; it contains an extensive list of CRM resources.)

CAA (2002) *Flight Crew Training: Cockpit Resource Management (CRM) and Line-oriented Flight Training (LOFT)*. London: Author. (This report can be downloaded from www.caa.co.uk.)

FAA (2004) Advisory Circular No 120-51E: Crew resource management training. Washington, DC: Author.

Macleod, N. (2005) *Building Safe Systems in Aviation. A CRM Developers Handbook*. Aldershot: Ashgate.

McAllister, B. (1997) *Crew Resource Management: The Improvement of Awareness, Self-discipline Cockpit Efficiency and Safety*. London: Crowood Press.

Okray, R. and Lubnau, T. (2004) *Crew Resource Management for the Fire Service*. Tulsa: PennWell.

Salas, E., Wilson, K.A, Burke, C.S., Wightman, D.C. and Howse, W.R. (2006b) A checklist for Crew Resource Management training. *Ergonomics in Design*, 14, 6–15.

Salas, E., Wilson, K.A., Burke, C.S., Wightman, D.C., and Howse, W. (2006c) Crew resource management training research and practice: A review, lessons learned and needs. In R.C. Williges (ed.) *Review of Human Factors and Ergonomics Series, Volume 2* (pp. 30–55). Santa Monica, CA: Human Factors and Ergonomics Society.

Salas, E., Bowers C. and Edens, E. (eds.) (2001) *Improving Teamwork in Organizations: Applications of Resource Management Training*. Mahwah, NJ: Lawrence Erlbaum Associates.

Wiener, E.L., Kanki, B.G. and Helmreich. R.L. (eds.) (1993) *Cockpit Resource Management*. San Diego: Academic Press.

Websites

Civil aviation authority: www.caa.co.uk

Federal Aviation Administration Human Factors and Engineering Group: www.hf.faa.gov/

Joint Aviation Authorities Human Factors Steering Group: www.hfstg.org

Neil Krey's CRM Developers: www.crm-devel.org

Royal Aeronautical Society Human Factors Group: www.raes-hfg.com/

University of Texas Human Factors Research Project homepage: psy.utexas.edu/homepage/group/HelmreichLAB/

References

Awad, S. et al. (2005) Bridging the communication gap in the operating room with medical team training. *American Journal of Surgery*, 190, 770–774.

Baker, D., Salas, E., Barach, P, Battles, J. and King, H. (2007) The relation between teamwork and patient safety. In P. Carayon (ed.) *Handbook of Human Factors and Ergonomics in Health Care and Patient Safety.* Mahwah: N.J. Lawrence Erlbaum.

Bleakley, A., Hobbs, A., Boyden, J. and Walsh, L. (2004) Safety in operating theatres. Improving teamwork through team resource management. *Journal of Workplace Learning*, 16, 83–91.

Blickensderfer, E., Cannon-Bowers, J.A. and Salas, E. (1998) Cross training and team performance. In J.A. Cannon-Bowers and E. Salas (eds.) *Making Decisions Under Stress: Implications for Individual and Team Training* (pp. 299–311). Washington, DC: American Psychological Association.

Blickensderfer, E., Cannon-Bowers, J.A. and Salas, E. (1997) Theoretical bases for team self-correction: fostering shared mental models. *Advances in Interdisciplinary Studies of Work Teams*, 4, 249–279.

Boehm-Davis, D.A., Holt, R.W. and Seamster, T.L. (2001) Airline resource management programs. In E. Salas, C. Bowers and E. Edens (eds.) *Improving Teamwork in Organizations* (pp. 191–217). Mahwah, NJ: Lawrence Erlbaum Associates.

CAA (2006a) *Crew Resource Management (CRM) Training*. London: Author.

CAA (2006b) *Guidance Notes for Accreditation Standards for CRM Instructors and CRM Instructor Examiners. Standards Doc. 29 Version 2*. London: Author.

CAA (2002) *Flight Crew Training: Cockpit Resource Management (CRM) and line-oriented Flight Training (LOFT)*. London: Author.

Chidester, T.R., Helmreich, R., Gregorich, S.E. and Geis, C.E. (1991) Pilot personality and crew coordination: Implications for training and selection. *The International Journal of Aviation Psychology*, 1, 25–44.

Crichton, M. and Flin, R. (2003) BP Drilling/Upstream technology: Enhanced teamwork project. Aberdeen: University of Aberdeen.

Crichton, M., Flin, R. and Rattray, W.A. (2000) Training decision makers – Tactical Decision Games. *Journal of Contingencies and Crisis Management*, 8, 208–217.

Dismukes, R.K. and Smith, G.M. (2000) *Facilitation in Aviation Training and Operations*. Aldershot: Ashgate.

Drury, C.G.T. (1998) Establishing a human factors/ergonomics program. In FAA (ed.) *The World Wide Web Edition of the Human Factors Issues in Aircraft Maintenance and Inspection 3.0 CD-ROM 1998* (chapter 2). Available: http://hfskyway.faa.gov/document.htm [2001, 3/5/01].

Dwyer, D.J., Oser, R.L., Salas, E. and Fowlkes, J. (1999) Performance measurement in distributed environments: Initial results and implications for training. *Military Psychology*, 11, 189–215.

Federal Aviation Administration (2004) *Advisory Circular No 120-51E: Crew Resource Management Training*. Washington, DC: Author.

Flin, R., O'Connor, P. and Mearns, K. (2002) Crew Resource Management: Improving safety in high reliability industries. *Team Performance Management*, 8, 68–78.

Flin, R., Yule, S., McKenzie, L., Paterson-Brown, S. and Maran, N. (2006) Attitudes to teamwork and safety in the operating theatre. *The Surgeon*, 4, 145–151.

Flin, R., Yule, S., Paterson-Brown, S., Maran, N., Rowley, D. and Youngson, G. (2007) Teaching surgeons about non-technical skills. *The Surgeon*, 5, 107–110.

Gliner, J.A., Haber, E. and Weise, J. (1997) Use of controlled vignettes in evaluation: Does type of response method make a difference? *Evaluation and Program Planning*, 22, 313–322.

Goldstein, I.L. (1993) *Training in Organizations*. Belmont, CA: Wadsworth.

Goldstein, I.L. and Ford, K.J. (2002) *Training in Organizations: Needs Assessment, Development, and Evaluation*. Belmont, CA: Wadsworth.

Gregorich, S.E., Helmreich, R.L. and Wilhelm, J.A. (1990) The structure of Cockpit Management Attitudes. *Journal of Applied Psychology*, 75, 682–690.

Gregorich, S.E. and Wilhelm, J.A. (1993) Crew resource management training assessment. In E.L. Wiener, B.G. Kanki and R.L. Helmreich (eds.) *Cockpit Resource Management* (pp. 173–196). San Diego: Academic Press.

Harrington, D.K. and Kello, J.E. (1992 June) Systematic evaluation of nuclear operator team skills training: A progress report. Paper presented at the STL conference on Human Factors and Power plants, Monterey, California.

Heiman, G.A. (1995) *Research Methods in Psychology*. Boston, MA: Houghton Mifflin.

Helmreich, R.L. (2000) On error management: Lessons from aviation. *British Medical Journal*, 320, 781–785.

Helmreich, R.L. (1984) Cockpit management attitudes. *Human Factors*, 26, 583–589.

Helmreich, R.L. and Merritt, A.C. (1998) *Culture at Work in Aviation and Medicine*. Aldershot: Ashgate.

Helmreich, R.L., Merritt, A.C. and Wilhelm, J.A. (1999) The evolution of crew resource management training in commercial aviation. *International Journal of Aviation Psychology*, 9, 19–32.

Helmreich, R.L. and Wilhelm, J.A. (1991) Outcomes of Crew Resource Management training. *International Journal of Aviation Psychology*, 1, 287–300.

Howard, S., Gaba, D., Fish, K., Yang, G. and Sarnquist, F. (1992) Anesthesia crisis resource management training: Teaching anesthesiologists to handle critical incidents. *Aviation, Space and Environmental Medicine*, 63, 763–770.

Howell, D.C. (1992) *Statistical Methods for Psychology*. Belmont, CA: Wadsworth Inc.

Ilgen, D.R., Fisher, C.D. and Taylor, M.S. (1979) Consequences of individual feedback on behaviour in organizations. *Journal of Applied Psychology*, 64, 531–545.

Irwin, C.M. (1991) The impact of initial and recurrent cockpit resource management training on attitudes. In R. Jensen (ed.) *Proceedings of the Sixth International Symposium on Aviation Psychology* (pp. 344–349). OH: Ohio State University.

JAA (2006) *JAR-OPS, 1 Subpart N, Crew Resource Management Flight Crew (Amendment 12).* Hoofddorp, Netherlands: Author.

Jensen, R. (1995) *Pilot Judgment and Crew Resource Management.* Aldershot: Ashgate.

Kerlinger, F. (1996) *Foundations of Behavioural Research.* New York: Holt, Rinehart and Winston.

Kirkpatrick, D.L. (1976) Evaluation of training. In R.L. Craig and L.R. Bittel (eds.) *Training and Development Handbook* (pp. 18.1–18.27). New York: McGraw Hill.

Kirkpatrick, D.L. (1998) *Evaluation Training Programs.* San Francisco: Berrett-Koehler.

Klinect, J.R., Wilhelm J.A. and Helmreich, R.L. (1999) Threat and error management: data from line operations safety audits. In R. Jensen (ed.) *Proceedings of the Tenth International Symposium on Aviation Psychology* (pp. 683–688). OH: Ohio State University.

Kozlowski, S.W.J., Gully, S.M., McHugh, P.P., Salas, E. and Cannon-Bowers, J.A. (1996) A dynamic theory of leadership and team leader effectiveness: Developmental and task contingent leader roles. In G.R. Ferris (ed.) *Research in Personnel and Human Resource Management* (Vol. 14, pp. 253–305). Greenwich, C.T.: JAI Press.

Lambert, T. (1993) *Key Management Tools*. London: Pitman.

Lauber, J.K. (1984) Resource Management in the cockpit. *Air Line Pilot*, 53, 20–23.

Macleod, N. (2005) *Building Safe Systems in Aviation. A CRM Developers Handbook.* Aldershot: Ashgate.

McAllister, B. (1997) *Crew Resource Management: The Improvement of Awareness, Self-discipline Cockpit Efficiency and Safety.* London: Crowood Press.

McCann, C., Baranski, J.V., Thompson, M.M. and Pigeau, R.A. (2000) On the utility of experiential cross-training for team decision making under time stress, *Ergonomics*, 43, 1095–1110.

Njaa, O. (2000, June) On the use of accident scenarios. Paper presented at the SPE conference on Health, Safety, and the Environment in Oil and Gas Exploration, Stavanger, Norway.

Norton, S.M. (1992) Peer assessment of performance and ability: An exploratory meta-analysis of statistical artefacts and contextual moderators. *Journal of Business and Psychology*, 6, 387–399.

O'Connor, P. (2005) An investigation of the non-technical skills required to maximize the safety and productivity of U.S. Navy divers. Panama City, FL: Navy Experimental Diving Unit, Research Report 05-03.

O'Connor, P. and Flin, R. (2003) Crew Resource Management training for offshore teams. *Safety Science*, 41, 591–609.

O'Connor, P., Flin, R. and Fletcher, G. (2002) Methods used to evaluate the effectiveness of CRM training: A literature review. *Human Factors and Aerospace Safety*, 2, 217–234.

Okray, R. and Lubnau, T. (2004) *Crew Resource Management for the Fire Service.* Tulsa: PennWell.

Paris, C.R., Salas, E. and Cannon-Bowers, J.A. (1999) Human performance in multi-operator systems. In P.A. Hancock (ed.) *Human Performance and Ergonomics* (pp. 329–386). San Diego, CA: Academic Press.

Rall, M. and Gaba, D. (2005) Patient simulators. In R. Miller (ed.) *Miller's Anesthesia.* Philadelphia: Elsevier.

Reason, J. (1990) *Human Error.* Cambridge: Cambridge University Press.

Riley, R. (ed.) (in press) *Simulation in Healthcare. Oxford: Oxford University Press.*

Robertson, M.M. (2001) Resource management for aviation maintenance teams. In E. Salas, C.A. Bowers and E. Edens (eds.) *Improving Teamwork in Organizations: Applications of Resource Management Training* (pp. 235–264). Mahwah, NJ: Lawrence Erlbaum Associates.

Royal Aeronautical Society (1999) Discussion document: crew resource management. London: Royal Society.

Salas, E. and Cannon-Bowers, J.A. (1997) Methods, tools and strategies for team training. In M. Quinones and E. Ehrestein (eds.) *Training for a Rapidly Changing Workplace: Applications in Psychological Research* (pp. 291–322). Washington, DC: American Psychological Association Press.

Salas, E. and Cannon-Bowers, J.A. (2000) The anatomy of team training. In L. Tobias and D. Fletcher (eds.) *Training and Retraining: A Handbook for Business, Industry, Government, and the Military* (pp. 312–335). New York: Macmillan.

Salas, E., Burke, C., Bowers, C. and Wilson, K. (2001) Team training in the skies: Does crew resource management (CRM) training work? *Human Factors*, 41, 161–172.

Salas, E., Wilson, K., Burke, C. and Wightman, D. (2006a). Does CRM training work? An update, extension and some critical needs. *Human Factors*, 14, 392–412.

Salas, E., Wilson, K., Burke, C., Wightman, D. and Howse, W. (2006b) A checklist for Crew Resource Management Training. *Ergonomics in Design*, 14, 6–15.

Salas, E., Wilson, K., Burke, C., Wightman, D. and Howse, W. (2006c) Crew resource management training research and practice: A review, lessons learned and needs. In R. Williges (ed.) *Review of Human Factors and Ergonomics Series, Volume 2* (pp. 30–55). Santa Monica, CA: Human Factors and Ergonomics Society.

Smith-Jentsch, K., Zeisig, R., Acton, B. and McPherson, J.(1998) Team dimensional training: A strategy for guided team self-correction. In J.A. Cannon-Bowers and E. Salas (eds.), *Making Decisions Under Stress: Implications for Individual and Team Training* (pp. 271–298). Washington, DC: American Psychological Association.

Tannenbaum, S.I., Smith-Jentsch, K.A. and Behson, S.J. (1998) Training team leaders to facilitate team learning and performance. In J.A. Cannon-Bowers and E. Salas (eds.) *Making Decisions Under Stress: Implications for Individual and Team Training* (pp. 247–270). American Psychological Association.

Taylor, J.C. (1998 May) Evaluating the effectiveness of Maintenance Resource Management (MRM). Paper presented at the 12th International Symposium on Human Factors in Aviation Maintenance.

Taylor, J.C. (2000 April) A new model for measuring return on investment (ROI) for safety programs in aviation: An example from airline maintenance resource management (MRM). Paper presented at the Advances in Aviation Safety Conference, Paper number 2000-01-2090, Daytona Beach, Florida.

Truelove, S. (1997) *Training in Practice*. Oxford: Blackwell.

Volpe, C.E., Cannon-Bowers, J.A. and Salas, E. (1996) The impact of cross-training on team functioning: An empirical investigation. *Human Factors*, 38, 87–100.

Walters, A. (2002) *Crew Resource Management is No Accident.* Wallingford, UK: Aries.

Wiener, E., Kanki, B. and Helmreich, R. (eds.) (1993) *Cockpit Resource Management.* San Diego: Academic Press.

Woldring, M. and Isaac, A. (1999) *Team Resource Management Test and Evaluation* (HUM.ET1.ST10.2000 -REP-01): EUROCONT. Brussels: European Air Traffic Management Programme.

Chapter 11

Assessing Non-Technical Skills

Introduction

As a growing number of professions begin to identify and train non-technical skills for safety-critical positions, the issue of assessment becomes increasingly relevant. In the context of this book, assessment means the processes of observing, recording, interpreting and evaluating individual performance, usually against a standard defined by a professional body, a company or a safety regulator. For example, the UK Civil Aviation Authority (2006a) state, 'In order to ascertain whether CRM training has been effective, it would be necessary to assess the CRM skills of flight crew members from time to time. CRM assessment is inevitable and essential if standards that address this major threat to safety are to be maintained and improved' (ch7, p1).

Evaluation or assessment of non-technical skills can be for several different purposes:

a. to give trainees feedback on their skill development (Patey et al., 2005; Yule et al., 2007a)
b. for testing skills in a competence-assurance or licensing programme (Flin, 2006)
c. to ascertain whether a non-technical skills or crew resource management (CRM) training programme is effective and transferring skills to the workplace (Goeters, 2002; O'Connor et al., 2002a; 2002b; Salas et al., 2001; 2006), and
d. to audit the level of skill demonstrated in a work unit (e.g. aircraft fleets) (FAA, 2006b; Helmreich et al., 2003). For instance, to build performance databases to identify norms and prioritise training needs for individuals, teams, departments.

This chapter presents an overview of how to evaluate non-technical skills using ratings of observed work behaviour (sometimes called behavioural marker systems). Behaviour rating systems in the workplace are mainly used for performance appraisal, selection and task analysis (Cook and Cripps, 2005; Fletcher, 2004; Woodruffe, 2000) and by psychologists as research tools. These systems are expensive to develop and utilise given the level of training and calibration required for users. Consequently, they have mainly been developed for occupations where safety is prime and simulators are used for training and assessment, e.g. in aviation, nuclear power generation, military settings and increasingly in medicine. By the late 1990s, non-technical/CRM skill sets and associated rating scales had been introduced to measure pilots' performance in several airlines (Antersijn and Verhoef, 1995; Flin and Martin, 2001; Helmreich, 1999). Behaviour rating systems can look deceptively simple. Before describing how they are designed, it is important to emphasise that any such system has a number of limitations:

- They can never capture every aspect of performance.
- There will be limited opportunity to observe some behaviours, such as important but infrequent behaviours (e.g. conflict management).
- The human observers have their own limitations – bias, distraction, overload (e.g. for complex situations, large teams).

Formal assessment using behavioural rating systems is most advanced for pilots in the civil aviation industry (CAA, 2006a; MacLeod, 2005)[1] and much of the material in this chapter has been drawn from aviation researchers (Baker et al., 2001, Baker and Dismukes, 2002; Beaubien et al., 2004; Holt et al., 2001). The chapter also reflects some general guidance devised by psychologists and practitioners who have designed behaviour rating systems in nuclear power and acute medicine, as well as in aviation (Klampfer et al., 2001, reproduced in Appendix 11 of CAA, 2006a). The key features of behaviour rating systems are described, along with recommendations for their design and use in the workplace (or simulated work environment). In the appendix of this chapter, a description is given of the development of NOTECHS (Flin et al., 2003; van Avaermate and Kruijsen, 1998), a non-technical skills taxonomy and behaviour rating system designed in Europe for pilots.

The focus of this final chapter is mainly on the assessment of an individual's non-technical skills by observing his or her behaviour at work or in a simulator, and usually in a team setting. There are also group-level assessment systems, typically used by researchers for rating observed aspects of a crew's or a team's performance. These are not reviewed in detail, with the exception of LOSA (Line Oriented Safety Audit, Helmreich, 1999), which is widely used in aviation to record ratings of a flight crew. The LOSA method evolved from the first behaviour rating method, University of Texas Behavioural Markers, which was devised for evaluating airline pilots' non-technical skills. This formed the basis of many airlines' rating systems (Flin and Martin, 2001) as well as systems devised for health care (Gaba et al., 1998) and the nuclear industry (O'Connor et al., 2002d).

Origins and evolution of the University of Texas (UT) Behavioural Markers

The first behavioural marker system developed for pilots (the NASA/UT Behavioural Markers) originated in the University of Texas Human Factors Research Project in the late 1980s. This project had two goals: the first was to evaluate the effectiveness of CRM training as measured by observable behaviours of flight deck crew, while the second was to aid in defining the scope of CRM programmes. This research produced the first manual to assist examiners in assessing the interpersonal component of flying (Helmreich and Wilhelm, 1987; Helmreich et al., 1990). Ratings of crew performance were made by observers assessing a complete flight from initial briefing to landing, taxi in and shutdown of engines (see Table 11.1).

1 In aviation, this is sometimes part of an assessment called line operational evaluation (LOE) (FAA, 2006a).

Table 11.1 University of Texas (UT) behavioural markers

Key to phase: P=Pre-departure/taxi; T=Take-off/climb; D=Descent/approach/land; G=Global

Markers	Definition	Anchors (examples)	Phase
Sop Briefing	The required briefing was interactive and operationally thorough	- Concise, not rushed, and met SOP requirements - Bottom lines were established	P–D
Plans Stated	Operational plans and decisions were communicated and acknowledged	- Shared understanding about plans – 'Everybody on the same page'	P–D
Workload Assignment	Roles and responsibilities were defined for normal and non-normal situations	- Workload assignments were communicated and acknowledged	P–D
Contingency Management	Crew members developed effective strategies to manage threats to safety	- Threats and their consequences were anticipated - Used all available resources to manage threats	P–D
Monitor/ Cross-check	Crew members actively monitored and cross-checked systems and other crew members	- Aircraft position, settings and crew actions were verified	P–T–D
Workload Management	Operational tasks were prioritised and properly managed to handle primary flight duties	- Avoided task fixation - Did not allow work overload	P–T–D
Vigilance	Crew members remained alert of the environment and position of the aircraft	- Crew members maintained situational awareness	P–T–D
Automation Management	Automation was properly managed to balance situational and/or workload requirements	- Automation set-up was briefed to other members - Effective recovery techniques from automation anomalies	P–T–D
Evaluation Of Plans	Existing plans were reviewed and modified when necessary	- Crew decisions and actions were openly analysed to make sure the existing plan was the best plan	P–T
Inquiry	Crew members asked questions to investigate and/or clarify current plans of action	- Crew members not afraid to express a lack of knowledge - 'Nothing taken for granted' attitude	P–T
Assertiveness	Crew members stated critical information and/or solutions with appropriate persistence	- Crew members spoke up without hesitation	P–T
Communication Environment	Environment for open communication was established and maintained	- Good cross-talk – flow of information was fluid, clear and direct	G
Leadership	Captain showed leadership and co-ordinated flight deck activities	- In command, decisive and encouraged crew participation	G

Rating scale

1	2	3	4
Poor Observed performance had safety implications	**Marginal** Observed performance was barely adequate	**Good** Observed performance was effective	**Outstanding** Observed performance was truly noteworthy

The first set of crew behavioural markers developed by Helmreich's group was included by the Federal Aviation Administration (FAA) as an appendix to its Advisory Circular on CRM (AC 120-51A, FAA, 1993). Systematic use of the markers grew as airlines enhanced their assessment of crew performance and as more detailed data were collected on an airline's operations. The markers themselves were incorporated in a rating form for systematic observations known as the Line/LOS Checklist (LOS refers to line operational or full mission simulation). As experience and the database of observations grew, it became apparent that there was significant variability in crew behaviour during flights that needed to be captured. Accordingly, the rating form was modified to assess the markers for each phase of flight (Helmreich et al., 1995a).

A validation of the markers was undertaken by classifying their impact (positive and negative) in analyses of aviation accidents and incidents (Helmreich et al., 1995b). The results provided strong support for the utility of the markers as indicators of crew performance and their value as components of CRM training. The UT markers, showing phase of flight where rated, are listed in Table 11.1. They are used in non-jeopardy observations of crews conducting normal line flights (see FAA, 2006b). The rating scale is shown above and has four scale points.

Each of these markers has been validated as relating to either threat and error avoidance or management. With the exception of two global ratings, specific markers are rated (if observed) during particular phases of flight. It should be noted that these scales are used to rate the non-technical skills of a crew (normally of two pilots) rather than an individual pilot. Moreover, the focus is mainly on interpersonal skills and tasks such as briefing, rather than on cognitive skills such as situation awareness or decision-making.

Designing a behaviour rating system

As Table 11.1 illustrates, the two components of a behavioural rating system are (i) the list of specified behaviours or categories of behaviour to be rated and (ii) the rating scale. These have to be combined into a scoring form that practitioners can easily use in the workplace to record their ratings. Moreover, the whole system needs to meet a set of design requirements for the environment where it is to be used.

i) System content: specified skills/behaviours

The skills that are to be assessed have to be specified and this is normally done by providing skill definitions and behavioural examples. In Chapter 9, we described how the non-technical skills (NTS), which contribute to superior or substandard performance in a given occupation, can be identified. These are usually then structured into a set of categories, containing sub-components, and examples of good and poor behaviours. These levels are labelled differently in various systems. For example, in ANTS they are called 'categories', 'elements' and 'markers', and in the University of Texas system shown above, the target behaviours are called 'anchors'. Figure 11.1 shows the basic structure of the ANTS system (Fletcher et al., 2003), with behavioural markers (examples of good and poor practice) for two elements.

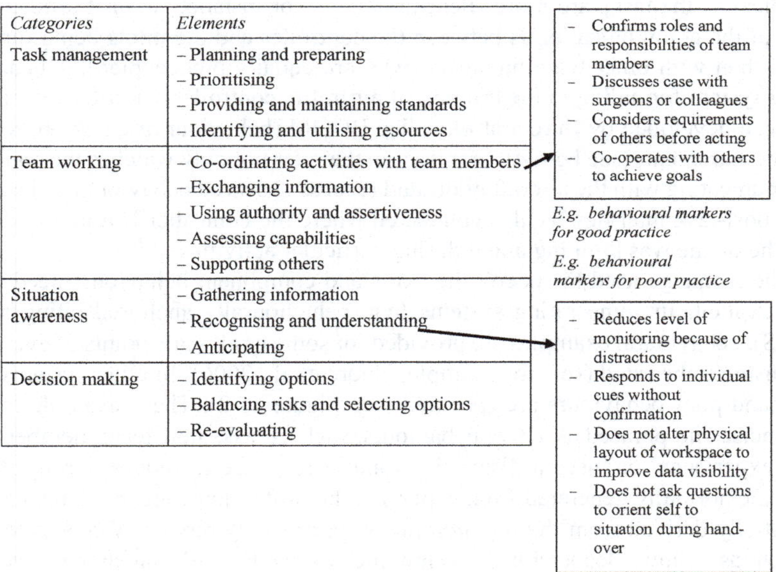

Figure 11.1 The structure of ANTS (Anaesthetists' Non-technical Skills)

The categories of skill present in a NTS taxonomy or CRM training syllabus (see Chapter 10) will not all necessarily be used in the associated evaluation method because a behaviour rating system that relies on observers has to focus only on observable behaviours. Thus, while stress and fatigue management are usually included as topics in CRM training, they can be difficult to rate unless extreme symptoms are displayed (in which case other skills will be affected) and so will not necessarily be included as a category in a behaviour rating system.

Not all systems include communication as an explicit category, on the basis that communication is the main medium by which almost all the behaviours relevant to all the categories will be conveyed. There is also some debate as to whether cognitive skills should be included in these rating systems, as they cannot be directly observed. It is generally agreed that it is acceptable to include cognitive skills if there are a set of observable behaviours from which they could be inferred. So, for example, we cannot see into a surgeon's head to determine what decision is being made. But the rater should be able to observe the surgeon communicating about her thinking, such as voicing the need for a decision, consulting other team members, agreeing with the anaesthetist what the best option is, discussing how long it will take to complete the case. Therefore a number of systems do include cognitive components, usually situation awareness and decision-making, when there are behaviours that reflect these mental activities. In fact, in many high-risk environments practitioners are trained to verbalise to other team members task-related thoughts, and behavioural intentions, in order to maximise shared understanding across the team.

There are, however, workplaces where rather less of this kind of communication is typical as the tasks are more silent in nature. For instance, in air traffic control, most of the communication is between the controller and the pilots being directed, rather than with other team members (who are equally preoccupied). A behaviour rating system for aiding in the training of air traffic controllers' non-technical skills has been developed by Eurocontrol, called BOOM (behaviour oriented observation and measurement) (see below). In this case, the rater not only watches the controller communicating with the aircraft pilots and records observed behaviours and context, but a post-task interview is also conducted where the controller is asked to discuss what he or she was thinking about during particular activities.

The language used to specify the skills and component behaviours needs to be very explicit. In some rating systems (e.g. behaviourally anchored rating scales; BARS), behavioural examples are provided for some or all scale points of every skill dimension to be rated (see, for example, Undre et al., 2006). In other systems, only good and poor behaviours are specified (see Figure 11.2). These examples should be concise and phrased as active behaviours, such as 'asks new team member about their experience' or 'uses a dismissive tone in response to requests from others'. They are normally generated from experts in the profession concerned, but can also be initially derived from the research literature or safety reports. Vague statements – such as 'shows leadership', 'communicates well', and statements based on unobservable processes, e.g. 'thinks about options', should be avoided. The need to present and describe the system in simple, precise language is particularly important when it may be used by practitioners in their non-native language. The number of skills or behaviours to be rated will depend on what the system is being used for and the skill of the raters. Beaubien et al. (2004) advise evaluating as few skills per event set as possible because this reduces workload and gives the rater more time for observation. They suggest 'including only those skills that are mission-critical' (p8).

ii) The rating scale and rating form

Rating scales come in a wide range of formats, for recording the quality and/or frequency of a given behaviour. They tend to be linear scales and may be specified only at the end points (anchors), such as low to high (a bipolar scale) or each point on the scale can be defined by words or numerals. The number of points on the scale (granularity) may be as few as two (e.g. pass/fail) or as many as 100. There may be an even number of scale points – which means there is no mid-point, or an uneven number. If the scale is too long, raters may have difficulty in making the fine-grained judgements that are required. If the scale is too short, then it may not allow the required degree of discrimination. In terms of choosing one scale format or length in preference to another, there are no fixed rules; only that they need to be fit for purpose. For rating behaviours related to non-technical/CRM skills, the most common scale format is a four- or five-point Likert-type scale (Beaubien et al., 2004; Holt et al., 2001), which shows each skill on the left and the rating scale on the right, with the scale points defined (see Figure 11.2). Macleod (2005) gives an example of four- and five-point rating scales used in the same airline to assess CRM skills:

> Five-point scale: Poor – Below Average – Average – Above Average – Outstanding
> Four-point scale: Poor – Minimum Expectation – Standard – Outstanding

The international airline Emirates assesses eight aspects of technical (e.g. aircraft handling) and non-technical (e.g. situation awareness) skills demonstrated by their pilots. These are rated with reference to the five-point scale shown below in Table 11.2, and for each category, the five performance levels are also specified in relation to that particular skill.

Table 11.2 Example of an airline's rating system with definitions of scale points

5. Very good	Performance of a very good standard. Errors may be made, but are trapped or managed appropriately. Minor points may still be discussed.
4. Good	Performance to a good standard. Inconsequential errors may be made that are not either trapped or managed appropriately. Points may require debrief, however displays sound underlying proficiency.
3. Satisfactory	Performance is satisfactory. Consequential errors may not be trapped or may only be managed with difficulty, however safety margin retained. No concerns raised about underlying proficiency. Could display opportunities to improve performance.
2. Minimum acceptable	Performance is acceptable. Errors are mismanaged or ignored inappropriately and safety margin reduced. Although competent, displays some weaknesses. Would benefit from some additional training.
1. Unacceptable	Performance is unacceptable. Errors are mismanaged or ignored inappropriately with potential for adverse outcomes. Safety compromised or considerable concerns over ability to operate effectively. Retraining is required.

Reproduced with permission of Emirates.

The question of establishing the criterion or threshold for successful performance can be extremely important when behavioural ratings are being used for formal assessment. There has to be a method for determining whether the overall performance observed was acceptable or not. In some systems, an additional judgement of 'acceptable/not acceptable' can be recorded. When used for formal assessment, this may have to be based on a judgement of both technical, as well as non-technical skills. An overall performance rating might be rephrased as 'pass/fail' or 'pass/refer for further training' or 'competent/not yet competent', depending on the operational requirements for the system. Organisations will usually devise their own company standard and procedures for this threshold judgement. In some rating systems, there may be a specific decision rule that has to be applied. For example a fail/low score in any one category of non-technical skill will produce an overall fail or there may even be particular behaviours, single actions or specific omissions that constitute a failure (see Rall and Gaba, 2005).

Most rating forms have a space for the rater to note down any observations about a particular skill component. These notes can be made during the observation or afterwards. In practice, many rating forms do not have much space to write comments and some raters prefer to record their own notes in a notebook and then later transfer their notes to the rating form. The guidance provided for the operational use of the NOTECHS system advises that a written reason should always be recorded when a rating of unacceptable is given. Beaubien et al. (2004) recommend that all skill ratings should be given a reason or justification. They give the example of airlines inadvertently rewarding their instructors for rating most crews as average or better, when the pilot instructors (i.e. the raters) only have to record a reason code or handwritten note next to a rating if it is less than average.

An example from the NOTSS system for rating the behaviour of surgeons is shown in Figure 11.2. This uses a four-point scale (1 is unsatisfactory and 4 is good) and the rater's notes for each judgement are recorded on the form as well as the scores.

Hospital: **Date:**		**Trainer name:** **Trainee name:**		**Operation:** Right inguinal hernia repair
Category	**Category rating***	**Element**	**Element rating***	**Feedback on performance and debriefing notes**
Situation Awareness	3	Gathering information	3	Didn't mark side/ arrived in theatre late
		Understanding information	4	Aware of INR importance and checked
		Projecting and anticipating future state	3	Take more of a lead in op – i.e. requesting retractions
Decision Making	3	Considering options	3	Be more explicit about relative merits of options
		Selecting and communicating option	2	Not sure about sutures/ mesh sizes etc...
		Implementing and reviewing decisions	4	Readily vocalised concerns
Communication and Teamwork	2	Exchanging information	2	Did not relate well to anaesthetist
		Establishing a shared understanding	3	Waited for trainer to take the lead
		Co-ordinating team activities	3	Did not enquire about pt condition from anaesthetist
Leadership	3	Setting and maintaining standards	4	Followed theatre protocol but didn't mark side
		Supporting others	4	Good rapport with trainee OPD and scrub nurse
		Coping with pressure	3	At times seemed to carry on regardless – oblivious to important anatomy – too focused on other things

Figure 11.2 Completed NOTSS rating form for a trainee general surgeon's performance on a hernia repair (from Yule et al., 2007a)

Box 11.1 Key properties for a behavioural rating system

- **Sensitivity**: The system should be based on detectable behaviours that differentiate performance. So, for example, raters can distinguish between behaviours indicating poor leadership from behaviours indicative of good leadership.

- **Reliability**: This relates to the consistency or stability of the measurement. Three aspects of reliability that are of particular relevance here are:
  a. *test–retest* – assesses stability over time. Raters would be asked to make the same judgements on two occasions and these would be compared (correlated).
  b. *internal reliability or consistency* – tests the level of inter-correlation between a set of items intended to be measuring the same construct (e.g. the inter-correlation of scores on elements of a skill category called decision-making).
  c. *inter-rater reliability* – measures whether the raters using the system are applying it in the same way and are showing agreement in their ratings.

- **Validity**: Refers to the extent to which a measure really assesses the construct. The behaviour ratings should accurately reflect real differences in the skills being measured. The skills and behaviours being assessed should also be related to the performance outcome of interest (e.g. safety). Relevant aspects of validity include:
  a. *face validity* – is whether the items look to practitioners as if they are measuring the appropriate construct. This is not a true measure of validity but if face validity is low (i.e. content of scale does not look relevant), then this can influence practitioner acceptance of a measure.
  b. *construct validity* – is whether the rating scale is actually measuring what it claims to measure. This can be assessed in different ways, such as comparing the new test with an established measure of the same construct (convergent validity) or by testing whether scores on the test actually relate to the key outcome measure (criterion validity). So for non-technical skills, this would be whether higher ratings actually relate to better safety and efficiency of practice.

- **Structure**: minimal overlap between components (e.g. categories).

- **Transparency**: those being rated understand the performance criteria against which they are being rated. The reliability and validity data should be available to show the system properties.

- **Usability**: the system needs to be usable – i.e. the framework is simple, easy to understand, has domain-appropriate language, is sensitive to rater workload, the target behaviours are easy to observe, and raters can be trained to use it.

- **Baselines** for performance criteria are used appropriately for the experience level of ratee (i.e. *ab initio*/trainees vs. more experienced practitioners).

It is also possible to use a behaviour checklist rather than a rating scale, where the observer simply notes whether a particular behaviour was demonstrated or not, without giving it a qualitative or frequency rating. Beaubien et al. (2004) suggest that a checklist can help to reduce the rater's workload, although it probably does not provide such useful data for assessment and feedback. For more detailed advice on rating scales, see Aiken (1996) or De Vellis (2003).

Behaviour rating systems: psychometric properties

Underlying the design of behaviour rating systems are a set of psychometric properties. A more carefully designed instrument will produce a higher quality of measurement. The most relevant psychometric characteristics are shown above in Box 11.1.

This is a specialised field in psychology and is not discussed in depth here but does need to be covered when training raters. Holt et al. (2001) give more information on quality of measurement (when evaluating CRM training) relating to sensitivity, validity and reliability. Nunnally and Bernstein (1993) or AERA (1999) are comprehensive texts on psychological measurement.

Design requirements

A critical step in designing any measurement instrument, such as a behaviour rating system, is to ascertain the basic design requirements for function and usability: what is this system to be used for, under what conditions and who will be using it (Noyes, 2001). When practitioners want a system for rating non-technical skills, they usually require an assessment method that is going to be simple to understand and easy to use while they are running a simulation or observing a real event. For example, the design requirements for the NOTECHS system for rating pilots' non-technical skills included:

1. The system was to be used to assess the skills of an individual pilot, rather than a flight deck crew; it was to be suitable for use across Europe, by both large and small airlines, i.e. it was to be culturally-robust.
2. The instructors/examiners who would be making the non-technical skills rating would usually be making a technical rating of the pilot(s) at the same time. If this was in the simulator, they would also be required to operate the simulator as well as acting roles from the scenario, such as air traffic controllers or cabin crew. With such heavy task demands, it was therefore important that the rating scale was as simple as possible, with clear, concise terminology and that the rating form could be fitted onto one side of a single sheet of paper.

The above design guidelines are typical of what practitioners require; usually the whole rating system and accompanying guidance needs to be provided in a single, readable document. The ANTS and NOTSS systems, designed for anaesthetists and surgeons respectively (see below), are normally used for rating trainees when the consultant is observing, and may also be assisting in, the operation. Consequently, their rating forms were also designed to be as simple as possible and to fit onto one side of paper. The resulting booklets, which may be taken into the operating theatre, have 'wipeable' covers and spiral binding (so that the booklet could be laid open, allowing two pages to be read at once). These booklets explain the non-technical skills taxonomy and the rating system, as well as giving instructions and advice for use. These are produced in A5 size with about 10 pages because practitioners found this size useful, as it would fit easily into a bag or a pocket. The booklets have been made available via websites so that they can be downloaded and printed as required.

Using a behaviour rating system

This section discusses key aspects of using a behaviour rating system for evaluation. It outlines a number of points concerning the training and qualification of raters, then considers the practicalities of undertaking assessments on the job or in a simulator. Several important considerations should be taken into account when thinking of using a behavioural rating system to assess NTS, such as:

- Raters require extensive training (initial and recurrent). They also need to be 'calibrated' – that is, their ratings need to be anchored onto the rating scale in the same way.
- The rating systems do not transfer across domains and cultures without adaptation (e.g. from aviation to medicine or western behaviours in eastern cultures).
- The rating systems need proper implementation into an organisation, and need management and workforce support. It is recommended that a phased introduction is required to build confidence and expertise in raters and ratees. Consideration needs to be given to how rating information will be stored and accessed (as with any other performance data held on file).
- Application of the rating system must be sensitive to the stage of professional development of the individual, and to the maturity of the organisational and professional culture (e.g. whether used as a diagnostic, training and/or assessment tool).
- Any use of the system must consider contextual factors when rating behaviour (e.g. crew experience, workload, operating environment, operational complexity).

Conducting the assessment – online or in the simulator

Performance ratings can be made either when real performance is taking place (live/online/*in vivo*) or in a simulator, if one is available. For some professions, simulators are not widely available or, at their present stage of evolution, they may not provide high-fidelity task conditions. In this case, ratings of non-technical skills are more likely to be made during real workplace operations. There are some advantages of using real tasks, as fidelity is guaranteed. The disadvantages are that the trainer or rater has limited or no control over the events about to unfold during the task.

Usually the whole task from start to finish is observed and one overall set of ratings recorded, but this does not have to be the case. Ratings could be based on particular events (e.g. TARGETS, targeted acceptable responses to generated events or tasks; Fowlkes et al., 1994). Alternatively, specific phases of a task – such as stages of an operation for surgical teams (Undre et al., 2006) or stages of flight (UT/LOSA markers, see Table 11.1) – can be rated separately and there is some evidence that this improves the validity of the ratings (Beaubien et al., 2004). However, in both these examples, the ratings are recorded by specially trained observers who are not involved in the work tasks. In environments where the raters are also engaged in the task (or are monitoring it with a view to intervening if necessary), then an event or stage-based rating may be too demanding in terms of workload.

In safety-critical environments, such as operating theatres or flight decks, additional concerns arise when conducting ratings during live operations. The guidance given to anaesthetists for using the ANTS rating system in the operating theatre includes: carefully selecting suitable cases, not allowing teaching or the assessment to interfere with the case, and when circumstances dictate, abandoning the assessment. In many busy work environments, there may also be limited time for any debrief following task completion, depending on workloads and shift patterns. However, the feedback is an essential part of the process and should be scheduled as close to task completion as possible.

In the simulator, there is obviously far greater control for the raters, as the scenarios can be carefully designed to include specific events and to require the ratee to demonstrate particular skills and behaviours in response. More than one scenario may be given to test performance in different situations. If necessary, the simulation can be stopped to allow feedback. Ratings can be made of complete tasks or of certain phases or even individual events. Ample time can be scheduled into the session to allow proper feedback and discussion with those being rated. However, the rater may also be required to operate the simulator, which can significantly increase workload and interfere with the observation and evaluation. For advice on simulation, including designing scenarios and event sets, see Riley (in press) or CAA (2002).

Videotaping is normally available in simulators and the tape can be observed at a later point and the ratings made then, with the tape available for viewing in the feedback and discussion session. This is particularly useful for the rater to explain to the ratee which behaviours contributed to a particular rating and why. However, videotaping is not possible or permitted in all work environments, due to practical and ethical issues. (See Mackenzie and Xiao, 2007 for a recent review of videotaping in health care settings.) Software systems for coding video-captured data, such as OBSERVER (www.noldus.com) or RATE (Guerlain et al., 2005) can be easily adapted for non-technical skills assessment.

Formal assessment

When using a behaviour rating system for summative or high-stakes assessment, there are additional considerations because the formal assessment of non-technical aspects of performance presents significant challenges. Where professionals are being formally rated they may produce 'angel performance' that may not reflect their normal practice; however this applies to any formal assessment and not just to non-technical skills. The rating system should capture the context in which that assessment is made (e.g. crew dynamics and experience, operating environment, operational complexity, current conditions). For example, in a team, the behaviour of one crew member can be adversely or positively impacted by another, which could result in a substandard or inflated performance rating. Marker systems should be designed to detect and record such effects. In an assessment designed for competence assurance, then an overall pass/fail judgement has to be reached, as discussed above. There also need to be remedial mechanisms in place for dealing with candidates who fail or score poorly on particular skills. Where formal assessment of non-technical skills is required for licensing, such as in UK civil aviation, then clear procedures and policies have to be put in place. Specific guidelines may be developed, as shown in Box 11.2.

Box 11.2 Guidance for assessment of pilots' CRM skills (CAA, 2006a: Chapter 7, p2)

'CRM assessment should only be tied to the assessment of technical issues, and not carried out as a stand alone assessment. Suitable methods of assessment should be established, together with selection criteria and training requirements of the assessors, and their relevant qualifications, knowledge and skills.

A crew member should not fail a licence or type-rating evaluation unless this is associated with a technical failure... However, an Operator Proficiency Check should not be considered as being satisfactorily completed unless the CRM performance of the pilot meets with company requirements...

For individual CRM skills assessment, the following methodology is considered appropriate:

- An operator should establish the CRM training programme including an agreed terminology. This should be evaluated with regard to methods, length of training, depth of subjects and effectiveness.
- The CRM standards to be used (e.g. NOTECHS) have been agreed by crews, operators and regulators, and reflect best practice.
- The standards are clear, briefed, and published (in the Operations Manual).
- The methodology for assessing, recording and feeding back has been agreed and validated.
- Training courses are provided to ensure that crews can achieve the agreed standards.
- Procedures are in place for individuals who do not achieve the agreed standards to have access to additional training, and independent appeal.
- Instructors and examiners are qualified to standards agreed by all parties, and are required to demonstrate their competency to the CAA or such persons as the CAA may nominate.
- A training and standardisation programme for training personnel should be established.'

Similar guidance is available in other countries for the evaluation of pilots' CRM skills (e.g. FAA, 2006a). As indicated above, one of the main considerations for introducing a behaviour rating system is whether there are competent raters available to make the assessments.

Selecting and training raters

The basic entry requirements for personnel to be trained as raters are as follows:

- commitment to human factors principles (e.g. appreciates the significance of non-technical skills for safety and efficiency)
- domain knowledge
- formal training in applicable aspects of human factors or non-technical skills (e.g. crew resource management).

The training requirements for raters should not be underestimated. 'Any airline that chooses to introduce CRM assessment without first training its cadre of line training captains or simulator instructors is simply ensuring that the process will be conducted in a haphazard and unreliable fashion' (MacLeod, 2005: p157). The danger is that behaviour rating systems look deceptively simple. Considerable skill is required to make accurate observations and ratings, and to give constructive feedback to those being rated. The following are recommended by Klampfer et al. (2001) for a training course in behaviour rating and feedback.

- minimum 2–5 consecutive days' training (depending on prior experience)
- ideal group size of 8–12 people
- half-day training follow-up (e.g. meetings, feedback via telephone) after use of rating system.

The training for raters should deliver the following:

- Make explicit the goals for use of the rating system (e.g. formal assessment, developmental feedback, organisational audit).
- Explain the design of the rating system, as well as content and guidelines for its use.
- Review main sources of rater biases (e.g. halo, recency, primacy; see Box 11.3) with techniques to be used for minimisation of these influences.
- Present the concept of inter-rater reliability and the methods that will be used to maximise it.
- Illustrate and define each point of the rating scale and different levels of situational complexity with video examples, discussions and hands-on exercises.
- Provide practical training with multiple examples.
- Include calibration with iterative feedback on inter-rater reliability score.
- Teach debriefing skills as appropriate.
- Conclude with a formal assessment of rater competence.

A typical course might cover basic theory, practice in observation and rating using videotapes, role play with small groups of trainees, followed by instructor-observed rating sessions at the worksite or in the simulator. The purpose of rater training is to improve *observation* accuracy (i.e. correctly identifying and recording the behavioural information) and *rating* accuracy (i.e. assigning the correct rating (e.g. from a scale) to the particular behaviour that was observed). Baker et al. (2001) suggest that the former can best be addressed by behavioural observation training (BOT), which will deal with detection, recall and recording of behaviours. This can include training raters how to take notes during the evaluation and to restrict these to the actual behaviours observed rather than premature evaluations of them. Issues covered can include systematic errors of observation, avoiding contamination from prior knowledge and from over-reliance on single sources of information.

To improve rater accuracy, rater error training (RET) can be used. This is designed to familiarise raters with common sources of error in behaviour ratings, with the aim of minimising these by increasing awareness of the risks of error. The most common rating errors are due partly to the way in which we perceive and remember information and also to our skill in recognising our own prejudices and in discriminating and judging behaviours (see Box 11.3). MacLeod (2005), discussing the measurement of pilots' CRM skills, also mentions that raters need to be aware of their own personal standards or preferences of what constitutes good or bad behaviour in the workplace. For a more extensive discussion of the errors that humans can make when judging each other, see Cook and Cripps (2005).

Box 11.3 Common rating errors

Halo effect – one particular positive aspect is overemphasised and enhances the ratings on other dimensions.
Horns effect – one particular negative aspect is overemphasised and diminishes the ratings on other dimensions.
Central tendency – ratings mainly given around the mid-point of the scale.
Leniency – tendency to give favourable (higher) ratings.
Severity – tendency to give unfavourable (lower) ratings.
Primacy – remembering better/over-weighting behaviours that were observed first.
Recency – remembering better/over-weighting behaviours that were observed last.

The training materials also need to include information about the background of the particular behaviour rating system with full reference documentation.

Difficulties in rating

There are a number of situations that pose difficulties for raters and these should be illustrated, practised and discussed in the training programme. The first of these is where the performance level of a particular skill varies during the observed scenario or task. When a global judgement has to be made, rather than separate ratings for different phases of the task, then the overall rating will obviously depend on the particular balance of behaviours and their significance. The implications for safety should be regarded as paramount when determining the rating. A second issue, mentioned earlier, is the effect of context, which includes the behaviour of others. For instance, a more junior team member may realistically only be able to challenge a domineering senior up to a certain point, or an unusually high level of distraction may be interrupting task concentration. These conditional influences have to be taken into account when rating performance and should be noted with the rating. However, the same problems apply to the rating of technical performance in which variability and context can be equally salient.

Recording scenarios for rater training

Video recordings of scenarios for illustration and for practising rating are not always easy to find; sometimes they can be obtained from research teams or from other organisations in the same sector. It is usually preferable for organisations to film their own in-house scenarios, but in small companies the cost can be prohibitive, especially if no simulator facilities are available. Videotapes of scenarios should ideally have professional sound and visual quality. They need to demonstrate various levels of performance and show different standards of behaviour relating to the component elements of the system. The scenarios should illustrate a range of task conditions and different degrees of complexity. It is important to include scenarios where some behaviours are ambiguous (i.e. open to interpretation) and where the actor demonstrates good and poor performance in the same scenario. For instance, the target actor responds too harshly to an error made by another team member but then leads the team in a very clear and supportive manner through the error recovery phase. The scenarios can depict different lengths of task segment (e.g. from two-minute vignette of a specific behaviour to an entire flight/surgical procedure). The length of clips can be increased as trainee raters become more proficient.

The actors in the target roles in the scenario must be very experienced in that role and real actors are rarely used for this. For the rating systems we have been involved in developing (for pilots, anaesthetists and surgeons), the training tapes for raters were all made in simulators with experienced professionals playing the key parts. The scenarios were only loosely scripted, as it is far easier for practitioners to use their own words than to learn full lines for a part. This also increases the level of local realism, as they can talk in their own normal technical jargon and the social conversations are authentic. It is surprising just how good practitioners can be at playing out the requested behaviours (roles involving bad behaviour being particularly popular).

Debriefing and facilitation skills

Apart from learning to observe and judge the behaviours demonstrated by the ratee, the rater will also need to develop good communication skills for debriefing the ratee and giving feedback on the behaviour ratings. This is not a skill that comes naturally to most people and giving feedback on non-technical skills has to be handled even more sensitively than debriefing a technical performance. If a videotape of the ratee's performance is available, then this should be used to provide examples of behaviour that led to a particular rating. If there is no video recording, then the rater should take care to document particular examples of observed behaviour that contributed to the final score. Dismukes and Smith (2000) give a good description of the skills required to facilitate and debrief training and assessment sessions designed to provide feedback on pilots' non-technical, as well as technical, skills. They emphasise the need to make this an individual or crew-centred experience in order to maximise learning (see also CAA, 2006a, Appendix 9, on CRM facilitation skills). A new Facilitation Skills Assessment Tool (FSAT) has been developed by a group of aviation psychologists and pilots (Smith et al., 2007). This measures 16 skills, broadly categorized into Introduction, Discussion, Trainer's Guidance, Closing and can be adapted for training and assessing facilitators (website given below).

In aviation, the debriefing of non-technical skills is usually conducted as part of line oriented flight training (LOFT) – see Chapter 10. With the increasing use of simulators in health care, debriefing has also been recognised as an important component of medical training. Rudolph et al. (2006) and Flanagan (in press) describe their debriefing techniques for instructors to disclose their evaluations of behaviour while eliciting trainees' assumptions about the situation and their reasons for acting the way they did.

Qualifying raters as competent

Once the raters have been trained, but before they undertake behaviour rating as part of a training or assessment programme, they should be formally assessed to ensure their competence as raters. In UK aviation, the examination of a pilot's CRM skills must be undertaken by a qualified CRM instructor/examiner (CRMIE) who has been formally assessed as competent for this role (see CAA, 2006b, for details). It is suggested that raters are qualified as competent in relation to the following:

- complete initial rater training on the behavioural rating system
- formal assessment as competent and calibrated following rater training in classroom
- calibration in operational environment (e.g. training, simulator, work environment)
- regular re-calibration for continuing use of the behavioural rating system.

Calibration is the process of ensuring that there is sufficiently high agreement between raters and that they are all aligned with company and/or regulatory standards. This is carried out by asking multiple raters to judge the same performance (usually from videotapes) and checking the level of correlation between their scores (Holt et al., 2001; Rall and Gaba, 2005).

Self assessment

In some operational environments, one person works alone (e.g. fighter pilots), and there may be limited opportunities to observe their behaviour, unless simulators or workplace video recordings are available. In this case, self-assessment of non-technical skills is an option (see MOSA below) and there is some evidence that these judgements correlate with observer assessments (Burdekin, 2004).

When non-technical skills assessment is being used for training purposes, self-assessment can be a valuable method in any environment – namely, asking trainees to rate themselves and then their ratings can be compared and discussed with the trainer's ratings. This exercise helps to build skills in self-critique and self-reflection, which are a key part of professional competence (De Cossart and Fish, 2005).

Organisational issues

Where performance evaluations are conducted as part of an organisation's safety management system, then the aggregated data can be used to diagnose emerging problems and to estimate the effectiveness of training and other safety interventions. There are obviously a whole set of organisational/legal/human resource management issues that relate to any kind of performance monitoring system; these would include validity of rating method, qualification of raters, preparation of ratees, data protection, consequences of failure. Discussion of these issues is beyond the scope of this chapter (see Fletcher, 2004, for a useful review). The aviation industry has very well-developed procedures and policies, given their history of regulatory requirements (Birnbach and Longridge, 1993; CAA, 2006a; FAA, 2006a) for both flight operations and pilot licensing. Yet for any profession, appropriate care needs to be taken when introducing the assessment of non-technical skills, especially if this is part of a competence assurance programme. This will require negotiation with key stakeholders, such as professional organisations and unions, to successfully implement the new evaluation practices.

Non-technical skills assessment in practice

The final section discusses how behavioural rating systems are used in three different industries: aviation, health care and nuclear power.

Aviation

As discussed earlier, the aviation community has put considerable emphasis on training flight crew members' non-technical skills as a crucial factor for enhanced safety. The international aviation regulators have generally mandated crew resource management (CRM) courses (see Chapter 10), designed to teach pilots about the essential 'cognitive and interpersonal skills needed to manage the flight within an organized aviation system' (CAA, 2006a: p1). Consequently, the aviation industry has led the field in relation to the assessment of non-technical skills. In the US, the Federal Aviation Administration (FAA) introduced the Advanced Qualification Program (AQP) in the 1990s (FAA, 2006a). This enabled airlines to develop their own training programmes for CRM but they also had to demonstrate to the regulator that these were evaluated. More recently, mandatory regulations have appeared in Britain (CAA, 2006a; 2006b) that require a more formal incorporation of non-technical (CRM) skills evaluation into all levels of training and checking of flight crew members' performance. This builds on earlier regulations and research developments instigated across Europe. The European Joint Aviation Authorities (JAA)[2] require the training and assessment of flight crews' crew resource management (CRM) skills: 'the flight crew must be assessed on their CRM skills in accordance with a methodology acceptable to the Authority and published in the Operations Manual. The purpose of such an assessment is to: provide feedback to the crew collectively and individually and serve to identify retraining; and be used to improve the CRM training system' (JAA, 2001: 1.965).

i) Pilots: NOTECHS This system resulted from a desire by the JAA to achieve a generic method of evaluation of pilots' non-technical skills throughout Europe. They intended to minimise cultural and corporate differences, and maximise practicability and effectiveness for airline instructors and examiners. As a consequence, in 1996, a research project was sponsored to work on what was called the NOTECHS (non-technical skills) project. Appendix 1 presents a summary of the development of the resulting taxonomy of pilots' non-technical skills (NOTECHS). For a more comprehensive account, see Flin et al. (2003) or van Avermaete and Kruijsen (1998).

In response to the JAA requirements on evaluation of CRM skills, many European airlines have now developed their own systems, such as the KLM SHAPE system, the Lufthansa System 'Basic Competence for Optimum Performance' (Burger et al., 2002), and the Alitalia PENTAPERF system (Polo, 2002). These systems have made use of the basic NOTECHS framework in the design of their own customised systems (Hörmann et al., 2002). Several other airlines have used NOTECHS or their own versions of it to complement their proficiency evaluation methods both in Europe (e.g. Finnair, Eastern Airways, Iberia) and beyond (e.g. Gulf Air). The NOTECHS system offers one method of ascertaining whether the CRM training provided to pilots is actually enhancing effectiveness of overall crew performance on the flight deck (Goeters, 2002). A research project in Switzerland used independent judgements from both the NOTECHS and LOSA systems to rate the behaviour of pilots in 46 crews filmed in an Airbus-320 simulator (Hausler et al., 2004). They concluded that 'NOTECHS and LOSA give a very similar picture of the sample regarding the overall performance of crews that were rated' (Klampfer et al., 2003: p133). Thomas (2004) carried out an observational study of crews flying 323 flight segments (on Boeing-737 and Airbus-330 aircraft) for a south-east Asian airline. He used behavioural ratings based on both LOSA and NOTECHS to evaluate the crews' non-technical performance and to compare them against errors and threat management.

2 JAR OPS (2001) 1.940, 1.945, 1.955, and 1.965.

ii) Pilots: LOSA In the line oriented safety audit (LOSA) system independent expert observers sit on the flight deck and watch the pilots' behaviour as they fly a sector on a regular operation. This mainly records behavioural observations at the team level (see Table 11.1). The assessment is generally reported back to the organisation at a fleet level, where a number of crews from a given fleet have been observed. LOSA observers need to be familiar with the airline's procedures and operations, be respected by pilots, and be able to capture data without being obtrusive or interfering with the crew's performance. They use a structured observation form that captures multiple aspects of normal operations. Categories and codes to streamline observations are used, but a written description of the flight is also included to capture the full context. A LOSA observation form is based on the threat and error management model (Helmreich et al., 2003; Merritt and Klinect, 2006), which defines errors as: communication errors, procedural errors (knowing what to do but getting it wrong), proficiency errors (not knowing what to do), decision errors, violations of formal policies or procedures.

A distinctive feature of LOSA is that contextual factors (threats) are taken into account. Threats are defined as situations that can encourage the occurrence of errors. The 12 elements of threat include: policies, standards and procedures; work preparation; job factors; person factors; competence and training; communication; teamwork; supervision; organisational and safety culture; work environment; system–equipment interface; tools and equipment. How air crew manage threat and error situations to maintain safety are recorded by observers. Safety performance strengths and weaknesses are gathered using LOSA as well as structured interviews with pilots to acquire suggestions for safety improvement. Confidential data collection and non-jeopardy assurance for pilots are fundamental to this system. The purpose of conducting a LOSA is to acquire data about an airline's defences and vulnerabilities in their flight operations. A LOSA can complement other safety data collection techniques by revealing pilots' workaround behaviours, usability of procedures, emerging threats to safety, identifying clusters of error events, allowing the comparison of fleets and operations. For more information, see Helmreich et al. (2003), Merritt and Klinect (2006) or FAA (2006b).

iii) Military pilots: MOSA In military aviation, non-technical skills are also recognised as critical and the pilots are trained on CRM courses. However, military pilots are often flying single-seat aircraft and it is not possible for an observer to be present on the flight deck to observe their behaviour to rate their non-technical skills. Burdekin (2004), working with the Royal Australian Air Force, reviewed both LOSA and NOTECHS and concluded that neither was suitable for this type of operation. So she developed a new method – MOSA (mission operations safety audits) – in which pilots self-assess their non-technical skills. In a study involving an F/A-18 simulator, pilots' self-assessments were found to correlate with observer assessments.

iv) Air traffic controllers: BOOM The importance of non-technical skills in aviation is not restricted to pilots; the air traffic controllers also need high-level cognitive and social skills to maintain safe air traffic management. In Europe their non-technical (CRM) skills training is usually called TRM (team resource management). These skills can be evaluated with a behavioural rating system known as behaviour oriented observation method (BOOM) (www.eurocontrol.int/humanfactors). This is a non-jeopardy evaluation where the observer records behaviours, identifies non-technical skills and notes the situation context. These observations are then discussed with the controller during an interview, where the controller can provide explanation and the rater can give feedback.

Health care

As concern about the rates of adverse events to patients caused by medical error grew in the late 1990s (Vincent, 2006), medical professionals began to look at safety management techniques being used in the high-risk industries. Drawing on the behaviour-based/CRM approach adopted by the aviation sector to enhance safety, the acute medical specialities in particular began to identify the non-technical skills contributing to safe and efficient performance, so that these could be trained and evaluated.

i) Anaesthetists Anaesthesia has been at the forefront of these developments, with Gaba and colleagues developing anaesthesia crisis resource management (ACRM) courses (Howard et al., 1992; Gaba et al., 1994) and the first behavioural rating systems for anaesthetists (Gaba et al., 1998; Rall and Gaba, 2005). These were based on UT markers from aviation (see Table 11.1) and were used to rate 14 anaesthesia teams managing complex critical events in a simulator. No ratings were recorded of individual team members, but the 'overall non-technical performance of the primary anaesthetist' was rated. They noted the difficulty in achieving agreement across five raters due to the variability of the teams' behaviour over the scenario, although the relative inexperience of the raters in making the judgements may also have been a factor.

In Scotland, a research project was instigated in 1999 to develop a taxonomy of non-technical skills and method of rating them from behavioural observations of individual anaesthetists working in an operating theatre. The ANTS (anaesthetists' non-technical skills) system was developed using a similar design and evaluation process as was used for NOTECHS. The content was derived from the research literature on anaesthetists' behaviour, observations, interviews, surveys and incident analysis (Fletcher et al., 2002; 2004; Flin et al., 2003). The basic framework is shown in Figure 11.1. An evaluation of the rating method was carried out with 50 consultant anaesthetists who were given basic training on the system and were then asked to rate the non-technical skills of consultant anaesthetists shown in eight videotaped scenarios. The levels of rater accuracy were acceptable and inter-rater reliability (across all 50 raters) was found to approach an acceptable level. Given that the raters had no previous experience of behaviour rating and minimal training (four hours) in the ANTS system, it was concluded that these findings were sufficient to move on to usability trials (Fletcher et al., 2003). The first measures of usability and acceptability from consultants and trainees were promising (Patey et al., 2005). The ANTS system has now had some preliminary trials in the UK through the Royal College of Anaesthetists (Patey, 2007) and the Australia/New Zealand College is also about to conduct an evaluation study. It has been translated into German and Hebrew and has been used to evaluate simulator training for anaesthetists in Canada (Yee et al., 2005) and in Denmark (Rosenstock et al., 2006). Rall and Gaba (2005: p3088) discuss the ANTS system and concluded, 'On the whole, the ANTS system appears to be a useful tool to further enhance assessment of nontechnical skills in anaesthesia, and its careful derivation from a current system of nontechnical assessment in aviation (NOTECHS) may allow for some interdomain comparisons.' They also outline some of the general issues inherent in both technical and non-technical performance assessment, including criterion thresholds, rating fluctuating performance and inter-rater reliability.

The process of introducing the ANTS system has revealed some of the difficulties of bringing a novel type of assessment system to a profession where there is no formal assessment of competence post-qualification. Not only is the notion of workplace assessment new (apart from trainees), it quickly became apparent in the early training courses for raters that the basic psychological language was unfamiliar to most anaesthetists. For example, the term 'situation awareness' was not known, although there was good conceptual understanding of the need to maintain attention and vigilance. Therefore, there was actually a need for basic awareness training courses in non-technical skills for both ratees and raters. In aviation, pilots are taught and examined in the psychological and physiological factors influencing cognitive and physical performance from the start of their training (performance limitations courses). They then undertake crew resource management training provided by their employing airline on a regular basis. Consequently they are very familiar with the cognitive and social skills influencing performance. So there appears to be a need for *ab initio* courses in healthcare introducing the concept of non-technical skills and describing their importance for patient safety.

ii) Surgeons and surgical teams The first studies recording the behaviour of surgeons and other operating theatre team members were carried out in Switzerland. Helmreich and Schaeffer (1994) observed operations using a set of nine categories of 'specific behaviours that can be evaluated in terms of their presence or absence and quality… that are essential for safe and efficient function'. Their goal was 'to develop a rating methodology that can be employed reliably by trained observers' (p242). Using this behaviour categorisation for the observations, they identified a number of instances of error relating to inadequate teamwork, failures in preparation, briefings, communication and workload distribution. According to the authors, this did not provide a quantitative database to provide a baseline against which to measure interventions, such as training or organisational change. So Helmreich, Schaeffer and Sexton (1995) designed the operating room checklist (ORCL), which consisted of a similar list of behaviour categories with rating scales that could be used to assess the non-technical performance of operating teams. This was adapted from the University of Texas marker system used in aviation (Table 11.1). The ORCL system was principally designed to measure team behaviours rather than to rate the non-technical skills of individual surgeons.

In one of the first research studies observing and rating the behaviour of paediatric cardiac surgeons in the UK, Carthey et al. (2003) developed a 'framework of the individual, team and organisational factors that underpin excellence in paediatric surgery'. Seven non-technical skills were rated in their taxonomy at the individual (surgeon) level, in addition to 'technical' skill. Leadership, communication and co-ordination were listed as team-level behavioural markers. This tool does not appear to have been developed further, but the authors concluded that 'behavioural markers developed to explain aviation crew performance can be applied to cardiac surgery to explain differences in process excellence between surgical teams' (p422).

A recent project in Scotland has produced a taxonomy and behavioural rating system for individual surgeons' non-technical skills while in the operating theatre (Yule et al., 2006b). As with ANTS for anaesthetists, this was designed from reviewing the literature, observations, interviews and survey data, assisted by expert surgeons (Yule et al., 2006a; Flin et al., 2006). The NOTSS system has recently been tested in an experimental study where 42 consultant surgeons received some basic training and then rated the behaviour of consultant surgeons shown on videotape. The scores for accuracy were acceptable and inter-rater reliability was better for some skills than others, but the surgeons were inexperienced in rating and had not been calibrated (Yule et al., 2007b). Early usability trials are being run with surgeons rating the non-technical skills of their trainees. Non-technical skill sets can also be used for debriefing and reviewing participants' behaviour during a particular event, such as a critical incident or a 'problem flight' in aviation (Klair, 2000). An example of using NOTSS for this purpose with trainee surgeons is given in Box 11.4.

Box 11.4 Case report from a trainee surgeon using the NOTSS system to debrief after an adverse event during an operation (laparoscopic appendicectomy)

'Following an operative complication I was recently afforded the opportunity to conduct a structured analysis and review of the operative episode based upon the NOTSS framework. The procedure concerned was an appendicectomy commenced initially by means of a laparoscopic approach without the presence of the patient's consultant but with another consultant colleague observing in theatre. The appendix was found behind the colon in a higher than normal location increasing the complexity of the procedure. After reaching a stage where I felt I could not continue, I dissected further under my senior colleague's instruction but still did not make progress. My colleague therefore scrubbed and dissected during which an iatrogenic injury to the wall of the colon occurred. There was intra-operative recognition of this complication and appropriate management following conversion to an open procedure. The patient made an unremarkable recovery and was discharged without significant delay.

NOTSS based feedback was conducted with my senior consultant colleague several days following the event, after the patient's discharge from hospital. Reflection upon previous mixed experience of feedback situations made me adopt a defensive approach to the process. My colleague judiciously selected the neutral venue of a comfortable quiet seating area to conduct the session rather than perhaps the more intimidating surround of his office or the less confidential setting of my shared office.

The operative process, which starts during the pre-operative preparation rather than in theatre, was analysed by both parties in relation to each NOTSS category; situation awareness, decision making, communication and teamwork then leadership. Facilitated by the NOTSS framework we formulated a qualitative, formative and constructive analysis of the operative complication. Precise use of the tool with quantitative application of ratings to all elements was superfluous and by contradistinction may have been inhibitory to the actual discussion that transpired. Regarding situation awareness discussions identified that performance in this area was satisfactory as it was recognised before and during the procedure that this would not be an easy appendicectomy for technical reasons. The overall technical approach was satisfactory with the procedure conducted in daytime hours, with senior anaesthetic staff, consultant surgical presence, sufficient equipment available and the patient adequately prepared. Situation awareness remained satisfactory following the complication with projection of the future state and appropriate action taken. Regarding decision making, I considered conversion to an open procedure when I was no longer making progress with the dissection. However since a senior colleague was present who offered to attempt to continue the procedure laparoscopically, I decided to transfer the role of operating surgeon to my colleague, recognising that completion of the procedure laparoscopically would be in the patient's interest. We concluded that appropriate decision making was being undertaken but that the timing of decision making was affected by communication and leadership. There was a lack of information exchange when I felt we should convert to an open procedure when dissection was undertaken under the instruction of and then by my consultant colleague. There was not a shared understanding of the apparent danger of continued dissection without more accurate appreciation of the anatomical planes. Lack of vocalisation of my concern at continued laparoscopic dissection also represented an issue of leadership particularly with regard to the element of setting and maintaining standards. Our analysis of elements of communication, teamwork and leadership did however also consider other satisfactory aspects such as good coordination of team activities during the process of conversion to the open procedure, rectification of the injury and completion of the procedure.

The feedback process therefore reached far deeper in analysis of situation, causation and process than the superficial technical evaluation I had anticipated. Indeed it was enlightening to dissect the events of the procedure to reveal the causation of the problem could be traced in effect to issues of communication and leadership rather than resulting from technical deficiency. The outcome was to enable me to extract positive constructive learning value from the process with wider application to my general operative approach therefore enhancing the development of my non-technical skills.'

From Yule et al. (2007a)

For more details of the NOTSS system see www.abdn.ac.uk/iprc/notss. As with the anaesthetists (see above), it was found during the development of NOTSS that surgeons do not share a common vocabulary for discussing non-technical skills nor are they knowledgeable about the psychological factors influencing individual and team performance. Consequently, a new introductory course on safety and non-technical skills (Safer Operative Surgery) is being delivered at the Royal College of Surgeons of Edinburgh (Flin et al., 2007). The model for this training is shown below in Figure 11.3.

A similar course, SLIPS (safety and leadership for interventional procedures and surgery), is taught at the Royal College of Surgeons of England.

Prior to the release of NOTSS, adaptations of NOTECHS were used in two studies to rate surgeons' behaviour (Catchpole et al., 2007; Moorthy et al., 2005). There are also rating systems being developed to rate a whole operating theatre team or sub-teams (e.g. nurses). The observational team assessment for surgery (OTAS) measures five team-level skills and is based on a model of teamwork, as well as behaviours derived from observations and from the ANTS system (Healey et al., 2004; Undre et al., 2006; see also www.csru.org.uk).

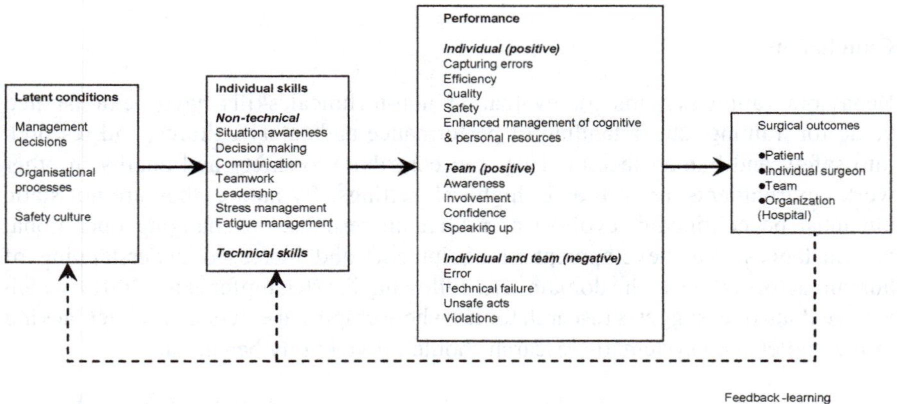

Figure 11.3 Surgical safety model (Flin and Yule, 2007)

iii) Other medical domains There is a body of research examining the performance of medical trauma teams, some of which has used behavioural ratings (Mackenzie and Xiao, 2007; Marsch et al., 2004). To date, most of this work has been for research purposes rather than to develop behaviour rating systems to be used by practitioners. Moreover, the systems are usually designed to rate the team, rather than individual specialists. Thomas et al. (2004) devised a team rating system based on the University of Texas/LOSA behavioural markers for neonatal resuscitation and compare the content of their system to several other medical teamwork rating scales. They used this scale to rate the team behaviours of 132 resuscitation teams caring for infants born by caesarean section (Thomas et al., 2006). They found that the behaviours reflected three underlying teamwork constructs – communication, management and leadership – and that these were related to independent ratings of the teams' quality of care (compliance with resuscitation guidelines).

Nuclear power industry

The nuclear power industry also became interested in non-technical skills and they developed versions of crew resource management training, especially for control room staff who operate the reactors and other areas of the plant. Nuclear power companies have well-established competence assurance programmes and, in some countries, nuclear regulators require control room staff to be licensed and revalidated on a regular basis. These assessments can include an evaluation of non-technical, as well as technical, skills. For example, British Energy has developed an assessment system that tests operators' performance managing routine and emergency scenarios in the control room simulator. This measures technical and non-technical skills (e.g. situation awareness, decision-making, co-operation).

Conclusion

Behaviour rating systems for evaluating non-technical skills have demonstrated value for training, understanding of performance and safety cultures, and research into safety and human factors. They can contribute to safety and quality in other work environments, as well as in high-risk settings. By nature, they are not static but must be continually evolved or refined in response to changing operational circumstances (e.g. development of equipment) and increased understanding of human factors issues in the domain. The following list (Klampfer et al., 2001), which is not exhaustive, suggests research topics where empirical evidence is either lacking or incomplete and systematic research should prove highly beneficial:

- developing empirical evidence for the relative merits of global vs. phase or event-specific ratings and individual vs. team ratings
- defining context effects on crew behaviour and developing a systematic system of integrating these measures with behavioural markers to provide a more comprehensive system
- investigating the distribution of ratings of markers taken in different data collection environments (i.e. training including technical and non-jeopardy, full mission simulation (LOFT), non-jeopardy assessment of system performance (LOSA), formal evaluations in both line operations and recurrent proficiency checks)
- integrating knowledge from incident analyses, especially coping/recovery strategies and translating them into behavioural markers
- providing practical guidance for the transfer of behaviour rating systems and/or their components across domains and cultures (national, professional and organisational).

Key points

- Non-technical skills are typically assessed using ratings of behaviour during task performance (actual task or in a simulator). This assessment should complement the evaluation of technical skills.
- Behavioural rating systems need to be developed carefully for a particular occupation and task set.
- A number of behavioural rating systems have been designed for safety-critical occupations, in aviation, transportation, the energy industry and acute medicine.
- Raters need to be properly trained and qualified to make assessments of non-technical skills, especially if these are for summative or jeopardy evaluation.

Resources

ANTS (anaesthetists' non-technical skills) www.abdn.ac.uk/iprc/ants
NOTSS (non-technical skills for surgeons (NOTSS) www.abdn.ac.uk/iprc/notss
University of Aberdeen Industrial Psychology Research Centre www.abdn.ac.uk/iprc
European Aviation Safety Agency www.easa.eu.int
Facilitation Skills Assessment Tool (FSAT) www.ypsilonassociates.com
Line Operations Safety Audit (LOSA) University of Texas Human Factors Aerospace Group www.psy.utexas.edu/psy/helmreich/nasaut.htm
Observational Teamwork Assessment for Surgery (OTAS) Imperial College, London www.csru.org.uk
UK Civil Aviation Authority www.caa.co.uk
US Federal Aviation Administration www.faa.gov

References

AERA (1999) *Standards for Educational and Psychological Testing.* Washington: American Educational Research Association/American Psychological Association.

Aiken, L. (1996) *Rating Scales and Checklists.* Chichester: Wiley.

Andlauer, E. and the JARTEL group (2001) *Joint Aviation Requirements – Translation and Elaboration.* JARTEL Project Report to DG-TREN European Commission. Paris: Sofreavia.

Antersijn, P. and Verhoef, M. (1995) Assessment of non-technical skills: is it possible? In N. McDonald, N. Johnston and R. Fuller (eds.) *Applications of Psychology to the Aviation System.* Aldershot: Avebury.

Baker, D. and Dismukes, K. (2002) Special issue on training instructors to evaluate aircrew performance. *International Journal of Aviation Psychology*, 12, 203–222.

Baker, D., Mulqueen, C. and Dismukes, K. (2001) Training raters to assess resource management skills. In E. Salas, C. Bowers and E. Edens (eds.) *Improving Teamwork in Organizations. Applications of Resource Management Training.* Mahwah, NJ: LEA.

Beaubien, M., Baker, D. and Salvaggio, A. (2004) Improving the construct validity of line operational simulation ratings: Lessons learned from the assessment center. *International Journal of Aviation Psychology*, 14, 1–17.

Birnbach, R. and Longridge, T. (1993) The regulatory perspective. In E. Wiener, B. Kanki and R. Helmreich (eds.) *Cockpit Resource Management.* San Diego: Academic Press.

Boehm-Davis, D., Holt, R. and Seamster, T. (2001) Airline resource management programs. In E. Salas, C. Bowers and E. Edens (eds.) *Improving Teamwork in Organizations. Applications of Resource Management Training.* Mahwah, NJ: LEA.

Burdekin, S. (2004) Mission Operations Safety Audits (MOSA). *Aviation Safety Spotlight*, 3, 21–29.

Burger, K., Neb, H. and Hörmann, H. (2002) Basic performance of flight crew – A concept of competence based markers for defining pilots' performance profile. *Proceedings of the 25th European Aviation Psychology Conference, Warsaw.* Warsaw: Polish Airforce.

CAA (2002) *Flight Crew Training: Cockpit Resource Management (CRM) and Line-Oriented Flight Training (LOFT).* CAP 720 – Reprint of ICAO Digest 2 – 217-AN/132 (1989). Gatwick: Civil Aviation Authority.

CAA (2006a) *Crew Resource Management (CRM) Training. Guidance for Flight Crew, CRM Instructors (CRMIs) and CRM Instructor-Examiners (CRMIEs).* CAP 737. Version 2. Gatwick: Safety Regulation Group, Civil Aviation Authority. Available on www.caa.co.uk

CAA (2006b) *Guidance Notes for Accreditation Standards for CRM Instructors and CRM Instructor Examiners. Standards Doc. 29 Version 2.* Gatwick: Civil Aviation Authority.

Carthey, J., de Leval, M., Wright, D., Farewell, D. and Reason, J. (2003) Behavioural markers of surgical excellence. *Safety Science*, 41, 409–425.

Catchpole, K., Giddings, A., Wilkinson, M., Hirst, G., Dale, T. and de Leval, M. (2007) Improving patient safety by identifying latent failures in successful operations. *Surgery*, 142, 102–110.

Cook, M. and Cripps, M. (2005) *Psychological Assessment in the Workplace. A Manager's Guide.* Chichester: Wiley.

De Cossart, L. and Fish, D. (2005) *Cultivating a Thinking Surgeon.* Shrewsbury: tfm.

DeVellis, R. (2003) *Scale Development.* (2nd ed.) London: Sage.

Dismukes, K. and Smith, G. (eds.) (2000) *Facilitation and Debriefing in Aviation Training and Operations.* Aldershot: Ashgate.

FAA (1993) *Advisory Circular 120-51A Crew Resource Management Training* Washington: Federal Aviation Administration. (current version 51E – 2004).

FAA (2006a) *Advisory Circular 120-54A. Advanced Qualification Program.* Washington: Federal Aviation Administration.

FAA (2006b) *Advisory Circular 120-90. Line Operations Safety Audit.* Washington: Federal Aviation Administration.

Flanagan, B. (in press) Debriefing – theory and techniques. In R. Riley (ed.) *A Manual of Simulation in Healthcare.* Oxford: Oxford University Press.

Fletcher, C. (2004) *Appraisal and Feedback. Making Performance Review Work.* (3rd ed.) London: Chartered Institute of Personnel and Development.

Fletcher, G., McGeorge, P., Flin, R., Glavin, R. and Maran, N. (2002) The role of non-technical skills in anaesthesia: A review of current literature. *British Journal of Anaesthesia*, 88, 418–429.

Fletcher, G., Flin, R. McGeorge, P. Glavin, R., Maran, N. and Patey, R. (2004) Rating non-technical skills. Developing a behavioural marker system for use in anaesthesia. *Cognition, Technology and Work*, 6, 165–171.

Fletcher, G., McGeorge, P, Flin, R., Glavin, R. and Maran, N. (2003) Anaesthetists' non-technical skills (ANTS). Evaluation of a behavioural marker system. *British Journal of Anaesthesia*, 90, 580–588.

Flin, R. (2006) *Safe in their Hands? Licensing and Competence Assurance of Safety Critical Roles in High Risk Industries.* Report to the Department of Health, London. University of Aberdeen. Available at www.abdn.ac.uk/iprc.

Flin, R., Fletcher, G., McGeorge, P., Sutherland, A. and Patey, R. (2003). Anaesthetists' attitudes to teamwork and safety. *Anaesthesia*, 58, 233–242.

Flin, R. and Martin, L. (2001) Behavioural markers for Crew Resource Management: A review of current practice. *International Journal of Aviation Psychology*, 11, 95–118.

Flin, R., Martin, L., Goeters, K., Höermann, J., Amalberti, R., Valot, C. and Nijhuis, H. (2003) Development of the NOTECHS (Non-Technical Skills) system for assessing pilots' CRM skills. *Human Factors and Aerospace Safety*, 3, 95–117.

Flin, R., Yule, S., McKenzie, L., Paterson-Brown, S. and Maran, N. (2006) Attitudes to teamwork and safety in the operating theatre. *The Surgeon*, 4, 145–151.

Flin, R., Yule, S., Maran, N., Paterson-Brown, S., Rowley, D. and Youngson, G. (2007) Teaching surgeons about non-technical skills. *The Surgeon*, 5, 107–110.

Fowlkes, J., Lane, N., Salas, E., Franz, T. and Oser, R. (1994) Improving the measurement of team performance: The TARGETS methodology. *Military Psychology*, 6, 47–61.

Gaba, D., Fish, K. and Howard, S. (1994) *Crisis Management in Anesthesiology.* New York: Churchill Livingstone.

Gaba, D., Howard, S., Flanagan, B., Smith, B., Fish, K. and Botney, R. (1998) Assessment of clinical performance during simulated crises using both technical and behavioural ratings. *Anesthesiology*, 89, 8–18.

Goeters, K.-M. (2002) Evaluation of the effects of CRM training by the assessment of non-technical skills under LOFT. *Human Factors and Aerospace Safety*, 2, 71– 86.

Guerlain, S., Adams, R., Turrentine, B., Shin, T., Guo, H., Collins, S. and Calland, F. (2005) Assessing team performance in the operating room: Development and use of a 'Black-Box' recorder and other tools for the intraoperative environment. *Journal of American College of Surgeons*, 200, 29–37.

Hausler, R., Klampfer, B., Amacher, A. and Naef, W. (2004) Behavioural markers in analysing team performance of cockpit crews. In R. Dietrich and T. Childress (eds.) *Group Interaction in High Risk Environments.* Aldershot: Ashgate.

Healey, A., Undre, S. and Vincent, C. (2004) Developing observational measures of performance in surgical teams. *Quality and Safety in Healthcare, 13 (Suppl 1),* i33–i40.

Helmreich, R. (1999) CRM training, primary line of defence against threats to flight safety, including human error. *ICAO Journal*, 54, 6–10.

Helmreich, R., Butler, R., Taggart, W. and Wilhelm, J. (1995a) The NASA/University of Texas/FAA Line/LOS checklist: A behavioral marker-based checklist for CRM skills assessment. Version 4. Technical Paper 94-02 (Revised 12/8/95). Austin, Texas: University of Texas Aerospace Research Project.

Helmreich, R., Butler, R., Taggart, W. and Wilhelm, J. (1995b) Behavioral markers in accidents and incidents: Reference list. NASA/UT/FAA Technical Report 95-1. Austin, TX: The University of Texas, Human Factors Research Project.

Helmreich, R., Klinect, J. and Wilhelm, J. (2003) Managing threat and error: Data from line operations. In G. Edkins and P. Pfister (eds.) *Innovation and Consolidation in Aviation.* Aldershot: Ashgate.

Helmreich, R. and Schaeffer, H. (1994) Team performance in the operating room. In M. Bogner (ed.) *Human Error in Medicine.* New Jersey: Lawrence Erlbaum.

Helmreich, R., Schaeffer, H. and Sexton, B. (1995) The Operating Room Checklist. Technical Report 95-10. University of Texas at Austin.

Helmreich, R. and Wilhelm, J. (1987) Reinforcing and measuring flightcrew resource management: Training Captain/Check Airman/Instructor Reference Manual. NASA/ University of Texas at Austin Technical Manual 87-1. Austin, TX: The University of Texas, Human Factors Research Project.

Helmreich, R., Wilhelm, J., Gregorich, S. and Chidester, T. (1990). Preliminary results from the evaluation of Cockpit Resource Management training: Performance ratings of flightcrews. *Aviation, Space, and Environmental Medicine*, 61, 576–579.

Holt, R., Boehm-Davis, D. and Beaubien, M. (2001) Evaluating resource management training. In E. Salas, C. Bowers and E. Edens (eds.) *Improving Teamwork in Organizations. Applications of Resource Management Training.* Mahwah, NJ: LEA.

Hörmann, J. (2001) Cultural variations of perceptions of crew behaviour in multi-pilot aircraft. *Le Travail Humain*, 64, 247–268.

Hörmann, J., Burger, K. and Neb, H. (2002) Integration of interpersonal skills into a pilot's proficiency reporting system. First results of a usability study at Lufthansa. In O. Truszczynski (ed.) *Proceedings of the 25th European Aviation Psychology Conference, Warsaw.* Warsaw: Polish Airforce.

Howard, S., Gaba, D., Fish, K., Yang, G. and Sarnquist, F. (1992) Anesthesia Crisis Resource Management training: teaching anesthesiologists to handle critical incidents. *Aviation, Space and Environmental Medicine*, 63, 763–770.

JAA (1999) *JAR-OPS (Joint Aviation Requirements for Flight Operations) 1 subpart N (NPA-OPS-16).* Hoofddorp, Netherlands: Joint Aviation Authorities.

JAA (2001) *JAR-OPS (Joint Aviation Requirements for Flight Operations – Crew Resource Management: Flight Crew).* Hoofddorp, Netherlands: Joint Aviation Authorities.

Klair, M. (2000) The mediated debrief of problem flights. In K. Dismukes and G. Smith (eds.) *Facilitation and Debriefing in Aviation Training and Operations.* Aldershot: Ashgate.

Klampfer, B., Flin. R., Helmreich R., et al. (2001) *Enhancing performance in high risk environments: Recommendations for the use of Behavioural Markers.* Daimler-Benz Foundation. Available to download as pdf file on www.abdn.ac.uk/iprc.

Klampfer, B., Haeusler, R. and Naef, W. (2003) CRM behaviour and team performance under high workload: Outline and implications of a simulator study. In G. Edkins and P. Pfister (eds.) *Innovation and Consolidation in Aviation.* Aldershot: Ashgate.

KLM (1996) Feedback and Appraisal System. Amsterdam: KLM Internal Paper.

Macleod, N. (2005) *Building Safe Systems in Aviation. A CRM Developer's Handbook.* Aldershot: Ashgate.

Mackenzie, C. and Xiao, Y. (2007) Video analysis in health care. In P. Carayon (ed.) *Handbook of Human Factors and Ergonomics in Health Care and Patient Safety.* Mahwah, NJ: Lawrence Erlbaum Associates.

Marsch, S., Muller, C., Marquardt, K., Conrad, G., Tscan, F. and Hunziker, P. (2004) Human factors affect the quality of cardiopulmonary resuscitation in simulated cardiac arrests. *Resuscitation*, 60, 51–56.

Merritt, A. and Klinect, J. (2006) Defensive flying for pilots: An introduction to threat and error management. Paper available on University of Texas LOSA website – see above.

Moorthy, K., Munz, Y., Adams, S. et al. (2005) A human factors analysis of technical and team skills among surgical trainees during procedural simulations in a simulated operating Theatre. *Annals of Surgery*, 242, 631–639.

Noyes, J. (2001) *Designing for Humans.* Hove: Taylor & Francis.

Nunnally, J. and Bernstein, I. (1993) *Psychometric Theory.* New York: McGraw Hill.

O'Connor, P., Flin, R. and Fletcher, G. (2002a) Techniques used to evaluate Crew Resource Management training: A literature review. *Human Factors and Aerospace Safety*, 2, 217–233.

O'Connor, P., Flin, R., Fletcher, G. and Hemsley, P. (2002b) Methods used to evaluate the effectiveness of flightcrew CRM training in the UK aviation industry. *Human Factors and Aerospace Safety*, 2, 235–255.

O'Connor, P., Hormann, H.-J., Flin, R., Lodge, M., Goeters, K.-M. and the JARTEL group (2002c) Developing a method for evaluating crew resource management skills: A European perspective. *International Journal of Aviation Psychology*, 12, 265–288.

O'Connor, P., O'Dea, A. and Flin, R. (2002d) Teamwork skills training for Nuclear Operations Personnel: Results (IMC project HF/GSNR/5064). University of Aberdeen/IMC report.

Patey, R. (2007) Non-technical skills and anaesthesia. In J. Cashman and R. Grounds (eds.) *Recent Advances in Anaesthesia and Intensive Care 24.* Cambridge: Cambridge University Press.

Patey, R., Flin, R., Fletcher, G., Maran, N. and Glavin, R. (2005) Developing a taxonomy of anaesthetists' non-technical skills (ANTS). In Hendriks, K. (ed.) *Advances in Patient Safety: From Research to Implementation.* Rockville, MD: Agency for Healthcare Research and Quality.

Polo, L. (2002) Evaluation of flight crew members' performance. Is evaluation a product or a tool? In O. Truszczynski (ed.) *Proceedings of the 25th European Aviation Psychology Conference, Warsaw.* Warsaw: Polish Airforce.

Rall, M. and Gaba, D. (2005) Simulation and anaesthesia. In R. Miller. *Anesthesia (6th Ed).* New York: Churchill Livingstone.

Riley, R. (ed.) (in press) *A Manual of Simulation in Healthcare.* Oxford: Oxford University Press.

Reader, T., Flin, R., Lauche, K. and Cuthbertson, B. (2006) Non-technical skills in the intensive care unit. *British Journal of Anaesthesia*, 96, 551–559.

Rosenstock, E., Kristensen, M., Rasmussen, L., Skak, C. and Ostergaard, D. (2006) Qualitative analysis of difficult airway management. *Acta Anaesthesiology, Scandinavia*, 50, 290–297.

Rudolph, J., Simon, R., Dufresne, R. and Raemer, D. (2006) There's no such thing as 'nonjudgmental' debriefing: A theory and method for good judgment. *Simulation in Healthcare*, 1, 49–55.

Salas, E., Burke, S., Bowers, C. and Wilson, K. (2001) Team training in the skies. Does Crew Resource Management (CRM) training work? *Human Factors*, 43, 641–674.

Salas, E., Wilson, K., Burke, S. and Wightman, D. (2006) Does Crew Resource Management training work? An update, an extension and some critical needs. *Human Factors*, 48, 392–412.

Smith, M., Smith, G., Budenberg, C., Willats, J. and Griffiths, P. (2007) An airline evaluation of the Facilitation Skills Assessment Tool (FSAT) as device for training and assessing facilitators. In *Proceedings of the International Symposium on Aviation Psychology*, Ohio State University, April.

Thomas, E., Sexton, J. and Helmreich, R. (2004) Translating teamwork behaviours from aviation to healthcare: development of behavioural markers for neonatal resuscitation. *Quality and Safety in Healthcare, 13 (Suppl. 1),* i57–i54.

Thomas, E., Sexton, J., Lasky, R., Helmreich, R., Crandell, D. and Tyson, J. (2006) Teamwork and quality during neonatal care in the delivery room. *Journal of Perinatology*, 26, 163–169.

Thomas, M. (2004) Predictors of threat and error management: Identification of core nontechnical skills and implications for training systems design. *International Journal of Aviation Psychology*, 14, 207–231.

Undre, S., Healey, A., Darzi, A. and Vincent, C. (2006) Observational assessment of teamwork: A feasibility study. *World Journal of Surgery*, 30, 1774–1783.

van Avermaete, J. and Kruijsen, E. (eds.) (1998) *NOTECHS. The Evaluation of Non-Technical Skills of Multi-Pilot Aircrew in Relation to the JAR-FCL Requirements.* Final Report NLR-CR-98443. Amsterdam: National Aerospace Laboratory (NLR).

Vincent, C. (2006) *Patient Safety.* London: Churchill Livingstone.

Woodruffe, C. (2000) *Development and Assessment Centres: Identifying and Developing Competencies.* London: Chartered Institute of Personnel and Development.

Yee, B., Naik, V., Joo, H., Savoldelli, G., Chung, D., Houston, P., Karatzogolou, B. and Hamstra, S. (2005) Nontechnical skills in anaesthesia crisis management with repeated exposure to simulation-based education. *Anesthesiology*, 103, 241–248.

Yule, S., Flin, R., Maran, N., Paterson-Brown, S. and Rowley, D. (2006a) Non-technical skills for surgeons in the operating room: A review of the literature. *Surgery*, 139, 140–149.

Yule, S. Flin, R., Maran, N., Paterson-Brown, S. and Rowley, D. (2006b) Development of a rating system for surgeons' non-technical skills. *Medical Education*, 40, 1098–1104.

Yule, S., Flin, R., Maran, N., Paterson-Brown, S., Rowley, D. and Youngson, G. (2007a) Observe one, rate one, debrief one. Using the NOTSS system to discuss non-technical skills with trainee surgeons. *Cognition, Technology and Work (in press).*

Yule, S., Flin, R., Maran, N., Paterson-Brown, S. and Rowley, D. (2007b) Evaluating surgeons' non-technical skills with NOTSS. *World Journal of Surgery (in press).*

Appendix 1

The development of the NOTECHS system

The European Joint Aviation Authorities (JAA) require the training and assessment of flight crews' crew resource management (CRM) skills: 'the flight crew must be assessed on their CRM skills in accordance with a methodology acceptable to the Authority and published in the Operations Manual. The purpose of such an assessment is to: provide feedback to the crew collectively and individually and serve to identify retraining; and be used to improve the CRM training system' (JAA 1999: 1.965).

This legislation resulted from a desire by the JAA to achieve a generic method of evaluation of pilots' non-technical skills throughout Europe. This was intended to minimise cultural and corporate differences, and maximise practicability and effectiveness for airline instructors and examiners. As a consequence, in 1996, the JAA Project Advisory Group on Human Factors initiated a research project that was sponsored by four European Civil Aviation Authorities (Germany, France, Netherlands, UK). A research consortium consisting of pilots and psychologists from DLR (Germany), IMASSA (France), NLR (Netherlands) and University of Aberdeen (UK) was established to work on what was called the NOTECHS (non-technical skills) project. This appendix describes the development of the resulting taxonomy of pilots' non-technical skills (NOTECHS). For a more comprehensive account, see Flin et al. (2003) or van Avermaete and Kruijsen (1998). The project team was required to identify or to develop a feasible and efficient methodology for assessing a pilot's non-technical (CRM) skills. (For the purpose of the project, these were defined as the cognitive and social skills of flight crew members in the cockpit, not directly related to aircraft control, system management, and standard operating procedures (SOPs).) The design requirements were (i) that the system was to be used to assess the skills of an individual pilot, rather than a crew, and (ii) it was to be suitable for use across Europe, by both large and small operators, i.e. it was to be culturally robust.

After reviewing alternative methods, it appeared that none of the existing behaviour rating systems being used by the larger European and American airlines (Flin and Martin, 2001) could be adopted in their original form. Nor did any single system provide a suitable basis for simple amendment that could be taken as an 'acceptable means of compliance' under the scope of the regulations. The reasons were that the existing systems were either too complex to be used on a pan-European basis, or too specific to a particular airline, or were designed to assess crews rather than individual pilots (e.g. University of Texas LLC (line/line-oriented simulation checklist system (Helmreich et al., 1995a)). Therefore, the NOTECHS group decided that it would have to design a new taxonomy of non-technical skills (see Chapter 9 for techniques that can be used for this). The four categories and elements of the NOTECHS system are shown below in Figure 11.4. They were presented in previous chapters when the component skills were discussed, therefore they are not discussed further here.

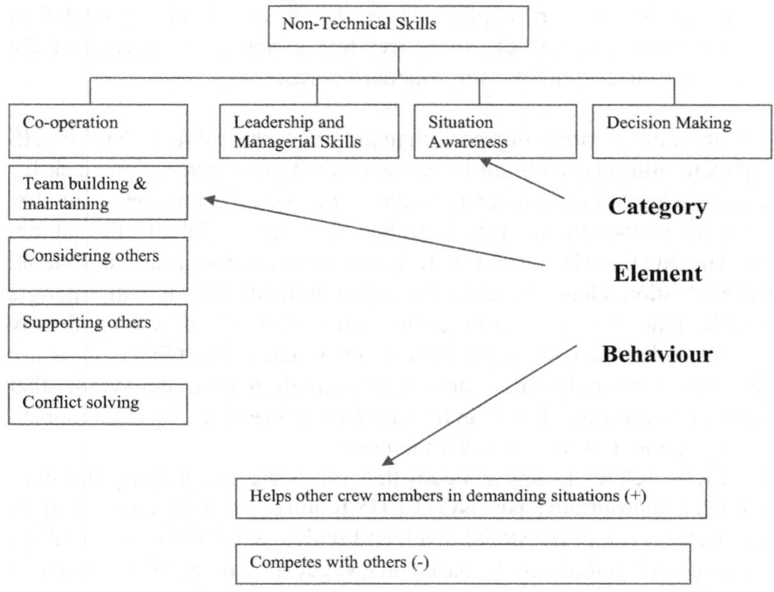

Figure 11.4 The NOTECHS system

Operational principles for using the NOTECHS system

Five operational principles were established to ensure that each crew member receives as fair and as objective an assessment as possible with the NOTECHS system:

1. *Only observable behaviour is to be assessed* – The evaluation must exclude reference to a crew member's personality or emotional attitude and should be based only on observable behaviour. Behavioural markers were designed to support an objective judgement.
2. *Need for technical consequence* – For a pilot's non-technical skills to be rated as unacceptable, flight safety must be actually (or potentially) compromised. This requires a related objective technical consequence.
3. *Acceptable or unacceptable rating required* – The JAR-OPS requires the airlines to indicate whether the observed non-technical skills are acceptable or unacceptable.
4. *Repetition required* – Repetition of unacceptable behaviour during the check must be observed to conclude that there is a significant problem. If, according to the JAR paragraph concerned, the nature of a technical failure allows for a second attempt, this should be granted, regardless of the non-technical rating.
5. *Explanation required* – For each category rated as unacceptable the examiner must: a) indicate the element(s) in that category where the unacceptable

behaviour was observed; b) explain where the observed NTS (potentially) led to safety consequences; c) give a free-text explanation on each of the categories rated unacceptable, using standard phraseology.

Judging behaviour is always more subjective than judging technical facts. NOTECHS has been designed to minimise ambiguities in the evaluation of non-technical skills. However, there are several factors that can occur in the evaluation process. The first relates to the unit of observation, i.e. who is evaluated: the crew globally, the captain or the co-pilot. The NOTECHS system is designed to be used to assess individual pilots. When an evaluation relates to individuals, a potential difficulty is to disentangle individual contributions to overall crew performance. But this difficulty already exists during checks when considering technical performance. NOTECHS does not magically solve this problem, but may serve to objectively point to behaviours that are related more to one crew member than the other, therefore allowing examiners to differentiate the judgement of the two crew members.

The second factor relates to any concern that raters are not judging the non-technical skills on an appropriate basis. NOTECHS requires the instructor/examiner to justify any criticisms at a professional level, and with a standardised vocabulary. Furthermore, a judgement should not be based on a vague global impression or on an isolated behaviour or action. Repetition of the behaviour during the flight is usually required to explicitly identify the nature of the problem.

The NOTECHS method is designed to be a guiding tool to look beyond failure during recurrent checks or training, and to help point out possible underlying deficiencies in CRM competence in relation to technical failures. The evaluation of non-technical skills in a check using NOTECHS should not provoke a failed (not acceptable) rating without a related objective technical consequence, leading to compromised flight safety in the short or long term. In the event of a crew member failing a check for any technical reason, NOTECHS can provide useful insights into the contributing individual human factors for the technical failure. Used in this way, the method can provide valuable assistance for debriefing and orienting tailored retraining.

Preliminary test of NOTECHS: the JARTEL project

The prototype NOTECHS system offered a systematic approach for assessing pilots' non-technical skills in simulator and flight missions. Testing of the basic usability and psychometric properties of the NOTECHS system was then required. The Directorate General Transportation (DGTREN) of the European Community, in co-ordination with the JAA research committee, tasked a consortium of five research centres (NLR (N), DLR (G), University of Aberdeen (UK), DERA (UK; now called QinetiQ) and IMASSA (F) and four aviation centres (Sofreavia (F), British Airways (UK), Alitalia (I) and Airbus) to test the NOTECHS method. A European research project, JARTEL (Joint Aviation Requirements – Translation and Elaboration of Legislation), began in January 1998 and was completed in 2001. The main goals of this project were to assess:

- usability of the NOTECHS system as an assessment tool
- reliability and validity of the assessment tool
- influence of cultural differences on the use of the NOTECHS system within Europe.

Experimental study

The experimental study was carried out using eight video scenarios filmed in a Boeing-757 simulator, with current pilots as actors. The scenarios showed simulated flight situations with predefined behaviours (from the NOTECHS elements) exhibited by the pilots at varying standards ('very poor' to 'very good'). The pilots' behaviours were rated using the NOTECHS systems by 105 instructors, recruited from 14 airlines in 12 European countries. The airlines represented large and smaller carriers within five European cultural groups.

Each of the experimental sessions was conducted within an airline training centre. It began in the morning with a standard briefing on the NOTECHS method (the participants had previously been supplied with background information on the NOTECHS system) and a practice session. Questionnaires were also completed by the instructors, providing data on their background and experience. During the subsequent afternoon session, the captain's and first officer's behaviour in each of the eight cockpit scenarios was rated by the instructors using the NOTECHS score forms. At the end of the session, a second questionnaire was given to the instructors for evaluating the NOTECHS rating process and material.

In summary, the results indicated that 80% of the instructors were consistent in their ratings and 88% of them were satisfied with the consistency of the method. On average, the difference between a reference rating (established for benchmarking by consensus ratings in a set of trained expert instructors) and the instructors' ratings was less than one point on the five-point scale, confirming an acceptable level of accuracy. In the evaluation questionnaire, the instructors were very satisfied with the NOTECHS rating system, especially with the five-point scale (98%). Cultural differences (relating to five European regions) were found to be less significant than other background variables, such as English language proficiency, experience with non-technical skills evaluation, and different role perceptions of captain and first officer (see Hörmann, 2001, for details of the cultural analysis). Full details of the experimental method and the results can be found in the JARTEL project reports for work packages 2 and 3 (Sofreavia website: sofreavia.com/jartel/) or see O'Connor et al. (2002c). A subsequent operational trial of NOTECHS was run with several airlines (see JARTEL work package 4 Report). It confirmed the applicability and feasibility of the system in real check events.

These first experimental and operational tests of the NOTECHS system showed that it was usable by instructors and appeared to have acceptable psychometric properties. These results were achieved with a minimal training period of half a day due to difficulties in recruiting experienced instructors to take part in the study, especially from the small companies. This level of training would be insufficient for using the NOTECHS system for regular training or assessment purposes. It is recommended that the basic training period is two full days or longer (depending on the level of previous experience of rating pilots' non-technical skills).

Training requirements for users of the NOTECHS system

Users of NOTECHS are certified flight instructors and authorised examiners. It is necessary to train all raters in the application of the method. NOTECHS presupposes sufficient knowledge of concepts included in the JAR-FCL theoretical programme on human performance and limitations (JAR-FCL1.125/1.160/1.165. Theoretical knowledge instruction PPL/ATPL). No additional theoretical knowledge is required. Being current in CRM training and recurrent CRM is required, at least as a participant. Experience of CRM instruction is a facilitating factor for standardisation, but is not a prerequisite (see CAA, 2006b, for the current UK position on CRM instructors and CRM instructor examiners). Most of the training effort should be devoted to the understanding of the NOTECHS methodology, the specific use of the evaluation grid, the calibration process of judgement and the debriefing phase. As the NOTECHS system is primarily used as a tool for debriefing and identification of training needs, it is important to ensure that in debriefing an emphasis is placed on skill components, rather than on more 'global' analyses of performance.

In summary, NOTECHS was designed as: (i) a professional pragmatic tool for instructors and authorised examiners; (ii) a tool to be used by non-psychologists; and (iii) a tool using common professional aviation language, with the primary intention of debriefing pilots and communicating concrete directions for improvements. NOTECHS was not primarily designed as: (i) a research tool (although it can be used for this purpose, see Goeters, 2002); (ii) a tool for judging flight crew personality on the basis of instructors' or authorised examiners' personal opinions; or (iii) a tool for introducing psychological jargon into the evaluation. The preliminary evaluation of the NOTECHS system from the experimental and operational trials indicated that the basic psychometric properties were acceptable and that the method was usable and accepted by practitioners. Clearly a more extensive test of the psychometric quality of NOTECHS would be desirable but this would require a large data set collected under standardised conditions.

Index

Page numbers in *italics* refer to tables, boxes and figures.